AF358820

Interaction between Sugarcane and Environmental Stressors: From Identification to Molecular Mechanism

Interaction between Sugarcane and Environmental Stressors: From Identification to Molecular Mechanism

San-Ji Gao

Basel • Beijing • Wuhan • Barcelona • Belgrade • Novi Sad • Cluj • Manchester

Editor
San-Ji Gao
Agronomy
Fujian Agriculture and
Forestry University
Fuzhou
China

Editorial Office
MDPI AG
Grosspeteranlage 5
4052 Basel, Switzerland

This is a reprint of articles from the Special Issue published online in the open access journal *Plants* (ISSN 2223-7747) (available at: www.mdpi.com/journal/plants/special_issues/Sugarcane_Environment).

For citation purposes, cite each article independently as indicated on the article page online and as indicated below:

Lastname, A.A.; Lastname, B.B. Article Title. *Journal Name* **Year**, *Volume Number*, Page Range.

ISBN 978-3-7258-2268-3 (Hbk)
ISBN 978-3-7258-2267-6 (PDF)
doi.org/10.3390/books978-3-7258-2267-6

Contents

Preface

Increased crop vulnerability to a wide range of abiotic (such as temperature, waterlogging, drought, oxidative, salinity, and ultraviolet) and biotic (such as viruses, bacteria, fungi, and insects) stresses can have a marked influence on the productivity of major crops. To cope with various stresses, plants have evolved complex rapid responses, but crop efficiency is still severely hampered. The prevailing, alarming climate change scenario and future changing weather provide challenges for researchers to better understand plant responses.

Sugarcane (*Saccharum* spp. hybrids) is a classic C_4 crop with the highest photosynthetic efficiency among other crops, which plays a crucial role in sugar and biofuel production. This Special Issue illustrated current research findings about the interaction between sugarcane and environmental stressors, which includes four main categories: (i) comparative analysis of agronomic characters (growth, yield, and quality) among sugarcane genotypes amidst different environmental conditions; (ii) microbial community analysis on sugarcane-growing soil and population genetic analysis on pathogens infecting sugarcane; (iii) transcriptome analysis of sugarcane defense responses to various pathogens; and (iv) transgenesis and functional analysis of sugarcane genes under different stressors. This reprint will help us to better understand the stress adaptation and tolerance mechanisms in sugarcane under climate change scenarios.

San-Ji Gao
Editor

MDPI

Editorial

Interaction between Sugarcane and Environmental Stressors: From Identification to Molecular Mechanism

San-Ji Gao [1],* and Talha Javed [2]

[1] National Engineering Research Center for Sugarcane, Fujian Agriculture and Forestry University, Fuzhou 350002, China

[2] Institute of Tropical Bioscience and Biotechnology, Chinese Academy of Tropical Agricultural Sciences, Haikou 571101, China; talhajaved@itbb.org.cn

* Correspondence: gaosanji@fafu.edu.cn

Citation: Gao, S.-J.; Javed, T. Interaction between Sugarcane and Environmental Stressors: From Identification to Molecular Mechanism. *Plants* **2024**, *13*, 1973. https://doi.org/10.3390/plants13141973

Received: 26 June 2024
Accepted: 18 July 2024
Published: 19 July 2024

An increase in the vulnerability of crops to a wide range of biotic and abiotic stresses can have a marked influence on the productivity and quality of major crops, especially sugarcane (*Saccharum* spp.). To cope with various stressful conditions, plants have evolved complex rapid response mechanisms [1,2]. The broader topic of this special issue emphasizes on the interaction between sugarcane and environmental stressors at genotypic, physio-chemical, and molecular levels. This special issue includes a total of 13 papers with 12 research papers and one review article. Papers were submitted from six countries, including Pakistan, Saudi Arabia, Egypt, Poland, Thailand, and China. These papers were grouped into four main categories: (1) comparative analysis of agronomic characters (growth, yield, and quality) among sugarcane genotypes amidst different environmental conditions; (2) microbial community analysis on sugarcane-growing soil and population genetic analysis on pathogens infecting sugarcane; (3) transcriptome analysis of sugarcane defense responses to various pathogens; and (4) transgenesis and functional analysis of sugarcane genes under different stressors. This editorial encompasses perspectives for the future vision of sugarcane improvement under climate change scenarios.

The study by Manzoor et al., investigated the influence of different ecological zones and irrigation conditions on sugarcane growth, yield, and quality in Pakistan. In general, the utilization of combined canal and tube well water as well as the combined application of potassium and boron improved sugarcane yield and quality [3]. Sajid et al. found that sugarcane genotypes S2006-US-658, HSF-240, and S2007-AUS-384 were better performers concerning growth and physiological attributes under sandy loam soil and water deficit conditions [4]. Similarly, Zhao et al., revealed that variance in sugarcane yields and sugar contents was caused by climate differences in the low-latitude plateau [5]. The observation by Wu et al. (2024) demonstrated that combined application of biochar and microplastics have potential to alter the soil nutrients and microbial community structure and function and alleviate the negative effects of micro-plastic accumulation on sugarcane biomass [6]. In general, the elite genotype breeding and soil improvement are important strategies for sugarcane adaptation under climate change scenarios.

Xu et al. analyzed viral population structure and evolutionary forces, demonstrating the phylogeny and genetic divergence among sorghum mosaic virus (SrMV) isolates infecting sugarcane. This work showed that the geographical isolation and host types are important factors promoting the genetic differentiation of SrMV populations [7]. Another study by Zhou et al., showed that significantly lower viral proliferation was observed in two cultivars ROC22 and Xuezhe under 1 mM exogenous salicylic acid (SA) application [8]. Additionally, Shan et al. revealed two non-synonymous mutation sites for amino acids located in the cap pocket of eIF4E, which might be related to the resistance to sugarcane streak mosaic virus (SCSMV) [9].

Transcriptome analysis based on the RNA-seq platform by Illumina NGS technology is commonly used for investigating sugarcane defense responses to pathogen infections.

Zhou et al., revealed that most of the differentially expressed genes (DEGs) were mainly concentrated in plant-pathogen interaction and phenylpropanoid biosynthesis, which involved in SA-induced disease resistance of sugarcane in response to SrMV infection [8]. The RNA-seq dataset and RT-qPCR assay showed that the transcript levels of genes related to SA biosynthesis and signal transduction pathways were significantly upregulated in sugarcane in response to SA treatment under SrMV infection [8]. A transcriptomic study by Lohmaneeratana et al., examines the increased expression of DEGs associated with the phenylpropanoid, plant hormones, calcium, and mitogen activated protein kinase (MAPK) cascades that positively promoted defense mechanisms against *Candidatus sacchari* (a causal agent of sugarcane white leaf disease) colonization by triggering calcium and accumulation in response to plant immunity [10]. Furthermore, Zhang et al., found that the *pglA* gene from *Leifsonia xyli* subsp. *xyli* (causing ratoon stunting disease) interacted with the host protein SoSnRK1β1 (an important protein in the ABA pathway), suggesting ABA pathway is likely triggered host resistance to this pathogen [11]. Khan and colleagues also conducted the comparative transcriptomic analysis to examine the functions of sucrose regulatory genes in high- and low-sucrose sister clones of sugarcane. According to this study, the enhanced expression of sucrose phosphate synthase (SPS) in the high-sucrose clone compared to the low sucrose-clone implies that SPS is mostly responsible for the greater accumulation of sucrose [12].

Recently, molecular breeding using transgenic and/or gene-edition technologies have contributed significantly to sugarcane improvement, but a simple, robust, and efficient transgenics is a crucial step in molecular breeding [2]. Thus, Wang et al., established an efficient sugarcane transformation system via herbicide-resistant CP4-EPSPS gene selection, as evidenced by high yield of transgenic lines from calli, an improved selection method, a higher transformation efficiency, and enhanced resistance against the Roundup herbicide in transgenic sugarcane plants [13]. Additionally, the illustrated mechanisms of sugarcane defense against environmental stressors provide important clues for sugarcane genetic improvement. Wei et al., showed that reactive oxygen species (ROS) production–scavenging system is involved in sugarcane response to *Xanthomonas albilineans* (*Xa*, causing leaf scald) infection under drought stress. This study also showed that, whereas drought stress had no discernible impact on resistant cultivars, it greatly enhanced the incidence of leaf scald and *Xa* populations in susceptible cultivars. Under coupled *Xa* infection and drought stress, sugarcane (particularly susceptible cultivars) showed a weakened defense response via the ROS-producing and scavenging system as compared to *Xa* infection alone [14].

In summary, this special issue illustrated current research findings about interaction between sugarcane and environmental stressors. Future research in this area will be driven by the integration of multi-omics tools (phenomics, transcriptomes, proteomics, and epigenetics) and utilization of modern biotechnological tools for understanding the trade-off between growth and defense in sugarcane, which will provide clues for the development of breeding strategies to generate elite cultivars for meeting rising global sugar and biofuel demand.

Author Contributions: Conceptualization, writing—original draft preparation, review, and editing by S.-J.G. and T.J. All authors have read and agreed to the published version of the manuscript.

Funding: This work was supported by the China Agriculture Research System of MOF and MARA (grant no. CARS-17).

References

1. Javed, T.; Shabbir, R.; Ali, A.; Afzal, I.; Zahoor, U.; Cao, S. J. Transcription Factors in Plant Stress Responses. Challenges and Potential for Sugarcane Improvement. *Plants* **2020**, *9*, 491. [CrossRef] [PubMed]
2. Kumar, T.; Wang, J.-G.; Xu, C.-H.; Lu, X.; Mao, J.; Lin, X.-Q.; Kong, C.-Y.; Li, C.-J.; Li, X.-J.; Tian, C.-Y.; et al. Genetic Engineering for Enhancing Sugarcane Tolerance to Biotic and Abiotic Stresses. *Plants* **2024**, *13*, 1739. [CrossRef]

3. Manzoor, M.; Khan, M.Z.; Ahmad, S.; Alqahtani, M.D.; Shabaan, M.; Sarwar, S.; Hameed, M.A.; Zulfiqar, U.; Hussain, S.; Ali, M.F.; et al. Optimizing Sugarcane Growth, Yield, and Quality in Different Ecological Zones and Irrigation Sources Amidst Environmental Stressors. *Plants* **2023**, *12*, 3526. [CrossRef] [PubMed]

4. Sajid, M.; Amjid, M.; Munir, H.; Ahmad, M.; Zulfiqar, U.; Ali, M.F.; Abul Farah, M.; Ahmed, M.A.A.; Artyszak, A. Comparative Analysis of Growth and Physiological Responses of Sugarcane Elite Genotypes to Water Stress and Sandy Loam Soils. *Plants* **2023**, *12*, 2759. [CrossRef] [PubMed]

5. Zhao, Y.; Yu, L.-X.; Ai, J.; Zhang, Z.-F.; Deng, J.; Zhang, Y.-B. Climate Variations in the Low-Latitude Plateau Contribute to Different Sugarcane (*Saccharum* spp.) Yields and Sugar Contents in China. *Plants* **2023**, *12*, 2712. [CrossRef] [PubMed]

6. Wu, Q.; Zhou, W.; Chen, D.; Tian, J.; Ao, J. Biochar Mitigates the Negative Effects of Microplastics on Sugarcane Growth by Altering Soil Nutrients and Microbial Community Structure and Function. *Plants* **2024**, *13*, 83. [CrossRef] [PubMed]

7. Xu, H.-M.; He, E.-Q.; Yang, Z.-L.; Bi, Z.-W.; Bao, W.-Q.; Sun, S.-R.; Lu, J.-J.; Gao, S.-J. Phylogeny and Genetic Divergence among Sorghum Mosaic Virus Isolates Infecting Sugarcane. *Plants* **2023**, *12*, 3759. [CrossRef] [PubMed]

8. Zhou, G.; Shabbir, R.; Sun, Z.; Chang, Y.; Liu, X.; Chen, P. Transcriptomic Analysis Reveals Candidate Genes in Response to Sorghum Mosaic Virus and Salicylic Acid in Sugarcane. *Plants* **2024**, *13*, 234. [CrossRef] [PubMed]

9. Shan, H.; Chen, D.; Zhang, R.; Wang, X.; Li, J.; Wang, C.; Li, Y.; Huang, Y. Relationship between Sugarcane *eIF4E* Gene and Resistance against Sugarcane Streak Mosaic Virus. *Plants* **2023**, *12*, 2805. [CrossRef] [PubMed]

10. Lohmaneeratana, K.; Leetanasaksakul, K.; Thamchaipenet, A. Transcriptomic Profiling of Sugarcane White Leaf (SCWL) Canes during Maturation Phase. *Plants* **2024**, *13*, 1551. [CrossRef] [PubMed]

11. Zhang, X.-Q.; Liang, Y.-J.; Zhang, B.-Q.; Yan, M.-X.; Wang, Z.-P.; Huang, D.-M.; Huang, Y.-X.; Lei, J.-C.; Song, X.-P.; Huang, D.-L. Screening of Sugarcane Proteins Associated with Defense against *Leifsonia xyli* subsp. *xyli*, Agent of Ratoon Stunting Disease. *Plants* **2024**, *13*, 448. [CrossRef] [PubMed]

12. Khan, Q.; Qin, Y.; Guo, D.-J.; Huang, Y.-Y.; Yang, L.-T.; Liang, Q.; Song, X.-P.; Xing, Y.-X.; Li, Y.-R. Comparative Analysis of Sucrose-Regulatory Genes in High- and Low-Sucrose Sister Clones of Sugarcane. *Plants* **2024**, *13*, 707. [CrossRef] [PubMed]

13. Wang, W.; Javed, T.; Shen, L.; Sun, T.; Yang, B.; Zhang, S. Establishment of an Efficient Sugarcane Transformation System via Herbicide-Resistant CP4-EPSPS Gene Selection. *Plants* **2024**, *13*, 852. [CrossRef] [PubMed]

14. Wei, Y.-S.; Zhao, J.-Y.; Javed, T.; Ali, A.; Huang, M.-T.; Fu, H.-Y.; Zhang, H.-L.; Gao, S.-J. Insights into Reactive Oxygen Species Production-Scavenging System Involved in Sugarcane Response to *Xanthomonas albilineans* Infection under Drought Stress. *Plants* **2024**, *13*, 862. [CrossRef] [PubMed]

Article

Optimizing Sugarcane Growth, Yield, and Quality in Different Ecological Zones and Irrigation Sources Amidst Environmental Stressors

Muhammad Manzoor [1], Muhammad Zameer Khan [1], Sagheer Ahmad [2], Mashael Daghash Alqahtani [3,*], Muhammad Shabaan [1], Sair Sarwar [1], Muhammad Asad Hameed [1], Usman Zulfiqar [4,*], Sadam Hussain [5], Muhammad Fraz Ali [5], Muhammad Ahmad [6] and Fasih Ullah Haider [7,8]

[1] Land Resources Research Institute, National Agricultural Research Centre, Islamabad 44000, Pakistan; manzoorm77@gmail.com (M.M.); zameerahmar@gmail.com (M.Z.K.); mshabaan@parc.gov.pk (M.S.); sardarsair@hotmail.com (S.S.); malik.asadhameed@gmail.com (M.A.H.)

[2] Pakistan Agricultural Research Council, Islamabad 45500, Pakistan; sagheersc@hotmail.com

[3] Department of Biology, College of Science, Princess Nourah bint Abdulrahman University, P.O. Box 84428, Riyadh 11671, Saudi Arabia

[4] Department of Agronomy, Faculty of Agriculture and Environment, The Islamia University of Bahawalpur, Bahawalpur 63100, Pakistan

[5] College of Agronomy, Northwest A&F University, Xianyang 712100, China; ch.sadam423@gmail.com (S.H.); frazali15@gmail.com (M.F.A.)

[6] Department of Agronomy, University of Agriculture Faisalabad, Faisalabad 38040, Pakistan; ahmadbajwa516@gmail.com

[7] Key Laboratory of Vegetation Restoration and Management of Degraded Ecosystems, South China Botanical Garden, Chinese Academy of Sciences, Guangzhou 510650, China; haider281@scbg.ac.cn

[8] University of Chinese Academy of Sciences, Beijing 100039, China

* Correspondence: mdalqahtani@pnu.edu.sa (M.D.A.); usman.zulfiqar@iub.edu.pk (U.Z.)

Citation: Manzoor, M.; Khan, M.Z.; Ahmad, S.; Alqahtani, M.D.; Shabaan, M.; Sarwar, S.; Hameed, M.A.; Zulfiqar, U.; Hussain, S.; Ali, M.F.; et al. Optimizing Sugarcane Growth, Yield, and Quality in Different Ecological Zones and Irrigation Sources Amidst Environmental Stressors. *Plants* **2023**, *12*, 3526. https://doi.org/10.3390/plants12203526

Academic Editor: San-Ji Gao

Received: 16 August 2023
Revised: 26 September 2023
Accepted: 27 September 2023
Published: 11 October 2023

Abstract: The imbalanced use of fertilizers and irrigation water, particularly supplied from groundwater, has adversely affected crop yield and harvest quality in sugarcane (*Saccharum officinarum* L.). In this experiment, we evaluated the impact of potassium (K) and micronutrients [viz. Zinc (Zn), Iron (Fe), and Boron (B)] application and irrigation water from two sources, viz. canal, and tube well water on sugarcane growth, yield, and cane quality under field trails. Water samples from Mardan (canal water) and Rahim Yar Khan (tube well water) were analyzed for chemical and nutritional attributes. The results revealed that tube well water's electrical conductivity (EC) was three-fold that of canal water. Based on the EC and total dissolved salts (TDS), 83.33% of the samples were suitable for irrigation, while the sodium adsorption ratio (SAR) indicated only a 4.76% fit and a 35.71% marginal fit compared with canal water. Furthermore, the application of K along with B, Fe, and Zn had led to a significant increase in cane height (12.8%, 9.8%, and 10.6%), cane girth (15.8%, 15.6%, and 11.6%), cane yield (13.7%, 12.3%, and 11.5%), brix contents (14%, 12.2%, and 13%), polarity (15.4%, 1.4%, and 14%), and sugar recovery (7.3%, 5.9%, and 6%) in the tube well irrigation system. For the canal water system, B, Fe, and Zn increased cane height by 15.3%, 13.42%, and 11.6%, cane girth by 13.9%, 9.9%, and 6.5%, cane yield by 42.9%, 43.5%, and 42%, brix content by 10.9%, 7.7%, and 8%, polarity by 33.4%, 28%, and 30%, and sugar recovery by 4.0%, 3.9%, and 2.0%, respectively, compared with sole NPK application. In conclusion, the utilization of tube well water in combination with canal water has shown better results in terms of yield and quality compared with the sole application of canal water. In addition, the combined application of K and B significantly improved sugarcane yields compared with Zn and Fe, even with marginally suitable irrigation water.

Keywords: sugarcane; water quality; balanced fertilizer use; climatic conditions; environmental stressors

---

1. Introduction

Water scarcity and quality have become serious issues for normal crop growth and sustainable agricultural development, which must be addressed through alternate water sources while ensuring soil fertility [1,2]. The long-term and excessive use of tube well water for irrigation, along with low precipitation and high evaporation, have increased salinity problems, resulting in low soil fertility and crop growth [2,3]. Applying highly saline irrigation water negatively impacts the soil-water-plant relationship, thereby restricting plants' normal physiological functions [4,5]. Furthermore, it causes osmotic effects, water scarcity, nutritional imbalances, and oxidative stress, which also hinder sugarcane growth, leaf surface expansion, and metabolic activities [6]. Collectively, low water quality is the major limiting factor for sugarcane crop production [7,8]. In Pakistan, a severe imbalance exists between nutrients application through fertilization and their use efficiency [9], resulting in stagnant crop yield and a deteriorated quality of harvested produce. Moreover, the generally adopted practice of extracting nutrients from soil through continuous cropping system has progressively degraded soil quality [10].

Sugarcane (*Saccharum officinarum* L.) is a major crop cultivated worldwide owing to its enormous dietary and commercial applications. It is one of the most imperative industrial crops based on its extensive production in multiple tropical and sub-tropical regions on the globe [11]. The sugarcane industry contributes nearly 80% of total sugar [12]. Hence, about 28.3 million hectares in 90 countries are cultivated with sugarcane, which produces about 1.69 billion tons globally [13]. Sugarcane requires tropical and subtropical climates for normal growth and production [14]. Different factors perform pivotal roles in ensuring optimum yield and sugar production, these including climatic conditions [15], varieties [16], agronomic management practices [17], and soil fertility status [5]. Metrological factors such as temperature, rainfall, humidity, and sunshine significantly impacted sugarcane yield and sugar recovery. It is found that high temperature favors excellent growth and yield potential, whereas a semi-humid climate produces better sucrose contents [18].

Potassium offers a significant role in cane production as it enhances the crop's resilience to intermittent drought conditions which commonly observed in sugarcane cultivated regions [19]. It also plays a pivotal role in controlling the stomatal aperture, thereby upholding turgor pressure during unfavorable moisture conditions. Under salinity stress, K application helps in preserving ion homeostasis and overseeing osmotic equilibrium [20]. Application of K has been reported to increase plant growth (by 19%), and count of millable cane attributable to the robust tillers developed in ratoon cane. The provision of sufficient K mitigates moisture stress [21] and salt stress in sugarcane [22]. It has been documented that distinct carrier proteins facilitate the transport processes, ensuring the crucial Na^+/K^+ equilibrium within cells, which is essential for plant survival in saline environments. Within plant cells, the vacuole serves as a reservoir of K^+, performing a vital role in upholding cellular turgor [23].

The impact of K and micronutrient fertilization, particularly Fe, Zn, manganese (Mn), and B, on sugarcane is very significant and area-specific [24]. Incorporating micronutrients into sugarcane cultivation enhances the yield and plays a vital role in elevating the quality of juice. Applying Zn and Fe minerals also becomes imperative for achieving superior cane production over the long term and superior juice characteristics [25,26]. Similarly, sugar transport and cell wall development heavily rely on the presence of B in the plant's system [27]. In this study, we hypothesized that applying K and different micronutrients with canal water irrigation could improve sugarcane growth, yield, and quality attributes. Despite a known fact regarding mixing tube well and canal waters for managing salinity, limited literature evidence exists about the combined application of K and different micronutrients under various irrigation schemes. In the current study, we evaluated the combined efficacy of K and other micronutrients in improving sugarcane growth, yield, and quality. Considering the importance of irrigation water with K and micronutrients on sugarcane growth, yield, and quality, the current study was conducted to investigate irrigation water characteristics, i.e., canal water from the Sawat River and tube well water from Rahim Yar Khan with varying climatic conditions.

2. Results

2.1. Characteristics of Irrigation Water and Nutrient Status

Evaluation of irrigation water from two sources showed a diverse variation regarding chemical characteristics and nutrient supply to crops. Results regarding water characteristics revealed that the EC of tube well water ranged from 0.218 to 5.34 dSm^{-1}, with an average of 1.09 dSm^{-1}, while canal water's EC varied between 0.08 to 1.08 dSm^{-1}, presenting an average of 0.24 dSm^{-1} showing high variability, i.e., 84% and 107%, respectively (Table 1). The EC of tube well water was almost three times higher than canal water. Based on EC, 83.33% of the total samples collected from tube well (Rahim Yar Khan) were suitable for sugarcane irrigation purposes, whereas 14.29% were marginally fit and 2.38% were unfit. In comparison, 100% of water samples from canal water (Mardan and Charsadda) were suitable for sugarcane irrigation. In the case of soluble salts, Na in water ranged from 27 to 398 mg L^{-1} for tube well sources compared with canal water (1.47 to 19.5 mg L^{-1}). Similarly, Ca^{2+}, Mg and $CO_3 + HCO_3$ contents varied between 74.56–885.4 and 61–406.9 mg L^{-1} in tube well water compared with canal water (42.6–160 and 40.9–447) with the mean values of 190.74–187.34 mg L^{-1} and 94.32–131.4 mg L^{-1} in tube well and canal water, respectively. Chloride contents in both water sources depicted that tube well water contained a high concentration (60–720 mg L^{-1}) compared with canal water (0.67–180 mg L^{-1}). Analysis of these water samples revealed that N, P, and K were 0.53, 0.88, and 25.05 mg L^{-1} in tube well water, whereas 1.36, 0.71, and 56.85 mg L^{-1} in canal water, respectively, indicating the suitability of canal water for sugarcane cultivation when compared with tube well water concerning soil health, crop growth, yield, and quality.

Table 1. Chemical characteristics and nutrient status of tube well and canal water surveyed from Districts Rahim Yar Khan (Punjab) and Mardan/Charsadda (Khyber Pakhtunkhwa).

Characters	Mean		Range		SD		SEM		CV	
	Tubewell	Canal	Tubewell	Canal	Tubewell	Canal	Tubewell	Canal	Tubewell	Canal
EC	1.09	0.24	0.218–5.34	0.08–1.08	0.912	0.256	0.14	0.036	84.14	107.3
Na	174.5	10.49	27–398	1.47–19.5	112.01	4.83	17.29	0.669	64.20	46.03
Ca + Mg	190.74	94.32	74.56–885.4	42.6–160	160.9	34.42	24.83	4.77	84.37	36.49
CO_3^+ HCO_3^-	187.34	131.4	61–406.9	40.9–447	77.44	92.1	11.95	12.76	41.34	70.01
Cl	172.38	78.64	60–720	0.67–180	132.5	33.41	20.45	4.632	76.87	42.48
TDS	714.4	153.2	139.5–4272	48.6–693	686.9	164.4	106	22.79	96.16	107.3
NO$_3$-N	0.532	1.36	0.39–0.94	0.33–3.39	0.148	0.893	0.023	0.123	27.9	65.51
P	0.881	0.71	0.58–2.79	0.37–1.26	0.361	0.167	0.056	0.023	40.9	23.8
K	25.05	56.85	13–46	29–87	7.37	13.8	1.14	1.91	29.4	24.3

Note: EC = Electrical conductivity, Na = Sodium, Ca + Mg = Calcium and magnesium, $CO_3 + HCO_3^-$ = Carbonates and bicarbonates, Cl = Chlorides, TDS = Total Issolved Salts, NO_3-N = Nitrate nitrogen, P = Phosphorus, K = Potassium

2.2. Growth and Yield of Sugarcane

It was also reported that applying soil K either alone or in combination with micronutrients, such as Zn, Fe, and B, led to a statistical significant ($p \leq 0.05$) inrement in sugarcane height, girth, and yield compared with conventional farming practices (Table 2). Applying K significantly increased cane height (8.4% and 9.74%) and girth (9.74% and 9.39%). Soil application of K and foliar application of B further increased cane height by 12.81% and 13.42% at Rahim Yar Khan and Mardan sites over conventional farming practices. Similarly, upon integrating K with micronutrients, both sites increased cane girth by 13.42% and 14.32%, respectively. Similarly, K application increased yield by 10.38% and 40.79% over conventional practices for Rahim Yar Khan and Mardan sites, respectively. The combined application of micronutrients and K increased cane yield by 12.49% and 42.79% compared with conventional farming practices. Soil applied K with foliar application of B showed maximum (197,767 and 198,826) millable canes (Figure 1) compared with sole NP (183,824 and 187,390) with resultant increases of 7.6% and 5.7% at both sites, respectively. The number of internodes in response to K application were improved by 12.2% and 9.3%, whereas the B application increased by 15.2% and 15.4% over conventional farming practices at Rahim Yar Khan and Mardan sites, respectively.

Table 2. Growth and yield of Sugarcane in response to potash (K) and micronutrient application at two locations with different irrigation water sources.

	Cane Height (cm)			Cane Girth (cm)			Cane Yield (Tonnes ha^{-1})		
	RYK	Mardan	Mean	RYK	Mardan	Mean	RYK	Mardan	Mean
NP	299.15 ab	227.03 d	263.09 B	7.7 ab	6.77 d	7.235 B	93.54 ab	59.86 c	76.71 B
NPK	324.27 a	249.15 cd	286.71 A	8.43 a	7.37 cd	7.90 A	103.25 a	84.32 b	93.79 A
NPK + Zn	330.82 a	253.28 bc	292.05 A	8.60 a	7.21 bc	7.905 A	104.29 a	85.03 b	94.66 A
NPK + Fe	328.49 a	257.49 bc	292.99 A	8.91 a	7.44 bc	8.175 A	105.07 a	85.88 b	95.47 A
NPK + B	337.48 a	261.71 bc	299.595 A	8.93 a	7.71 bc	8.32 A	106.32 a	85.53 b	95.92 A
LSD ($p \leq 0.05$)	48.25		28.78	1.89		1.13	14.21		8.48
Site effect	324 A	244 B	12.66	8.51	8.09	NS	102.49 A	80.12 B	3.73

Uppercase letters indicate grand means; lowercase letters indicate individual means. LSD, least significant difference at 5% level of significance; N, nitrogen; P, phosphorus; K, potassium; Zn, zinc; Fe, iron; B, boron.

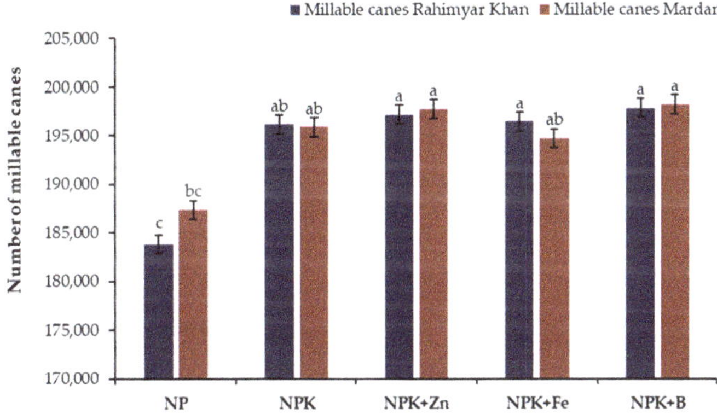

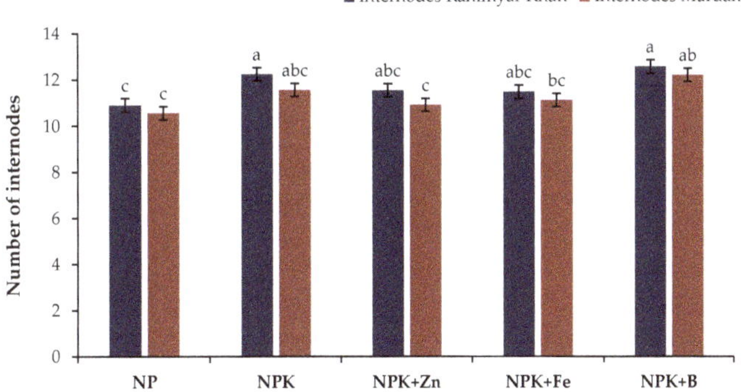

Figure 1. Effect of potash (K) and micronutrients on millable canes and number of internodes. Different lowercase letters indicate significant difference among nutrients application. N, nitrogen; P, phosphorus; K, potassium; Zn, zinc; Fe, iron; B, boron.

2.3. Quality Characteristics

Quality attributes of sugarcane in response to different sources of irrigation water along with K and micronutrient application showed that brix contents, polarity, and sugar recovery were significantly ($p \leq 0.05$) improved (Table 3). Brix contents in conventional farming practices were 15.75% and 19.24% for both sites. Soil application of K with Zn increased brix content by 20.78% with the canal water irrigation at the Mardan site compared with 17.67% at the Rahim Yar Khan Site irrigated with tube well and canal water in combination (Figure 2). In addition, K and B application increased brix content by 12.4% over conventional farming practices. Similarly, adding K with B at the Mardan site yielded the highest (18.67%) polarity by canal water irrigation. In sugar recovery, conventional farming practices resulted in the lowest (10.28%), which increased by 13.4% after adding K with Fe and B (Figure 1). Site effects revealed that sugar recovery was 17.5% higher at the Mardan site irrigated with canal water than Rahim Yar Khan Site irrigated with canal water and tube well water combined. For sugar production, K application in combination with B increased yield by 22% and 49%, respectively, at the Rahim Yar Khan and Mardan sites, with relative increases of 22% and 49% over conventional farming practices (Table 2). The response to K and micronutrients was inverted in the case of fiber contents where soil treated with NP as conventional farming practices showed the highest fiber contents, i.e., 12.29% and 12.19%, which reduced to 12.03% and 11.67% in soil applied K with Fe at Rahim Yar Khan and Mardan sites, respectively (Figure 3). Thus, results revealed that using K and micronutrients reduced the fiber contents and increased sugar production. Similarly, canal water decreased fiber contents and improved sugar production compared with tube well and canal water combined. It was found that chlorophyll contents were increased by 31.4 to 42.5% and 33.8 to 43.8% at both sites, respectively, in response to combined application of K and Fe, over conventional farming practices (Figure 3). Maximum juice purity (84.9%) was recorded for Rahim Yar Khan site followed by Mardan site with soil-applied K along with foliar Zn application and sole K application, respectively, where it was recorded that canal water irrigation was better in term of juice purity and sugar production than canal and tube well water (Figure 3).

Table 3. Quality characteristics of sugarcane in response to potash (K) and micronutrients application at two locations with different irrigation water sources.

	Brix (%)			Polarity (%)			Sugar Recovery (%)		
	RYK	Mardan	Mean	RYK	Mardan	Mean	RYK	Mardan	Mean
NP	15.75 b	19.24 ab	17.50 B	12.86 b	14.00 ab	13.43	10.28 AB	12.91 B	11.60 B
NPK	17.49 ab	20.65 ab	19.07 AB	14.63 ab	18.39 ab	16.51	10.83 AB	13.13 A	11.98 AB
NPK + Zn	17.79 ab	20.78 ab	19.29 AB	14.68 ab	18.20 ab	16.44	10.94 AB	13.16 AB	12.05 AB
NPK + Fe	17.67 ab	20.72 ab	19.20 AB	14.66 ab	17.97 ab	16.31	10.88 AB	13.41 A	12.15 AB
NPK + B	17.99 ab	21.33 a	19.66 A	14.84 a	18.67 a	16.75	11.03 AB	13.42 A	12.23 A
LSD ($p \leq$ 0.05)	5.99		NS	5.61		NS	2.93		1.75
Site effect	17.34 B	20.75 A	1.57	14.33 B	17.44 A	1.47	10.85 B	12.75 A	0.77

Uppercase letters indicate grand means; lowercase letters indicate individual means. LSD, least significant difference at 5% level of significance; N, nitrogen; P, phosphorus; K, potassium; Zn, zinc; Fe, iron; B, boron.

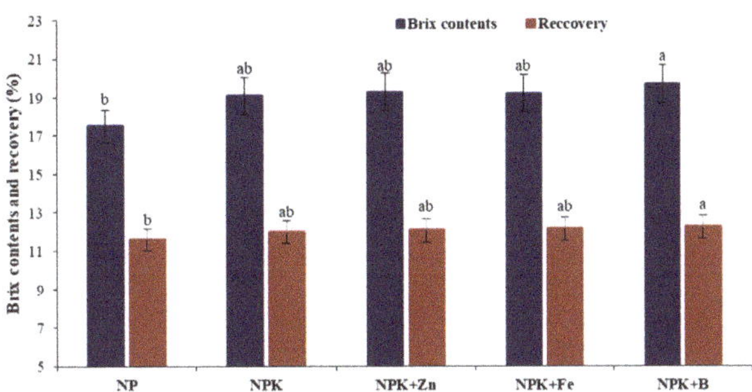

Figure 2. Effect of potash (K) and micronutrients on brix contents and sugar recovery average over two sites. Different lowercase letters indicate significant difference among nutrients application. N, nitrogen; P, phosphorus; K, potassium; Zn, zinc; Fe, iron; B, boron.

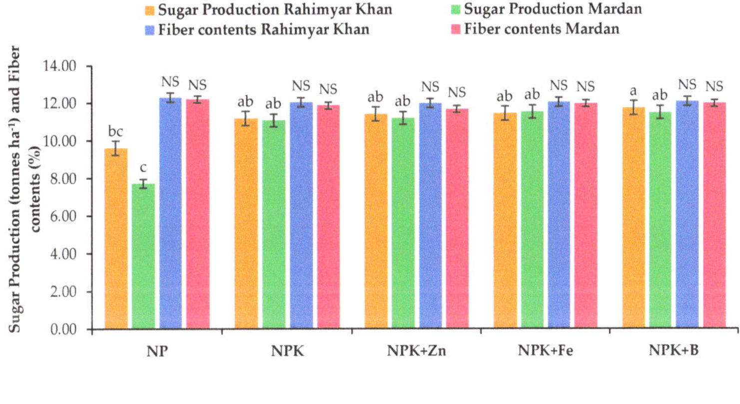

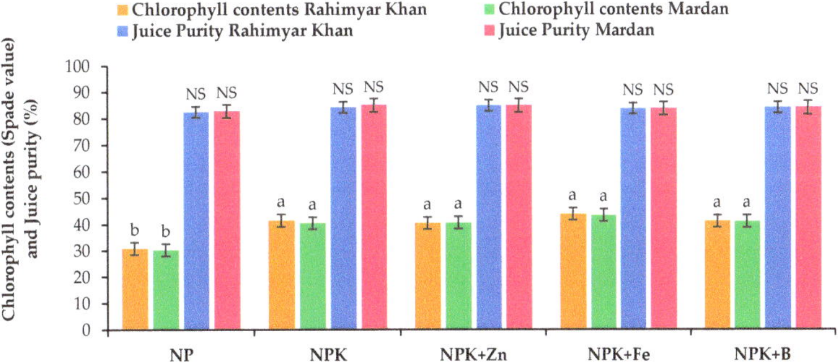

Figure 3. Effect of potash (K) and micronutrients on sugar production, fiber contents, chlorophyll contents, and juice purity of sugarcane at the Rahim Yar Khan and Mardan sites. Different lowercase letters indicate significant difference among nutrients application. N, nitrogen; P, phosphorus; K, potassium; Zn, zinc; Fe, iron; B, boron.

2.4. Potash and Micronutrients

Soil applied K and micronutrients showed a significant impact on K contents at both sites, resulting in a significant increase in K content in leaf blade (15.3%) and leaf sheath of sugarcane (2.7%) over conventional farming practices (Figure 4). The values were further increased by 16.3% and 4.0% after adding Fe to K in both plant tissues at the Rahim Yar Khan site. For Mardan site, the application of K alone showed a maximum increase of 41.6%

and 13.5% in leaf blade and leaf sheath, respectively. Sugarcane micronutrient contents such as Zn, Fe, and B in response to soil K application were remarkably improved by 26.3%, 59%, and 14.5% at the Rahim Yar Khan site, whereas at the Mardan site, values were increased by 123%, 104%, and 16.5% in leaf blade as well as leaf sheath, respectively, as compared with conventional farming practices (Figure 4). The combined application of K with Zn resulted in 34% and 37% increase at Rahim Yar Khan, while at the Mardan site, 144% and 160% in leaf blade and leaf sheath Zn contents, respectively. Similarly, Fe contents were increased by 70% and 50% at Rahim Yar Khan and 21% and 22.5% at the Mardan site in both plant parts. The combined application of K and B resulted in 247% and 152% increase in B content (Rahim Yar Khan) and 86% and 167% (Mardan) in leaf blade and leaf sheath of sugarcane, respectively, compared with conventional farming practices. Thus, it was found that soil application of sole K effectively improved K uptake, which, combined with Zn, Fe, and B, further improved the respective micronutrient uptake.

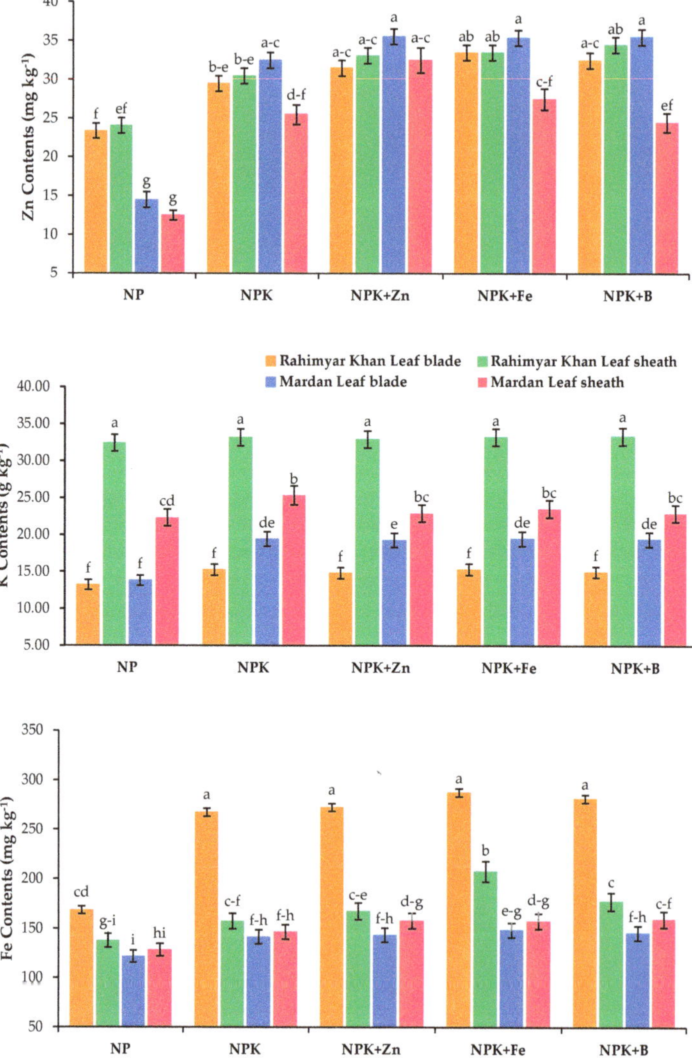

Figure 4. *Cont.*

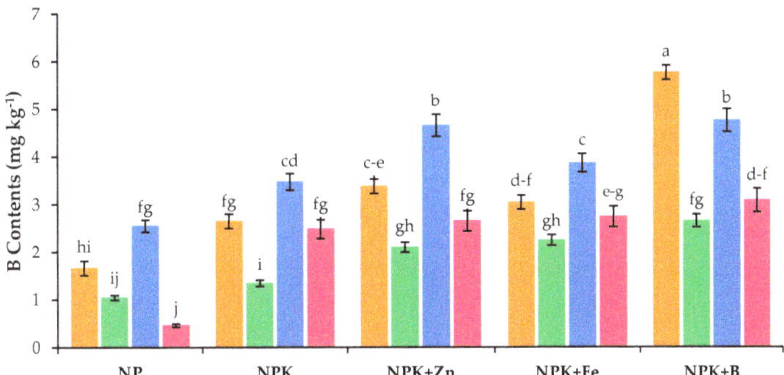

Figure 4. Effect of potash (K) and micronutrients on K, Zn, Fe, and B contents of sugarcane leaf blade and leaf sheath at Rahim Yar Khan and Mardan sites. Different lowercase letters indicate significant difference among nutrients application. N, nitrogen; P, phosphorus; K, potassium; Zn, zinc; Fe, iron; B, boron.

3. Discussion

Irrigation water is crucial in influencing soil health and crop quality and is categorized into suitable, marginal, and unsuitable classes based on its EC and TDS [27]. Our study showed that EC and TDS levels of tube well water were nearly three times higher than canal water. Consequently, tube well water was deemed unsuitable for irrigation purposes. These findings are similar with Feng et al. [28]. Furthermore, previous research has also indicated that using saline water for irrigation can lead to soil salinity and increased pH [27]. Similarly, our investigations revealed that the SAR of water samples from tube wells was substantially higher than canal water, rendering it unsuitable for irrigation purposes. Similar results were previously reported by Kumar et al. [27] who stated that water samples from tube wells water source exhibited substantially higher SAR. Irrigation with saline water that exceeds threshold levels for EC, SAR, and TDS has detrimental effects on plant growth and development, resulting in reduced crop yields. Previous studies have demonstrated that K is vital in supporting plants during stress by maintaining ion equilibrium and regulating osmotic balance [29–32]. Our study further recorded that K and micronutrient applications improved sugarcane growth and yield under canal water irrigation schemes compared with conventional farming practices.

Extreme environmental events associated with climate change are becoming more frequent worldwide and could push roughly one-third of global food crop and livestock production beyond the safe climatic space by the end of this century [33–44]. Salinity stress harms photosynthesis and the translocation of photoassimilates in sugarcane. Pabuayon et al. [45] stated that salinity negatively affects growth and development. However, K plays a crucial role in regulating stomatal aperture, thereby maintaining turgor pressure even under less favorable soil moisture conditions. It was worth noticing that even lower cane yield and higher sugar contents, as observed at the Mardan site, produced sugar at par with the higher cane yield at the Rahim Yar Khan site. Similar reports have highlighted that sustainable sugar production with improved sucrose content is achieved through suitable irrigation water and appropriate fertilizer management [46]. As reported earlier, changes in temperature and rainfall during critical stages resulted in a significant variation in crop yield and sugar recovery in three provinces of Pakistan, ultimately affecting overall production [47]. Jaiswal et al. [48] described that K application to sugarcane resulted in a 19% increase in overall plant growth compared with soil without K fertilization. Thus, adequate K supply to sugarcane reduces moisture stress, induces salt tolerance to sugarcane [16,21], and produces higher stalk and cane yields [47,48].

Furthermore, K application reduces N requirements through better N metabolism and transportation of nitrate to shoots [49,50]. A foliar spray of micronutrients combined with soil-applied K also improved cane yield (12.49% and 42.79% over conventional farming practices) but also yielded the higher brix contents and sugar recovery [51,52]. Similar was also reported for other crops [53]. Our results were supported by Ismail et al. [54], that Zn and Fe applied as foliar to the sugarcane effectively improved the growth characteristics of sugarcane, which ultimately increased its yield. Ghaffar et al. [25] described that sugarcane growth and yield traits were significantly improved due to applied Zn and Fe. Zhao et al. [55] reported that K application improved sugarcane growth under saline conditions. Franco et al. [56] stated that Zn fertilization enhanced stalk weight and cane quality. The brix contents in soil-applied K and B in combination with canal water showed high brix contents and sugar recovery. Although Watanabe et al. [57] correlated elevated K levels with a decrease in sucrose content. Zhao et al. [55] supported our results that K supplied either alone or in combination with Zn resulted in improved brix and sugar recovery. Potash with micronutrients is a balanced and complete package of sugarcane nutrition as, according to our findings. Thus, it was found that the soil application of K alone effectively improved K uptake, which, combined with Zn, Fe, and B, further improved the respective micronutrient uptake. The better uptake of micronutrients contributed to the heightened activity of sucrose-synthesizing enzymes, which also improved yield and sucrose content [57]. Iron is an essential micronutrient exhibiting a synergistic interaction with K [58,59]. Zinc is a critical component of numerous enzymes and proteins. As reported in the previous findings, K, Zn, and B uptake in sugarcane leaf tissues has effectively been improved by soil applied K and micronutrient integration [59].

4. Materials and Methods

A comprehensive study was conducted to survey sugarcane-growing districts such as Rahim Yar Khan in Punjab province and Mardan/Charsadda in Khyber Pakhtunkhwa (KPK) province. These districts are significant sugarcane-growing areas in their respective provinces and are known for utilizing different water sources for sugarcane irrigation. Subsequently, two specific fields were identified within these selected districts for conducting field experiments. These fields were situated on farmer-owned land and received irrigation water from distinct sources. In Mardan, the irrigation source was the Swat River canal, while in Rahim Yar Khan, the fields received a combination of canal and tube well water for irrigation, with alternating irrigation timings.

4.1. Collection of Water Samples and Characteristics Studied

Forty-two union councils were selected from District Rahim Yar Khan, and 52 were chosen from Mardan and Charsadda districts. To ensure the integrity of the water samples, plastic bottles were used, which were thoroughly rinsed five times with tap water, followed by distilled water. For tube well water, samples were collected after allowing the well to run for 15 min to stabilize. These collected samples were then transported to the Soil Fertility Laboratory of the Soil-Plant Nutrient Program at NARC, Islamabad. Here, the samples underwent processing and analysis for various parameters, including pH, electrical conductivity (EC), and levels of N, P, K, Ca^{2+}, Mg, CO_3, HCO_3, Na, B, and Cl^-, by following standard protocols [49]. Data obtained from these analyses were utilized to assess the nutrient status and determine the extent of different nutrient-related disorders.

4.2. Experimental Sites and Treatments

Field experiments were conducted during the 2021–2022 seasons at two selected sites on farmer fields where sugarcane was cultivated. In February 2021, sugarcane plantation was prepared before pre-plantation soil sampling was conducted to determine soil physicochemical traits and fertility status (Table 4). The treatment combinations were: T1 = local farmer practice; N and P application (200 and 100 kg ha^{-1}), T2 = N + P + K (200 + 100 + 150 kg ha^{-1}), T3 = N + P + K (200 + 100 + 150 kg ha^{-1}) + Zn (0.1%) 3 Foliar Spray, T4 = N + P + K (200 + 100 + 150 kg ha^{-1}) + Fe (0.1%) − 3 Foliar Spray and T5 = N + P + K (200 + 100 + 150 kg ha^{-1}) + B (0.05%) − 3 Foliar Spray. During the seeding process, fertilizers P and K were applied, while N was administered in three separate doses: during seeding, tillering, and knot formation stages. Micronutrients (Fe, Zn, and B) were applied through three foliar sprays, beginning at a height of

92 cm and continuing at one-month intervals after the first application. The experimental units were arranged using a randomized complete block design (RCBD) comprising five treatments, each with three replications. The plot size was maintained at 70×10 m for each treatment and replication. Planting was conducted at a spacing of 0.726 m (R $\times$ R) distance with four budded sets placed in a single row. The sugarcane variety 'CP-77400', approved uniformly for both provinces (Punjab and Khyber-Pakhtunkhwa), was selected as a test cultivar due to its adaptability and uniformity in agronomic, physiological, yield, and quality traits. One month after the completion of fertilizer application, chlorophyll contents were measured using a chlorophyll meter (SPAD-502), and plant tissue samples were collected in August 2022 to assess nutrient status at the grand growth stage. Tissue sampling involved collecting ten samples from each replication of one treatment, taken from the middle three rows. The fifth and sixth leaves from the top of the cane were collected, and leaf sheaths and leaf blades were separated, washed, air-dried, and oven-dried at 65 °C for 24 h. Data regarding growth and yield components, including cane height, cane girth, number of internodes, millable canes, and cane quality traits such as brix contents, polarity, juice purity, fiber contents, and sugar recovery, were recorded at the end of December upon weighing the fresh millable canes harvested from three inner rows of each plot. Brix contents (%), polarity (%), juice purity (%), and fiber contents (%) were quantified in a sugar mill cane quality laboratory. Sugar recovery (%) was calculated from brix, purity, polarity, and fiber contents, whereas sugar production (T ha^{-1}) was determined by multiplying cane yield by sugar recovery.

Table 4. Soil physico-chemical characteristics and nutrient status of selected sites.

Location	Depth	Texture	pH	EC	OM	NO$_3$-N	P	K	Fe	Zn	B
	(inches)		1:1	dSm^{-1}	(%)			(mg kg^{-1})			
Rahim Yar Khan	0–30	Silt loam	8.2	0.512	0.7	3.2	0.78	99	6.70	0.40	1.05
	30–45	-	8.5	0.410	0.4	4.6	0.77	102	5.03	0.60	1.30
Mardan	0–30	Silt	8.22	0.414	0.3	3.9	2.6	168	6.5	0.6	0.61
	30–45	-	8.18	0.885	0.20	1.3	2.4	198	3.4	0.7	0.6

4.3. Plant Tissue Analysis and Quality Characteristics

Plant tissue samples collected at the grand growth stage underwent analysis for their N, P, K, Zn, Fe, and B contents. Samples obtained from the field, specifically leaf sheath and leaf blade, were subjected to oven drying, followed by grinding and sieving. The total N content was determined using the Kjeldahl method, while P, K, Zn, and Fe were analyzed in a di-acid digestion mixture with a ratio of 2:1 (HNO$_3$:HClO$_4$). Boron content was determined through dry ashing and subsequent measurement [60].

4.4. Statistical Analysis

Data obtained were statistically analyzed using STATISTIX 8.1 software. For water characteristics, we analyzed samples to determine the range, calculate average values, and assess variability. For field experiments, data were analyzed for analysis of variance (ANOVA) in RCBD design with two-factor factorial, and a Least Significant Difference (LSD) was calculated at a 5% significance level.

5. Conclusions

Water quality has a direct impact on soil properties and its suitability for sugarcane cultivation. Water quality, soil conditions, and soil fertility directly affect plant growth and yield response. The present study investigated the relationship between K and different micronutrients instead of improving sugarcane growth, yield, and quality under two irrigation schemes (canal and tube well water). We concluded that canal water contributes to superior sugarcane quality with increased yield and improved cane quality over tube well water. Furthermore, using K and micronutrients (Zn, Fe, and B) may support sugarcane in enduring and yielding more effectively even under irrigation water's salt stress conditions. However, we need to expand our experimental area towards more heterogeneous agro-ecological areas to validate these results. Since sugarcane is one of the major crops worldwide, applying these treatments under diverse climate zones can help researchers enhance the sugarcane quality, yield, and even nutrient use efficiencies under those regions.

Author Contributions: Conceptualization, M.M. and M.Z.K.; methodology, M.M., M.Z.K. and S.A.; software, M.M.; validation, M.D.A., M.S. and S.S.; formal analysis, M.S.; investigation, M.A.H. and M.M.; resources, M.Z.K.; data curation, U.Z.; writing—original draft preparation, M.M., M.S. and U.Z.; writing—review and editing, S.H., F.U.H. and M.F.A.; visualization, M.A. and F.U.H.; supervision, M.Z.K.; project administration, M.M. and M.Z.K.; funding acquisition, M.D.A. All authors have read and agreed to the published version of the manuscript.

Funding: Princess Nourah bint Abdulrahman University Researchers Supporting Project number (PNURSP2023R355), Princess Nourah bint Abdulrahman University, Riyadh, Saudi Arabia.

Data Availability Statement: Not applicable.

Acknowledgments: The author extends their appreciation to Princess Nourah bint Abdulrahman University Researchers Supporting Project number (PNURSP2023R355), Princess Nourah bint Abdulrahman University, Riyadh, Saudi Arabia. The authors are also thankful to the Land Resources Research Institute, National Agricultural Research Center, Islamabad, for supporting this research work.

Conflicts of Interest: The authors declare no conflict of interest.

References

1. Hussain, S.; Wang, J.; Asad Naseer, M.; Saqib, M.; Siddiqui, M.H.; Ihsan, F.; Xiaoli, C.; Xiaolong, R.; Hussain, S.; Ramzan, H.N. Water stress memory in wheat/maize intercropping regulated photosynthetic and antioxidative responses under rainfed conditions. *Sci. Rep.* **2023**, *13*, 13688. [CrossRef]
2. Ramzan, T.; Shahbaz, M.; Maqsood, M.F.; Zulfiqar, U.; Saman, R.U.; Lili, N.; Irshad, M.; Maqsood, S.; Haider, A.; Shahzad, B.; et al. Phenylalanine supply alleviates the drought stress in mustard (*Brassica campestris*) by modulating plant growth, photosynthesis and antioxidant defense system. *Plant Physiol. Biochem.* **2023**, *201*, 107828. [CrossRef] [PubMed]
3. Abbas, A.; Huang, P.; Hussain, S.; Shen, F.; Wang, H.; Du, D. Mild Evidence for Local Adaptation of Solidago canadensis under Different Salinity, Drought, and Abscisic Acid Conditions. *Pol. J. Environ. Stud.* **2022**, *31*, 2987–2995. [CrossRef]
4. De Pascale, S.; Orsini, F.; Pardossi, A. Irrigation water quality for greenhouse horticulture. In *Good Agricultural Practices for Greenhouse Vegetable Crops*; Food and Agriculture Organization of the United Nations: Rome, Italy, 2013; p. 169.
5. Plaut, Z.; Edelstein, M.; Ben-Hur, M. Overcoming salinity barriers to crop production using traditional methods. *Crit. Rev. Plant Sci.* **2013**, *32*, 250–291. [CrossRef]
6. Zafar, S.; Afzal, H.; Ijaz, A.; Mahmood, A.; Ayub, A.; Nayab, A.; Hussain, S.; Hussan, M.U.; Sabir, M.A.; Zulfiqar, U.; et al. Cotton and drought stress: An updated overview for improving stress tolerance. *S. Afr. J. Bot.* **2023**, *161*, 258–268. [CrossRef]
7. Sajid, M.; Amjid, M.; Munir, H.; Valipour, M.; Rasul, F.; Khil, A.; Alqahtani, M.D.; Ahmad, M.; Zulfiqar, U.; Iqbal, R.; et al. Enhancing Sugarcane Yield and Sugar Quality through OptimalApplication of Polymer-Coated Single Super Phosphate and Irrigation Management. *Plants* **2023**, *12*, 3432. [CrossRef]
8. Luqman, M.; Shahbaz, M.; Maqsood, M.F.; Farhat, F.; Zulfiqar, U.; Siddiqui, M.H.; Masood, A.; Aqeel, M.; Haider, F.U. Effect of strigolactone on growth, photosynthetic efficiency, antioxidant activity, and osmolytes accumulation in different maize (*Zea mays* L.) hybrids grown under drought stress. *Plant Signal Behav.* **2023**, 2262795. [CrossRef]
9. Singhal, R.K.; Fahad, S.; Kumar, P.; Choyal, P.; Javed, T.; Jinger, D.; Singh, P.; Saha, D.; Md, P.; Bose, B.; et al. Beneficial elements: New Players in improving nutrient use efficiency and abiotic stress tolerance. *Plant Growth Regul.* **2023**, *100*, 237–265. [CrossRef]
10. Wang, J.; Hussain, S.; Sun, X.; Chen, X.; Ma, Z.; Zhang, Q.; Yu, X.; Zhang, P.; Ren, X.; Saqib, M.; et al. Nitrogen application at a lower rate reduce net field global warming potential and greenhouse gas intensity in winter wheat grown in semi-arid region of the Loess Plateau. *Field Crop Res.* **2022**, *280*, 108475. [CrossRef]
11. Hussain, S.; Khaliq, A.; Mehmood, U.; Qadir, T.; Saqib, M.; Iqbal, M.A.; Hussain, S. Sugarcane production under changing climate: Effects of environmental vulnerabilities on sugarcane diseases, insects and weeds. In *Sugarcane Production—Agronomic, Scientific and Industrial Perspectives*; IntechOpen: London, UK, 2018; pp. 1–17.
12. Ali, A.; Chu, N.; Ma, P.; Javed, T.; Zaheer, U.; Huang, M.T.; Fu, H.Y.; Gao, S.J. Genome-wide analysis of mitogen-activated protein (MAP) kinase gene family expression in response to biotic and abiotic stresses in sugarcane. *Physiol. Plant.* **2021**, *171*, 86–107. [CrossRef]

13. Thibane, Z.; Soni, S.; Phali, L.; Mdoda, L. Factors impacting sugarcane production by small-scale farmers in KwaZulu-Natal Province-South Africa. *Heliyon* **2023**, *9*, e13061. [CrossRef] [PubMed]

14. Tew, T.; Cobill, R.M. Genetic improvement of sugarcane (*Saccharum* spp.) as an energy crop. In *Genetic Improvement of Bioenergy Crops*; Springer: New York, NY, USA, 2008; pp. 249–272.

15. Shabbir, R.; Javed, T.; Afzal, I.; Sabagh, A.E.; Ali, A.; Vicente, O.; Chen, P. Modern biotechnologies: Innovative and sustainable approaches for the improvement of sugarcane tolerance to environmental stresses. *Agronomy* **2021**, *11*, 1042. [CrossRef]

16. Hu, Z.T.; Ntambo, M.S.; Zhao, J.Y.; Javed, T.; Shi, Y.; Fu, H.Y.; Huang, M.T.; Gao, S.J. Genetic divergence and population structure of Xanthomonas albilineans strains infecting *Saccharum* spp. Hybrid and *Saccharum officinarum*. *Plants* **2023**, *12*, 1937. [CrossRef] [PubMed]

17. Kandil, E.E.; Lamlom, S.F.; Gheith, E.S.M.; Javed, T.; Ghareeb, R.Y.; Abdelsalam, N.R.; Hussain, S. Biofortification of maize growth, productivity and quality using nano-silver, silicon and zinc particles with different irrigation intervals. *J. Agric. Sci.* **2023**, *161*, 339–355. [CrossRef]

18. Liao, Z.; He, Y.; Mo, S. The effect of meteorological factors on sugarcane yield and research progress of environmental interaction genes. *Chin. Agric. Sci. Bull.* **2022**, *38*, 82–87.

19. Bhatt, R.; Singh, J.; Laing, A.M.; Meena, R.S.; Alsanie, W.F.; Gaber, A.; Hossain, A. Potassium and Water-Deficient Conditions Influence the Growth, Yield and Quality of Ratoon Sugarcane (*Saccharum officinarum* L.) in a Semi-Arid Agroecosystem. *Agronomy* **2021**, *11*, 2257. [CrossRef]

20. Ahmad, N.; Virk, A.L.; Hussain, S.; Hafeez, M.B.; Haider, F.U.; Rehmani, M.I.A.; Yasir, T.A.; Asif, A. Integrated application of plant bioregulator and micronutrients improves crop physiology, productivity and grain biofortification of delayed sown wheat. *Environ. Sci. Pollut. Res.* **2022**, *29*, 52534–52543. [CrossRef]

21. Wang, M.; Zheng, Q.; Shen, Q.; Guo, S. The critical role of potassium in plant stress response. *Int. J. Mol. Sci.* **2013**, *14*, 7370–7390. [CrossRef]

22. Ma, L.; Zhang, H.; Sun, L.; Jiao, Y.; Zhang, G.; Miao, C.; Hao, F. NADPH oxidase AtrbohD and AtrbohF function in ROS-dependent regulation of Na^+/K^+ homeostasis in Arabidopsis under salt stress. *J. Exp. Bot.* **2012**, *63*, 305–317. [CrossRef]

23. Ragel, P.; Raddatz, N.; Leidi, E.O.; Quintero, F.J.; Pardo, J.M. Regulation of K^+ nutrition in plants. *Front. Plant Sci.* **2019**, *10*, 281. [CrossRef]

24. Naz, M.; Hussain, S.; Ashraf, I.; Farooq, M. Exogenous application of proline and phosphorus help improving maize performance under salt stress. *J. Plant Nutr.* **2023**, *46*, 2342–2350. [CrossRef]

25. Ghaffar, A.; Ehsanullah, N.A.; SH, K. Influence of zinc and iron on yield and quality of sugarcane planted under various trench spacings. *Pak. J. Agri. Sci.* **2011**, *48*, 25–33.

26. de Lima Vasconcelos, R.; Cremasco, C.P.; de Almeida, H.J.; Garcia, A.; Neto, A.B.; Mauad, M.; Gabriel Filho, L.R.A. Multivariate behavior of irrigated sugarcane with phosphate fertilizer and filter cake management: Nutritional state, biometry, and agroindustrial performance. *J. Soil Sci. Plant Nutr.* **2020**, *20*, 1625–1636. [CrossRef]

27. Kumar, A.; Babar, L.; Mohan, N.; Bansal, S.K. Effect of potassium application on yield, nutrient uptake and quality of sugarcane and soil health. *Indian J. Fertil.* **2019**, *15*, 782–786.

28. Feng, G.; Zhang, Z.; Wan, C.; Lu, P.; Bakour, A. Effects of saline water irrigation on soil salinity and yield of summer maize (*Zea mays* L.) in subsurface drainage system. *Agric. Water. Manag.* **2017**, *193*, 205–213. [CrossRef]

29. Hussain, S.; Hussain, S.; Ali, B.; Ren, X.; Chen, X.; Li, Q.; Saqib, M.; Ahmad, N. Recent progress in understanding salinity tolerance in plants: Story of Na+/K+ balance and beyond. *Plant Physiol. Biochem.* **2021**, *160*, 239–256. [CrossRef]

30. Maucieri, C.; Zhang, Y.; McDaniel, M.D.; Borin, M.; Adams, M.A. Short-term effects of biochar and salinity on soil greenhouse gas emissions from a semi-arid Australian soil after re-wetting. *Geoderma* **2017**, *307*, 267–276. [CrossRef]

31. Wei, C.; Li, F.; Yang, P.; Ren, S.; Wang, S.; Wang, Y.; Xu, Z.; Xu, Y.; Wei, R.; Zhang, Y. Effects of irrigation water salinity on soil properties, N2O emission and yield of spring maize under mulched drip irrigation. *Water* **2019**, *11*, 1548. [CrossRef]

32. Kumar, R.P.; Somashekar, R.K. A geochemical assessment of coastal groundwater quality in the Varahi river basin, Udupi District, Karnataka State, India. *Arab. J. Geosci.* **2013**, *6*, 1855–1870.

33. Zhou, J.R.; Sun, H.D.; Ali, A.; Rott, P.C.; Javed, T.; Fu, H.Y.; Gao, S.J. Quantitative proteomic analysis of the sugarcane defense responses incited by Acidovorax avenae subsp. avenae causing red stripe. *Ind. Crops Prod.* **2021**, *162*, 113275. [CrossRef]

34. Ahmed, J.; Qadir, G.; Ali, M.F.; Javed, T.; Jhanzab, H.M.; Wattoo, F.M.; Mahmood, I.; Ansar, M.; Khan, M.A.; Zulfiqar, U.; et al. Screening and growth assessment of indigenous and exotic sesame genotypes under osmotic Stress. *S. Afr. J. Bot.* **2023**, *158*, 203–213. [CrossRef]

35. Hussain, S.; Hafeez, M.B.; Azam, R.; Mehmood, K.; Aziz, M.; Ercisli, S.; Javed, T.; Raza, A.; Zahra, N.; Hussain, S. Deciphering the role of phytohormones and osmolytes in plant tolerance against salt stress: Implications, possible cross-talk, and prospects. *J. Plant Growth Regul.* **2023**, 1–22. [CrossRef]

36. Guo, R.; Qian, R.; Yang, L.; Khaliq, A.; Han, F.; Hussain, S.; Zhang, P.; Cai, T.; Jia, Z.; Chen, X.; et al. Interactive effects of maize straw-derived biochar and n fertilization on soil bulk density and porosity, maize productivity and nitrogen use efficiency in arid areas. *J. Soil Sci. Plant Nutr.* **2022**, *22*, 4566–4586. [CrossRef]

37. Shah, A.A.; Shah, A.N.; Bilal Tahir, M.; Abbas, A.; Javad, S.; Ali, S.; Rizwan, M.; Alotaibi, S.S.; Kalaji, H.M.; Telesinski, A.; et al. Harzianopyridone supplementation reduced chromium uptake and enhanced activity of antioxidant enzymes in vigna radiata seedlings exposed to chromium toxicity. *Front. Plant Sci.* **2022**, *13*, 881561. [CrossRef]

38. Ahmad, M.; Waraich, E.A.; Shahid, H.; Ahmad, Z.; Zulfiqar, U.; Mahmood, N.; Al-Ashkar, I.; Ditta, A.; El Sabagh, A. Exogenously Applied Potassium Enhanced Morpho-Physiological Growth and Drought Tolerance of Wheat by Alleviating Osmotic Imbalance and Oxidative Damage. *Pol. J. Environ. Stud.* **2023**, *32*, 4447–4459. [CrossRef]

39. Javeed, H.M.R.; Wang, X.; Ali, M.; Nawaz, F.; Qamar, R.; Rehman, A.U.; Shehzad, M.; Mubeen, M.; Shabbir, R.; Javed, T.; et al. Potential utilization of diluted seawater for the cultivation of some summer vegetable crops: Physiological and nutritional implications. *Agronomy* **2021**, *11*, 1826. [CrossRef]

40. Cui, D.; Huang, M.T.; Hu, C.Y.; Su, J.B.; Lin, L.H.; Javed, T.; Deng, Z.H.; Gao, S.J. First report of *Pantoea stewartii* subsp. stewartii causing bacterial leaf wilt of sugarcane in China. *Plant Dis.* **2021**, *105*, 1190. [CrossRef]

41. Zulfiqar, U.; Haider, F.U.; Maqsood, M.F.; Mohy-Ud-Din, W.; Shabaan, M.; Ahmad, M.; Kaleem, M.; Ishfaq, M.; Aslam, Z.; Shahzad, B. Recent Advances in MicrobialAssisted Remediation of Cadmium-Contaminated Soil. *Plants* **2023**, *12*, 3147. [CrossRef]

42. Shahid, S.; Shahbaz, M.; Maqsood, M.F.; Farhat, F.; Zulfiqar, U.; Javed, T.; Fraz Ali, M.; Alhomrani, M.; Alamri, A.S. Proline-induced modifications in morpho-physiological, biochemical and yield attributes of pea (*Pisum sativum* L.) cultivars under salt stress. *Sustainability* **2022**, *14*, 13579. [CrossRef]

43. Ali, A.; Javed, T.; Zaheer, U.; Zhou, J.R.; Huang, M.T.; Fu, H.Y.; Gao, S.J. Genome-wide identification and expression profiling of the bHLH transcription factor gene family in *Saccharum spontaneum* under bacterial pathogen stimuli. *Trop. Plant Biol.* **2021**, *14*, 283–294. [CrossRef]

44. Mustafa, A.; Zulfiqar, U.; Mumtaz, M.Z.; Radziemska, M.; Haider, F.U.; Holatko, J.; Hammershmiedt, T.; Naveed, M.; Ali, H.; Kintl, A.; et al. Nickel (Ni) phytotoxicity and detoxification mechanisms: A review. *Chemosphere* **2023**, *328*, 138574. [CrossRef] [PubMed]

45. Pabuayon, I.C.M.; Kitazumi, A.; Gregorio, G.B.; Singh, R.K.; De Los Reyes, B.G. Contributions of Adaptive Plant Architecture to Transgressive Salinity Tolerance in Recombinant Inbred Lines of Rice: Molecular Mechanisms Based on Transcriptional Networks. *Front. Genet.* **2020**, *11*, 1318. [CrossRef] [PubMed]

46. Ayub, S.; Huang, S.; Raza, S.H. Effect of climatic vulnerabilities on sugarcane production in punjab, sindh and kpk: A case study from Pakistan. *J. Agric. Res.* **2021**, *59*, 305–312.

47. Silva, W.; Neves, T.; Silva, C.; Carvalho, M.; Abraho, R. Sustainable enhancement of sugarcane fertilization for energy purposes in hot climates. *Renew. Energy* **2020**, *159*, 547–552. [CrossRef]

48. Jaiswal, V.P.; Shukla, S.K.; Sharma, L.; Singh, I.; Pathak, A.D.; Nagargade, M.; Ghosh, A.; Gupta, C.; Gaur, A.; Awasthi, S.K.; et al. Potassium influencing physiological parameters, photosynthesis and sugarcane yield in subtropical India. *Sugar Tech* **2021**, *23*, 343–359. [CrossRef]

49. Hasanuzzaman, M.; Bhuyan, M.H.M.; Nahar, K.; Hossain, M.; Mahmud, J.A.; Hossen, M.; Masud, A.A.C.; Fujita, M. Potassium: A vital regulator of plant responses and tolerance to abiotic stresses. *Agronomy* **2018**, *8*, 31. [CrossRef]

50. Han, F.; Guo, R.; Hussain, S.; Guo, S.; Cai, T.; Zhang, P.; Jia, Z.; Naseer, M.A.; Saqib, M.; Chen, X.; et al. Rotation of planting strips and reduction in nitrogen fertilizer application can reduce nitrogen loss and optimize its balance in maize–peanut intercropping. *Eur. J. Agron.* **2023**, *143*, 126707. [CrossRef]

51. Alves Flores, R.; de Mello Prado, R.; Pancelli, M.A.; Almeida, H.J.; Moda, L.R.; Montes Nogueira Borges, B.M.; Pereira de Souza Junior, J. Potassium nutrition in the first and second ratoon sugarcane grown in an Oxisol by a conservationist system. *Chil. J. Agric. Res.* **2014**, *74*, 83–88. [CrossRef]

52. Kumar, P.; Kumar, T.; Singh, S.; Tuteja, N.; Prasad, R.; Singh, J. Potassium: A key modulator for cell homeostasis. *J. Biotechnol.* **2020**, *324*, 198–210. [CrossRef]

53. Hussain, S.; Naseer, M.A.; Han, F.; Guo, R.; Saqib, M.; Farooq, M.; Chen, X.; Ren, X. Influence of nitrogen application on soil chemical properties, nutrient acquisition, and enzymatic activities in rainfed wheat/maize strip intercropping. *J. Soil Sci. Plant Nutr.* **2023**, 1–14. [CrossRef]

54. Ismail, M.; Ahmad, T.; Ali, A.; Nabi, G.; Haq, N.U.; Munsif, F. Response of sugarcane to different doses of Zn at various growth stages. *PAB* **2021**, *5*, 311–316. [CrossRef]

55. Zhao, D.; Zhu, K.; Momotaz, A.; Gao, X. Sugarcane plant growth and physiological responses to soil salinity during tillering and stalk elongation. *Agriculture* **2020**, *10*, 608. [CrossRef]

56. Franco, H.C.; Mariano, E.; Vitti, A.C.; Faroni, C.E.; Otto, R.; Trivelin, P.C. Sugarcane response to boron and zinc in Southeastern Brazil. *Sugar Tech* **2011**, *13*, 86–95. [CrossRef]

57. Watanabe, K.; Fukuzawa, Y.; Kawasaki, S.I.; Ueno, M.; Kawamitsu, Y. Effects of potassium chloride and potassium sulfate on sucrose concentration in sugarcane juice under pot conditions. *Sugar Tech* **2016**, *18*, 258–265. [CrossRef]

58. Mangrio, N.; Kandhro, M.N.; Soomro, A.A.; Mari, N.; Shah, Z.U.H. Growth, Yield and Sucrose Percent Response of Sugarcane to Zinc and Boron Application. *Sarhad J. Agric.* **2020**, *36*, 459–469. [CrossRef]

59. Tewari, R.K.; Hadacek, F.; Sassmann, S.; Lang, I. Iron deprivation-induced reactive oxygen species generation leads to non-autolytic PCD in *Brassica napus* leaves. *Environ. Exp. Bot.* **2013**, *91*, 74–83. [CrossRef]
60. Olsen, S.R.; Sommers, L.E. Phosphorus. In *Methods of Soil Analysis Part 2 Chemical and Microbiological Properties*; Page, A.L., Miller, R.H., Keeney, D.R., Eds.; SSSA: Madison, WI, USA, 1982; pp. 403–430.

Article

Comparative Analysis of Growth and Physiological Responses of Sugarcane Elite Genotypes to Water Stress and Sandy Loam Soils

Muhammad Sajid [1], Muhammad Amjid [1], Hassan Munir [1], Muhammad Ahmad [1], Usman Zulfiqar [2,*], Muhammad Fraz Ali [3], Mohammad Abul Farah [4], Mohamed A. A. Ahmed [5,6] and Arkadiusz Artyszak [7]

[1] Department of Agronomy, University of Agriculture, Faisalabad 38040, Pakistan; amjidm70@gmail.com (M.A.); hmbajwauaf@gmail.com (H.M.); ahmadbajwa516@gmail.com (M.A.)

[2] Department of Agronomy, Faculty of Agriculture and Environment, The Islamia University of Bahawalpur, Bahawalpur 63100, Pakistan

[3] College of Agronomy, Northwest A & F University, Xianyang 712100, China; frazali15@gmail.com

[4] Department of Zoology, College of Science, King Saud University, Riyadh 11451, Saudi Arabia; mfarah@ksu.edu.sa

[5] Plant Production Department (Horticulture—Medicinal and Aromatic Plants), Faculty of Agriculture (Saba Basha), Alexandria University, Alexandria 21531, Egypt; drmohamedmarey19@alexu.edu.eg

[6] School of Agriculture, Yunnan University, Kunming 650091, China

[7] Institute of Agriculture, Warsaw University of Life Sciences—SGGW, Nowoursynowska 159, 02-776 Warsaw, Poland; arkadiusz_artyszak@sggw.edu.pl

* Correspondence: usman.zulfiqar@iub.edu.pk

Citation: Sajid, M.; Amjid, M.; Munir, H.; Ahmad, M.; Zulfiqar, U.; Ali, M.F.; Abul Farah, M.; Ahmed, M.A.A.; Artyszak, A. Comparative Analysis of Growth and Physiological Responses of Sugarcane Elite Genotypes to Water Stress and Sandy Loam Soils. *Plants* 2023, *12*, 2759. https://doi.org/10.3390/plants12152759

Academic Editor: San-Ji Gao

Received: 13 June 2023
Revised: 23 July 2023
Accepted: 23 July 2023
Published: 25 July 2023

Abstract: Stumpy irrigation water availability is extremely important for sugarcane production in Pakistan today. This issue is rising inversely to river flow due to inadequate water distribution and an uneven rainfall pattern. Sugarcane growth faces a shortage of available water for plant uptake due to the low water–holding capacity of sandy loam soil, particularly under conventional flood irrigation methods. To address this problem, sugarcane clones were evaluated for their agronomic and physiological traits under conditions of low water availability in sandy loam soil. Ten cane genotypes, HSF–240, SPF–213, CPF–249, CP 77–400, S2008–FD–19, S2006–US–469, S2007–AUS–384, S2003–US–633, S2003–US–127, and S2006–US–658, were exposed to four levels of water deficit created through skip irrigations. These deficit levels occurred during the 9th, 11th, 13th, and 16th irrigations at alternate deficit levels between 2020 and 2022. Physiological data were collected during the tillering and grand growth stages (elongation) in response to the water deficit. The sugarcane clones S2006–US–658, S2007–AUS–384, and HSF–240 exhibited resistance to low water availability at both the tillering and grand growth stages. Following them, genotypes S2006–US–658, S2007–AUS–384, and HSF–240 performed better and were also found to be statistically significant. Clones susceptible to water deficit in terms of growth and development were identified as CP 77–400, S2008–FD–19, S2006–US–469, and S2003–US–633. These genotypes showed reduced photosynthetic rate, transpiration rate, stomatal conductance, relative water content, cane yield, and proline content under stressed conditions. Therefore, genotypes S2006–US–658, S2007–AUS–384, and HSF–240 were better performers concerning physiological traits under water deficit and sandy loam soil in both years. Moreover, a significant positive correlation was assessed between agronomic traits and photosynthetic rats. This study highlights that sugarcane can sustain its growth and development even with less irrigation frequency or moisture availability, albeit with certain specific variations.

Keywords: sugarcane; genotypes; drought; physiology; sandy loam

1. Introduction

Sugarcane holds significant global importance, accounting for approximately 85% of the world's sugar consumption. In recent years, there has been a rapid expansion of sugarcane planting for biofuel production, including the emergence of energy canes. Cultivating sugarcane is a vital endeavor worldwide due to its multifaceted impact on nutrition, the environment, society, and the economy. It offers the potential for productive diversification through co–products and by–products. With cultivation spanning over 105 countries, sugarcane has a long–standing history of enhancing socioeconomic conditions from its inception to its present–day production [1]. In recent years, there has been a notable expansion in sugarcane cultivation worldwide. However, data for 2019–2020 indicates a decline in global sugar production by approximately 3% to 1745 million tons. This decrease can be attributed to various factors, including a significant drop of 5 million tons in India's production. The decline in India's sugar production is primarily due to reduced cultivation area and lower expected yields. Despite this decline, Brazil and India remain in close competition as the top sugar–producing countries globally. Both countries make made significant contributions to global sugar production. Additionally, Pakistan holds the fifth position in terms of sugar production [2].

Drought is a critical threat to sugarcane production due to contagious water resources and uneven rainfall patterns in Pakistan. These factors limit the growth and development of sugarcane significantly [3]. Screening sugarcane clones is an authentic means for exploiting the genetic potential to a higher extent, particularly under stressful conditions, as adopted by Javed et al. [4]. Being a C_4 specie, sugarcane has four distinct growth phases, germination, tillering, grand growth, and maturity, which significantly impact growth and ultimate yield statistics [5].

Drought stress adversely impacts the different phenological stages of tillering and elongation, critical phases where internodal length depends highly on water availability and vice versa [6,7]. Physiologically, such delays in the growth and development of plants are expressed by three mechanisms, i.e., reduced absorption of photo–synthetically active radiations due to limited canopy size, poor radiation use efficiency due to sun–leeward leaf orientation, and narrow sizing of leaf in response to soil moisture deficit [8]. Furthermore, numerous ongoing physiological processes under moderate and severe water–stressed conditions, such as photosynthesis, transpiration, and stomatal conductance, are hindered. Photosynthesis is comparatively more sensitive among several physiological processes, which slows the turgor–based rate of stomatal conductance [9]. The reduction in stomatal conductance to conserve water under severely stressed conditions couples with a decreased CO_2 fixation rate in the diffusion site of bundle sheet cells in the surrounding areas of ribulose 1,5–bisphosphate carboxylase oxygenase (Rubisco), resulting in poor fresh weight accumulation in the sugarcane [10,11]. Moreover, leaf water status directly interacted with stomatal conductance in water–deficient environments. Hence, physiological traits such as water potential, leaf osmotic potential, and stomatal conductance have been used to assess the impact of water deficit on sugarcane, which can provide the basis for screening the sugarcane clones for drought tolerance [12]. Such physiological processes are interlinked with each other and expose possibilities of growth and development assurance under water–scarce conditions [13]. Sugarcane's aptitude for supporting various physiological processes during water stress leads to understanding its genetic flexibility for growth and yield performance in water scarcity regions [11,13]. Many other physiological traits, such as stomatal conductance, relative water contents, leaf osmotic potential, leaf temperature, and chlorophyll content, positively correlate with water stress [14]. These changes ultimately cause a reduction in such physiological processes as stomatal conductance, transpiration, and photosynthesis [13,15]. Reduced performance of these physiological parameters results due to moderate to severe water stress during cane growth and development [16]. Reduced relative water content, more negative leaf osmotic potential, and loss in turgor potential are the chief restrictions reflected in cane growth for resistance against water deficit [17–19]. Free proline accumulation in the leaves and roots is a response of the plant to avoid cell–level injuries due to severe drought by assisting and stabilizing the membrane, making osmotic adjustments and scavenging oxygen radicals [16].

Water deficit is a very common global issue, and it is getting harsher daily [20–23]. The hydrophilic association of water with sugarcane and the dependence of many traits on water availability demands research on identifying a smart number of responsible traits for developing drought–tolerant cane varieties [24]. The screening of water stress–tolerant cane clones remains the aim of many sugarcane genomic improvement programs, which require imperative physiological trait identification as principles for the assortment [1,25]. Such determined physiological and agronomic traits for sugarcane crops can further be used for the screening and development of new germplasm under biotic and abiotic stressors [1,13,26,27]. The drought and water deficit effect is much pronounced on millions of hectares of land, with special reference to sugarcane cultivation. Identifying and screening available germplasm is a valuable target to assure cane yields, even with reduced water consumption and irrigation frequency [3,27]. Physiological traits identified for screening the sugarcane clones under water deficit conditions were utilized in this study to assess the response of the locally available cane germplasm for its performance under sandy loam texture soils. The aim of the study was to identify cane clones for water–limited conditions through agronomic and physiological approaches, which are potentially applicable to water–scarce regions of Pakistan. These drought–resistant, drought–tolerant, and drought–associated physiological traits were analyzed for screening suitable cane clones based on their agronomic and physiological performance under water–stressed conditions.

2. Results

Two years of statistical analysis showed significant variation in all studied traits comprising cane length and girth, photosynthetic rate, transpiration rate, gas stomatal conductance, proline content, sugarcane yield, leaf temperature, relative water content, water use efficiency, and osmotic potential (Table 1). During 2021–2022, the interaction effect for leaf temperature was found insignificant ($p \geq 0.05$).

Table 1. F value of following traits of sugarcane genotypes under varying irrigation levels.

Traits	2020–2021			2021–2022		
	I	G	I × G	I	G	I × G
Cane length	406.84 **	96.9 **	7.43 **	833.64 **	100.17 **	6.1 **
Cane girth	4074.89 **	1078.04 **	2.08 **	4509.79 **	490.87 **	2.37 **
Photosynthetic rate	315.8 **	44.95 **	2.56 **	224.4 **	5.36 **	1.27 *
Transpiration rate	213.66 **	74.39 **	2.12 **	1196.16 **	112.38 **	5.13 **
Gas Stomatal conductance	227.84 **	55.64 **	0.05 *	913.81 **	73.92 **	2.12 **
Proline content	3873.95 **	22.43 **	3.73 **	8653.61 **	46.36 **	13.37 *
Yield	24.73 **	8.93 **	3.2 **	19.95 **	8.93 **	3.2 **
Leaf temperature	46.03 **	36.34 **	0.22 ns	18.3 **	5.53 **	0.08 ns
Relative water content	13.95 **	30.75 **	0.08 *	35.94 **	7.2 **	0.05 *
Water use efficiency	9.66 **	13.64 **	1.16 *	136.18 **	23.3 **	4.71 *
Osmotic Potential	36,267.3 **	2251.26 **	166.01 *	476.64 **	185.9 **	52.69 **

*, **, significant at $p \leq 0.05$, $p \leq 0.01$, respectively.

2.1. Yield Traits

Results regarding cane length exposed significant ($p \leq 0.05$) responses for different genotypes under varying irrigation regimes. However, the highest cane length was attained by the S2007–AUS–384 sugarcane genotype, whereas the lowest cane length was recorded in CP 77–400 genotypes that were statistically similar to the S2008–FD–19 genotype during both years of study under all irrigation regimes. For cane girth, the maximum cane girth was measured in HSF–240, statistically similar to the CPF–249 sugarcane genotype. Moreover, the minimum cane girth was exposed in S2006–US–658, followed by CP 77–400 and S2008–FD–19 under all irrigation levels (Figure 1).

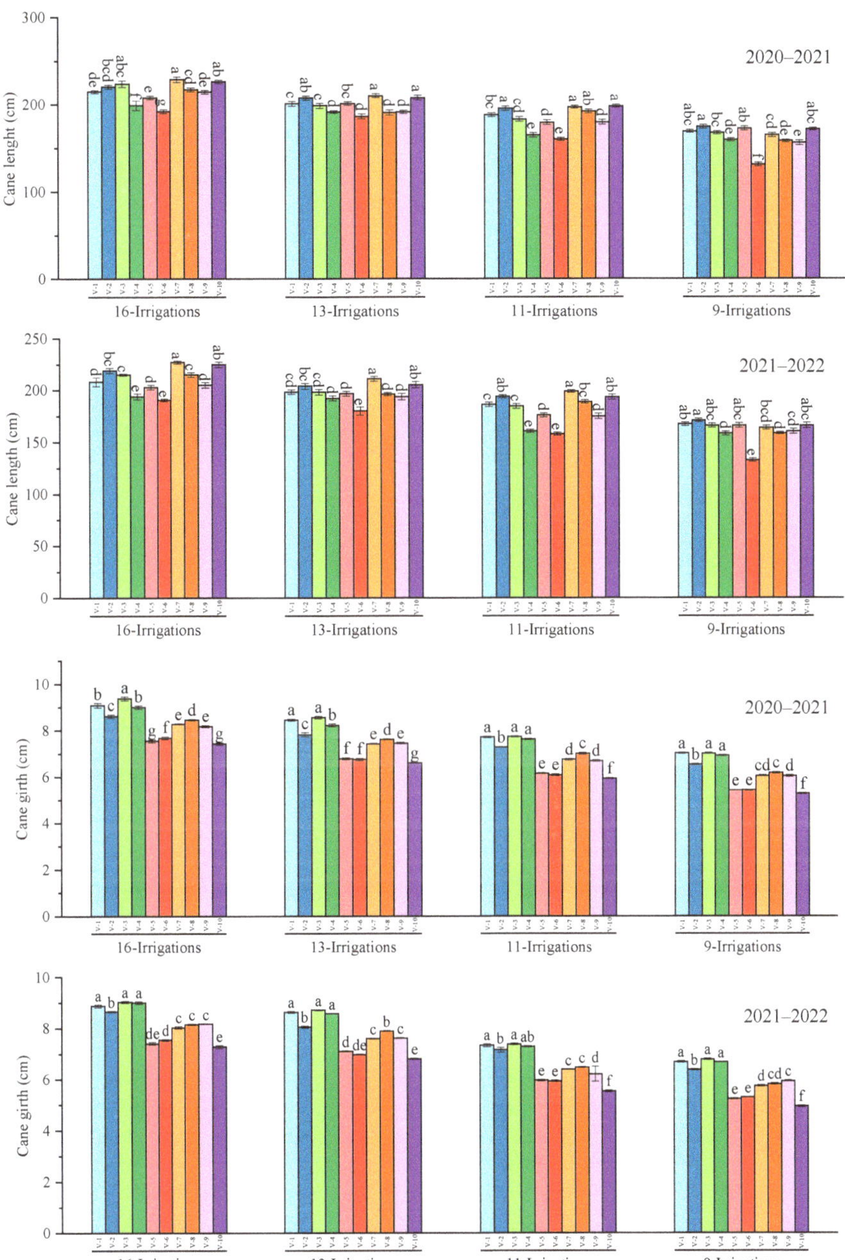

Figure 1. Performance of 10 cane varieties for cane length and girth under different water deficit levels over the years. V1 = HSF–240, V2 = SPF–213, V3 = CPF–249, V4 = CP 77–400, V5 = S2008–FD–19, V6 = S2006–US–469, V7 = S2007–AUS–384, V8 = S2003–US–633, V9 = S2003–US–127, and V10 = S2006–US–658; Variations in lowercase show significant difference between water deficit levels over the years at the $p \leq 0.05$.

Similarly, the highest sugarcane yield was recorded in HSF–240 and S2007–AUS–384 among all genotypes under all water deficit levels during both years, whereas the lowest cane yield was collected in CP 77–400 when 13 and 9 irrigations were applied in both years (Figure 2).

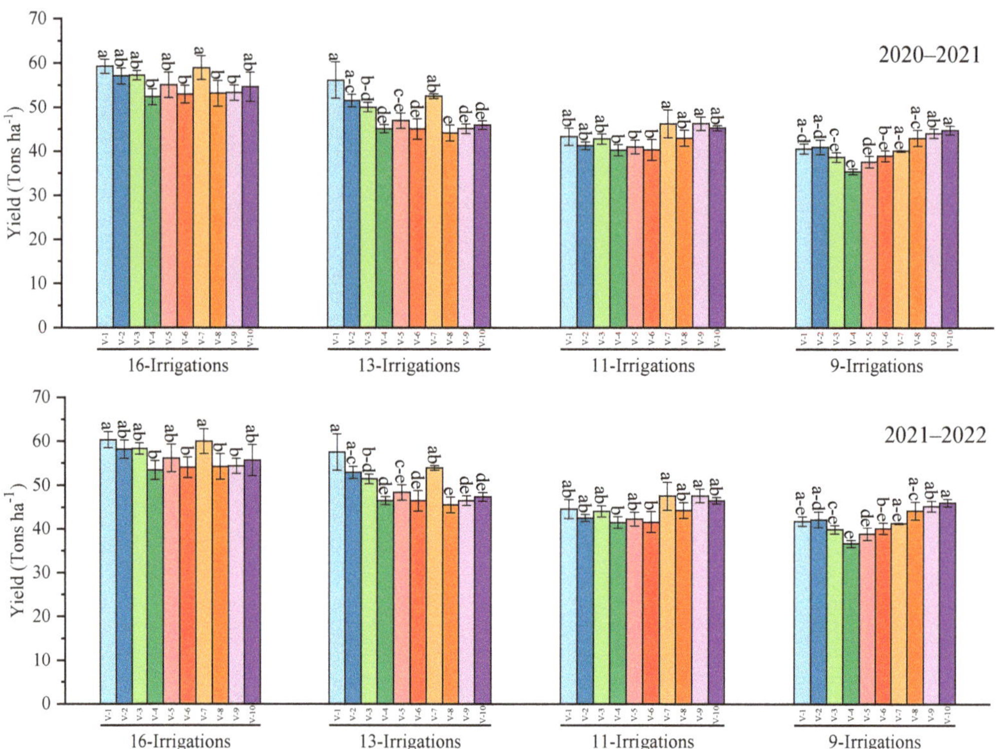

Figure 2. Performance of 10 cane varieties for cane yield under different water deficit levels over the years. V1 = HSF–240, V2 = SPF–213, V3 = CPF–249, V4 = CP 77–400, V5 = S2008–FD–19, V6 = S2006–US–469, V7 = S2007–AUS–384, V8 = S2003–US–633, V9 = S2003–US–127, and V10 = S2006–US–658. Variations in lowercase show significant difference between water deficit levels over the years at the $p \leq 0.05$.

2.2. Photosynthetic and Transpiration Rate

Data regarding net photosynthetic rate highlighted similar trends during 2020–2021 and 2021–2022 for different genotypes under differential irrigation levels. Results showed that the highest net photosynthetic rate was observed in S2006–US–658, statistically at parity with S2007–AUS–384, and the lowest rate was exhibited by CP 77–400, followed by S2006–US–469 and S2008–FD–19 under different irrigation regimes. Regarding transpiration rate, the maximum value was recorded in S2007–AUS–384, statistically at par with S2006–US–658, whereas the minimum transpiration was observed in CP–77400 during 2020–2021. During 2021–2022, the highest transpiration rate was recorded in S2007–AUS–384, at parity with SPF–213 and S2006–US–658, and the lowest transpiration rate was assimilated in CP–77400 under different irrigation regimes (Figure 3).

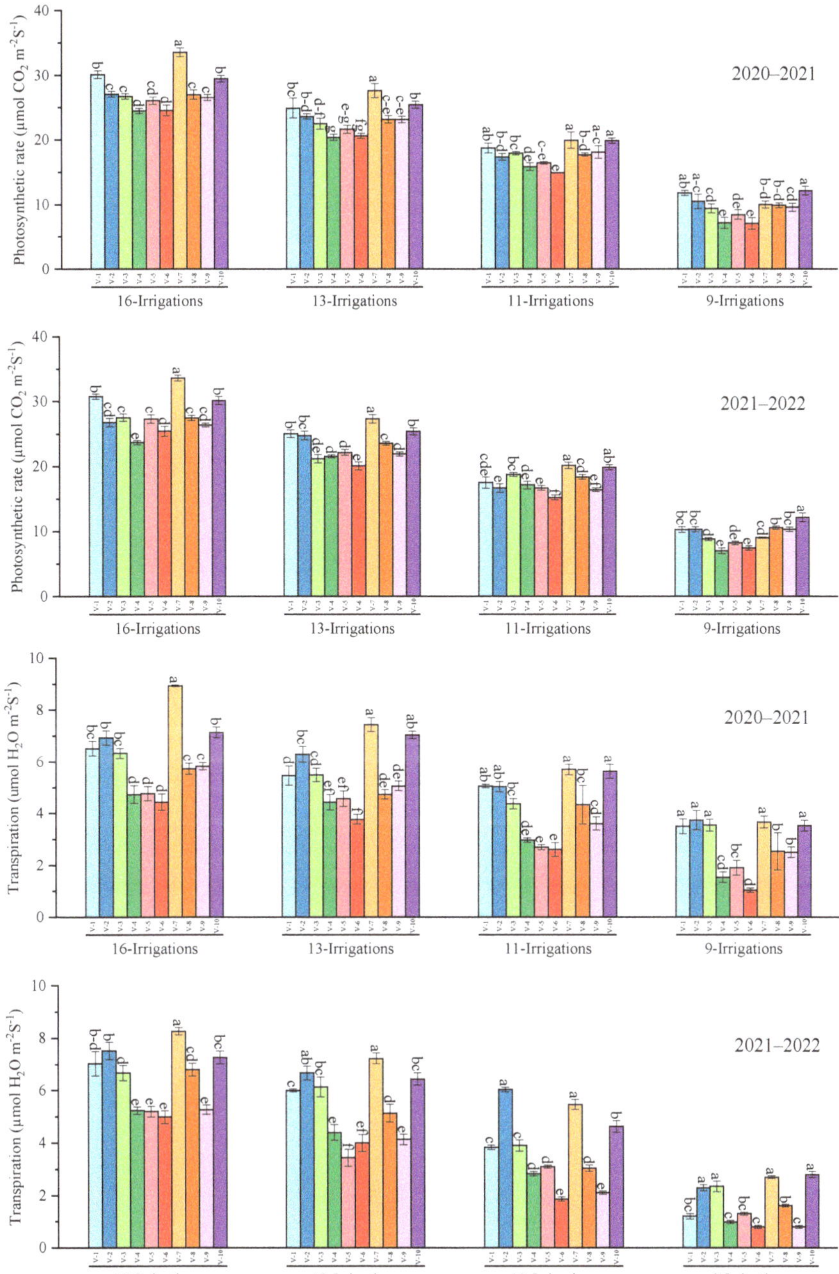

Figure 3. Performance of 10 cane varieties for photosynthetic and transpiration rate under different water deficit levels over the years. V1 = HSF–240, V2 = SPF–213, V3 = CPF–249, V4 = CP 77–400, V5 = S2008–FD–19, V6 = S2006–US–469, V7 = S2007–AUS–384, V8 = S2003–US–633, V9 = S2003–US–127, and V10 = S2006–US–658. Variations in lowercase show significant difference between water deficit levels over the years at the $p \leq 0.05$.

2.3. Description of Stomatal Conductance and Proline

Stomatal conductance data showed that S2007–AUS–384 and SPF–213 were the leading sugarcane genotypes, whereas the S2006–US–469 was the lowest performer under different irrigation regimes during 2020–2021. For 2021–2022, the highest gas stomatal conductance was measured in S2007–AUS–384, and the lowest was recorded in CP 77–400 under different irrigation levels. Moreover, concerning results regarding proline content, the highest was assessed in the S2007–AUS–384 genotype under 9 irrigation treatments during both years, which was statistically insignificant with SPF–213. In contrast, the lowest proline content was accumulated in S–2006–US–469 and CP 77–400 genotypes under all irrigation regimes during 2020–2021 and 2021–2022 (Figure 4).

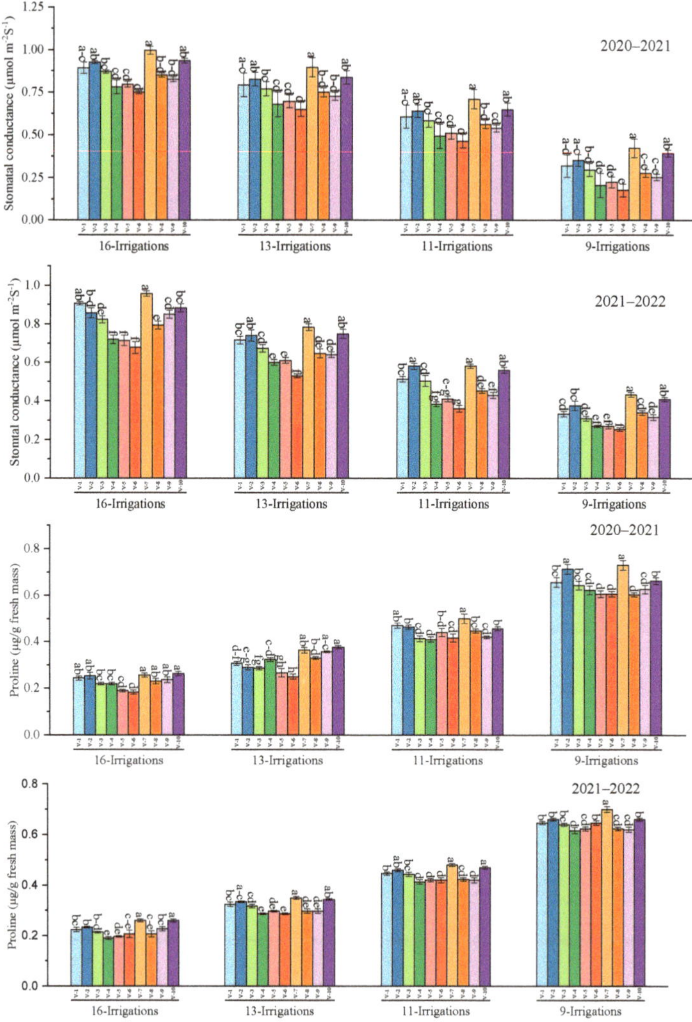

Figure 4. Performance of ten cane varieties for stomatal conductance and proline content under different water deficit levels over the years. V1 = HSF–240, V2 = SPF–213, V3 = CPF–249, V4 = CP 77–400, V5 = S2008–FD–19, V6 = S2006–US–469, V7 = S2007–AUS–384, V8 = S2003–US–633, V9 = S2003–US–127, and V10 = S2006–US–658. Variations in lowercase show significant difference between water deficit levels over the years at the $p \leq 0.05$.

2.4. Leaf Temperature and Osmotic Potential

Results regarding leaf temperature exhibited a significant response ($p \leq 0.05$) towards sugarcane yield and irrigation regime, and the maximum leaf temperature resulted in CP 77–400, followed by S2006–US–469 and S2008–FD–19. The minimum leaf temperature was recorded in HSF–240 and S2006–US–658 under all irrigation regimes in 2020–2021, whereas leaf temperature during 2021–2022 resulted in a non–significant variation. Significantly, the highest osmotic potential was determined in S2007–AUS–384, whereas the lowest potential was assessed in S2006–US–469 under all irrigation regimes during 2020–2021. Similarly, in 2021–2022, the maximum osmotic potential was recorded in HSF–240 sugarcane genotypes where 11 and 13 irrigations were applied, whereas under 9 and 16 irrigations, S2007–AUS–384 and S2006–US–658 had the highest osmotic potential, respectively. However, the least osmotic potential was assessed in S2006–US–469 and CP 77–400 by employing 9 and 16 irrigations, respectively (Figure 5).

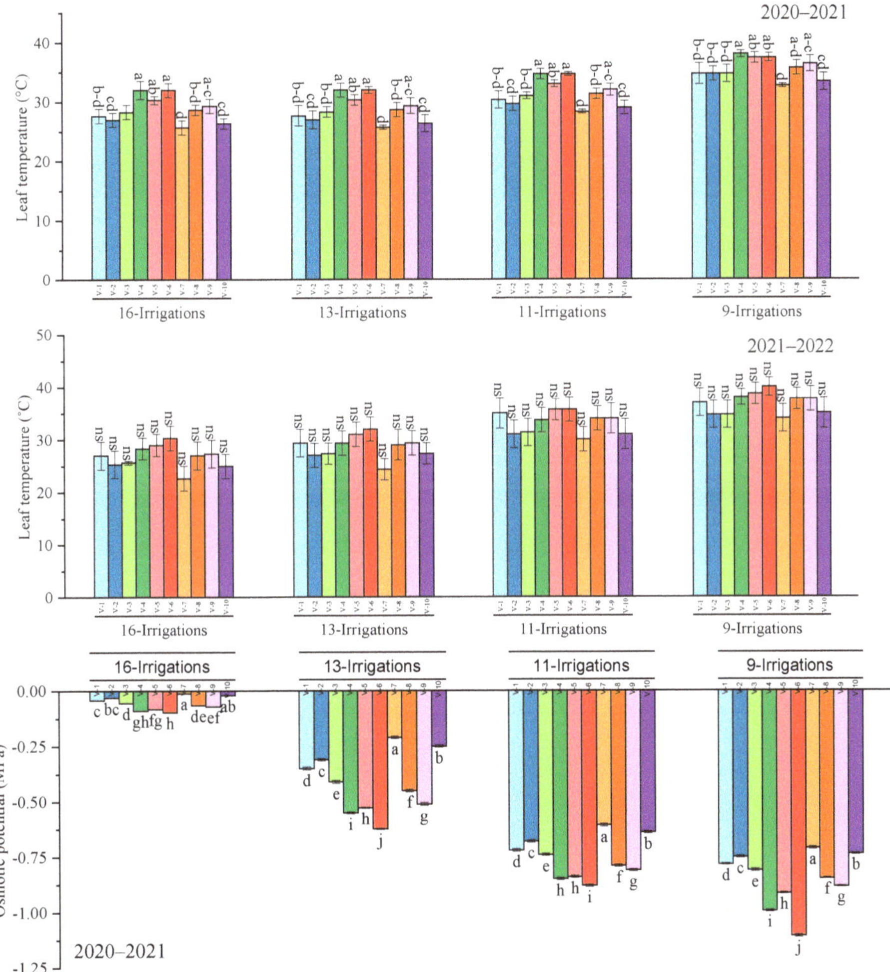

Figure 5. *Cont.*

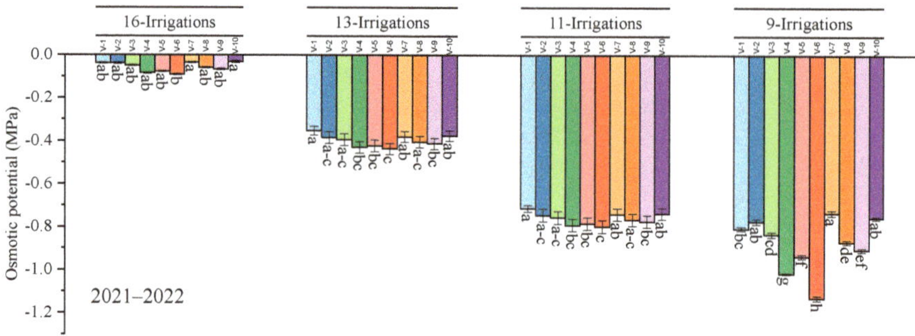

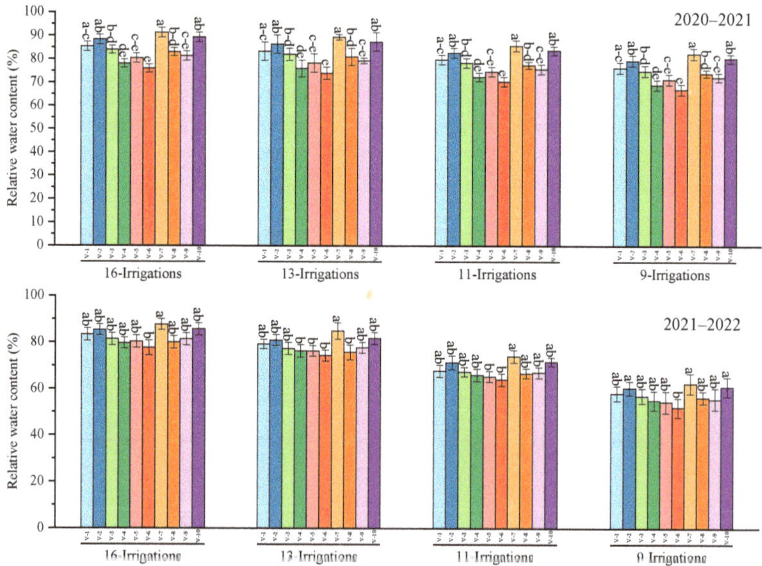

Figure 5. Performance of 10 cane varieties for leaf temperature and osmotic potential under different water deficit levels over the years. V1 = HSF–240, V2 = SPF–213, V3 = CPF–249, V4 = CP 77–400, V5 = S2008–FD–19, V6 = S2006–US–469, V7 = S2007–AUS–384, V8 = S2003–US–633, V9 = S2003–US–127, and V10 = S2006–US–658.

2.5. Relative Water Content and Water Use Efficiency

Data determined the highest relative water content was in S2007–AUS–384 and S2006–US–658 under all irrigation regimes during 2020–2021, while the minimum relative water content was calculated in S–2006–US–469, statistically on par with CP 77–400 and S2008–FD–19 in all irrigation regimes. The overall more relative water content was measured during 2021–2022. In the case of water use efficiency, the highest water use efficiency was calculated in S2006–US–469, similar to CP 77–400 and S2008–FD–19, whereas the least water use efficiency was recorded in SPF–13 under all irrigation treatments during 2020–2021. However, during 2021–2022, the highest water use and water use efficiency was determined in S2003–US–127 under 9 irrigations, and the least value was observed in SPF–213 through employing 11 irrigations. (Figure 6).

Figure 6. *Cont.*

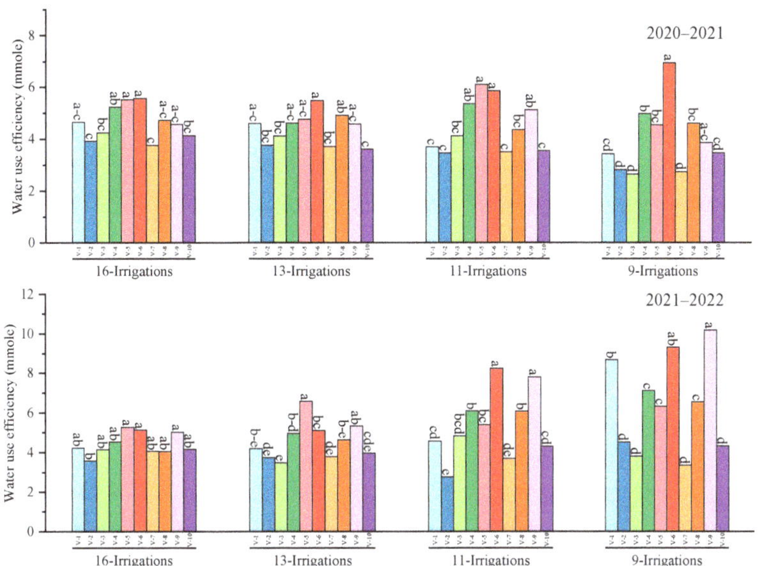

Figure 6. Performance of 10 cane varieties for relative water content and water use efficiency under different water deficit levels over the years. V1 = HSF–240, V2 = SPF–213, V3 = CPF–249, V4 = CP 77–400, V5 = S2008–FD–19, V6 = S2006–US–469, V7 = S2007–AUS–384, V8 = S2003–US–633, V9 = S2003–US–127, and V10 = S2006–US–658.

2.6. Pearson Correlation Analysis of Physiological and Agronomic Traits of Sugarcane

Two years (2020–2021 & 2021–2022) correlation analysis assessed the relationship of studied traits (Figure 7). During 2020–2021, sugarcane yield positively correlated with cane girth (CG) and cane length (CL), thus showing these traits had positively contributed to cane yield. Moreover, stomatal conductance (SC), transpiration (Trans), and photosynthetic rate (PSR) also positively correlated with CG, CL, and yield; hence these traits significantly influenced the cane yield. Osmotic potential (OP) and relative water content (*RWC*) showed a strong positive correlation with CL, CG, Yield, PSR, Trans, and SC traits. Similarly, proline and leaf temperature (LT) showed a strong negative correlation with all studied traits, excluding water use efficiency (WUE), but LT showed a positive correlation with proline content. However, 2021–2022 data also showed a similar correlation trend, except for WUE, which negatively correlated with CL, CG, yield, PSR, Trans, and SC.

2.7. Principal Component Analysis Description

Principal component analysis was presented in 3–D plots during both years (2020–2021 and 2021–2022) of the study (Figure 8). Two years of analyses exhibited that PC–1 had shown the highest variation, followed by PC–2 and PC–3. In 2020–2021, cane length (CL), photosynthetic rate (PSR), transpiration (Trans), and stomatal conductance (SC) were closely related. In addition, overall relative water content (*RWC*) was the maximum in treatment 11–Irri+V5 (11 Irrigations + V5 = S2008–FD–19), whereas transpiration was the highest in 16–Irri+V7 (16–Irirgations + V7 = S2007–AUS–384). Moreover, the highest cane girth (CG) was exhibited in 16–Irri+V4 & 16–Irri+V3 (16 Irrigations + V4 = CP 77–400 & V3 = CPF–249). Treatment 9–Irri + V7 had greater content of proline whereas, 9–Irri+V2 (9–Irrigations + V2 = SPF–213 & V3 = CPF–249) showed the maximum value of leaf temperature. Correspondingly, in the 2021–2022 study PSR, yield, SC, and CL are closely related, and similar findings were also recorded during this year. In this year, the highest water use efficiency was calculated in 9–Irri + V9 (9–irrigations + V9 = S2003–US–127).

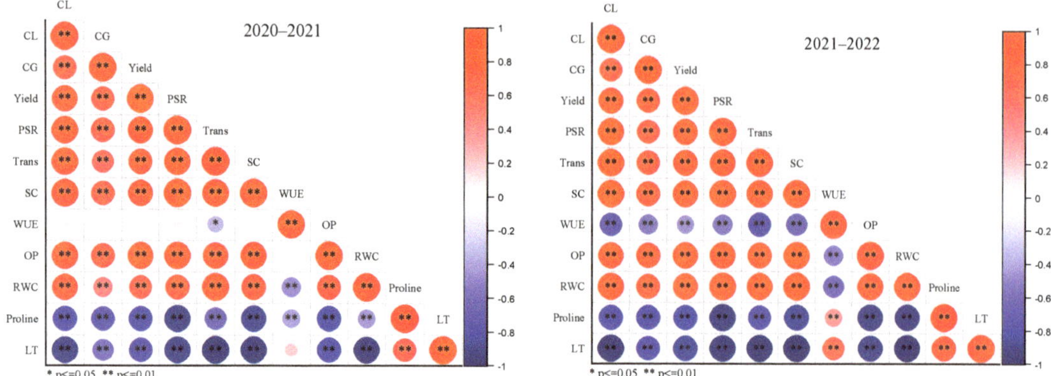

Figure 7. Pearson correlation analysis of studied traits: CL = cane length, CG = cane girth, PSR = photosynthetic rate, Trans = Transpiration, SC = stomatal conductance, WUE = water use efficiency, OP = osmotic potential, *RWC* = relative water content, proline, and LT = leaf temperature of sugarcane cultivars under differential irrigation levels.

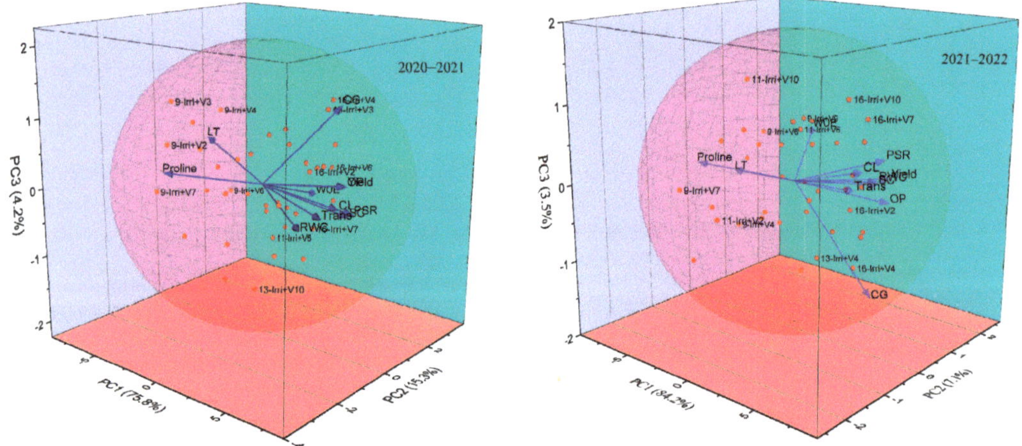

Figure 8. Principal component analysis of targeted sugarcane traits of different genotypes under differential irrigation regimes over the years.

3. Discussion

Low water availability has drastically impacted cane clones and reduced the growth and development of cane plantations. Similar findings were observed as a result of this experiment when considered based on the physiological traits of these cane clones. Cane clones analyzed through Biplot analysis underwent screening for water deficit tolerance using physiological markers. Water scarcity causes a retardation of plant growth. Plant growth decline during drought occurs due to changes in plant water relationships, reduced assimilation of CO_2, cellular oxidative stress, damage to membranes in affected tissues, and inhibition of enzyme activities [1]

Results confirmed significantly better growth and development of cane clones being grown under normal irrigation when compared with water–stressed conditions during both experimental years, as described by Silva et al. [13], who opined similarly regarding the productivity and physiology of sugarcane under sandy loam soil conditions. The variation in cane growth response to drought stress is contingent upon the genetic makeup of sugarcane genotypes [28,29]. Some clones were typically screened for their better performance under normal irrigation regimes, while others were found to be differentially better performers in moderate and severe water deficit conditions. Under sandy loam soils, the regenerative ability of cane was conspicuously observed, showing an immediately triggered growth response of clones to water availability in irrigation spells. While some of the clones were weak responders to water availability following water deficit exposures. Such clones having regenerative abilities were S2006–US–658, S2007–AUS–384, and HSF–240, whereas clone names were weak responders, as mentioned above. Similarly, cane clones group CP 77–400, S2008–FD–19, S2006–US–469, and S2003–US–633 showed a medium response or were insensitive to water deficit levels in sandy loam soil for physiological traits. On the other hand, cane clones S2003–US–127, CPF–249, and SPF–213 had better performance for photosynthesis rate, transpiration rate, stomatal conductance, chlorophyll content, osmotic potential, relative water content, proline content, and leaf temperature traits had high values under moderate and severe water deficit levels.

The varieties S2007–AUS–384 and S2006–US–658 were suitable under control and water deficit levels (D1) for those sandy loam texture soils during year I and year II. These clones show non–significant ($p \geq 0.05$) differences in their performance under mild water deficit compared to control in both years. Thus, clones may be successfully cultivated without any decline in sugarcane production with judicial use of water in sandy loam conditions. Similarly, clones S2003–US–127 and CPF–249 could be grown under medium to severe water deficit conditions, having higher production and survivability in sandy loam soil texture during years I and II. During years I and II, cane clones S2007–AUS–384, S2006–US–658, and SPF–213 had the best performance in control conditions due to good water availability, and these clones produced the highest biomass production under control conditions. Still, their performance reduced significantly due to decreased water quantity from mild, moderate, and severe stress conditions in a sandy loam environment. Cane clones S2003–US–127, CPF–249, and SPF–213 were retained and resistant to drought spells and had maximum values of physiology markers in moderate and severe water stress and sandy loam soil conditions. The performance of these cane clones during year II significantly differs from year I.

In the study, these cane clones were screened out for high rates of physiological mechanisms under drought spell duration of the crop growth and development in sandy loam soil conditions. The high water demand is a great concern for the sugarcane crop production system among agriculture policymakers, who often suggest substituting the sugarcane crop with other sugar crops that demand less irrigational water [30]. Therefore, the reduction in water mandate of the cane clones is an important part of the sustainable sugarcane crop system. The leakage of electrolytes from leaf samples is associated with membrane damage resulting from oxidative stress [31,32]. The oxidative stress induces injury to the cell membrane, which allows for the diffusion of electrolytes and ions. Under drought stress conditions, the destruction of thylakoid membranes directly or indirectly impacts the chlorophyll content [33]. Drought causes a decline in the SPAD value of sugarcane leaves after exposure to drought [9,34,35]. SPAD value was employed in screening drought tolerance in sugarcane, and under drought stress, it could be used for screening for drought–tolerant and drought–sensitive genotypes [3]. Soil texture also profoundly affected sugarcane productivity, and higher production could be obtained in sandy loam soil. Cane clones S2003–US–127, CPF–249, and SPF–213 had better performance under sandy soil and thus may be suitable for the specific sandy loam soil types. Based on studied cane clones, S2003–US–127, CPF–249, and SPF–213 had high production in most of the cane traits during year II, followed by agronomic intervention in a drought period of crop growth under sandy loam soil conditions in moderate (D2) and severe (D3) stress levels as compared to year I. Sandy loam soil had less water holding capacity and other media necessary for growth and development, the enactment of cane clones S2003–US–127, CPF–249, and SPF–213 proved that under drained water conditions during both years. So, cane clones S2003–US–127, CPF–249, and SPF–213 performed better regarding agronomic and physiological assessment under moderate to severe drought stress situations in sandy loam soil conditions. Drought has a significant impact on photosynthesis, primarily through the substantial decrease in stomatal conductance resulting from stomatal closure in response to drought [33]. The reduction in stomatal conductance, caused by drought stress, can adversely affect various photosynthetic processes. Several studies on sugarcane have reported that varieties considered more sensitive to drought stress exhibit greater stomatal closure and reduced transpiration [26,34]. In our study rate of photosynthesis, transpiration rate, and stomatal conductance were decreased after inducing water stress. These findings are similar to those of Medeiros et al. [16] and Natarajan et al. [36]. Numerous studies have demonstrated the significant impact of drought on photosynthetic traits, including the net photosynthetic rate. Drought conditions often coincide with low soil moisture and high air temperature, and this combination of factors leads to increased evapotranspiration and subsequently affects photosynthesis in field conditions da Graca et al., [37]. Additionally, when drought stress was imposed, it was observed that drought–sensitive sugarcane varieties experienced a significant decrease in transpiration rate. In certain sugarcane genotypes, however, the plants could sustain their photosynthetic activity and maintain their water status by developing deeper roots [38].

Water use efficiency is a crucial characteristic when selecting drought–resistant varieties [39]. Drought–tolerant sugarcane genotypes exhibit higher intrinsic instantaneous water use efficiency, along with the ability to maintain higher water potential and photosynthetic capacity during water deficit conditions [40]. Drought stress has a substantial impact on sugarcane yield. Water deficit stress results in stunted growth and limited tillering, leading to empty and low–quality millable stalks. Consequently, exposure to drought stress during the early growth stages and midseason leads to reductions in both cane and sugar yields [28,29]. Hemaprabha [41] reported that water stress treatments had a negative impact on cane yield and total dry weight when compared to irrigation treatments. Insufficient water availability during the formative phase resulted in significant changes in yield and its related parameters. Specifically, after withholding water for 90 days during the formative phase, there were reductions in single stalk weight, cane height, and internode length. Under prolonged drought stress conditions, there was a considerable reduction in cane yield of approximately 21%. Furthermore, significant variations were observed among different genotypes for nearly all the traits evaluated, except for stalk length and diameter, when subjected to moderate drought stress conditions [42].

The approach of either spending or conserving water depends on factors such as the timing, intensity of the drought period, and the specific sugarcane genotype in a given location. The findings indicate that maintaining optimal photosynthetic performance during drought and facilitating recovery after re–watering, particularly during the stalk growth stage, play a vital role in determining the final yield of sugarcane. In the future, climate–smart breeding combined with conventional breeding strategies can be employed to develop climate–resilient sugarcane cultivars [43–47].

4. Materials and Methods

4.1. Field Preparation

Optimum land preparation is an important segment regarding the physical anchorage of the plant, which help the plant to uptake soil water and nutrition. Better tillage operation for the planting of crops is good practice concerning water holding capacity as well as infiltration rate in the case of conservation agriculture; hence 500 m^2 sandy loam soil was selected at Ramzan Sugar Mills, Limited, for the experiment. For preparation, firstly a disc harrow was used to open the soil, and two ploughings were done with the help of a cultivator. After a good tilth, the trenches were made with the help of a sugarcane planter, the trenches being 120 cm wide. Soil sampling was done at 0–30 cm depth of soil by using soil auger and composite soil samples were analyzed by using standard procedures and protocol and analysis that is given in supplementary data (Table S1, Figure S1) as well as weather data of subject study is given in supplementary data.

4.2. Collection of Planting Material and Trial Execution

For experimentation, ten cane genotypes were collected from different Ramzan Sugar Mill farms, which are mentioned above. The proposed study was conducted at the University of Agriculture, Faisalabad, during spring 2020–2021. In both years, a fresh crop was planted. Experiments were replicated three times with four drought levels, i.e., 16 irrigations, 13 irrigations, 11 irrigations, and 9 irrigations upheld via alternate irrigation methods in sandy loam soil. Sugarcane varieties S2003–US–127, CP 77–400, SPF–213, S2006–US–469, CPF–249, S2003–US–633, S2006–US–658, S2007–AUS–384, S2008–FD–19, and HSF–240 were sown in 120 cm wide trenches with a net plot size of 2.4 m × 6.0 m and double budded cane setts were planted through hand placement in single row at 7500 setts ha^{-1} during spring 2021 and 2022. Recommended macronutrient–based fertilizer doses (NPK 168:112:112 kg ha^{-1}) and plant protection measures were adopted to ensure uniform growth. The experiment was laid out in a randomized complete block design (RCBD) with a split–plot arrangement. The crop was irrigated by imposing water deficit levels, i.e., D_0 (16 irrigations), D_1 (13 irrigations), D_2 (11 irrigations), and D_3 (9 irrigations).

4.3. Data Collection

Physiological traits were recorded at the tillering and grand growth phases of cane clones by using an infra–red gas analyzer (IRGA) for the determination of stomatal conductance, gaseous exchange, photosynthesis rate, and transpiration rate between 9:00 a.m. and 11:00 a.m. [48]. Chlorophyll content was recorded using SPAD model MC–100, Apogee Pvt. Leaf osmotic potential was also estimated using an osmometer model VAPRO, while agronomic traits, such as cane length and girth, were recorded at harvest maturity.

To determine relative water content, leaf disks were collected with a 2 mm diameter cork borer from the same leaf samples frozen on ice in a glass vessel. The fresh weight (W_f) of leaf disks was measured within 2 h of separation, which were then saturated in water, and reweighed after four hours to get their turgid/ saturated weight (W_t) at room temperature. Leaf discs were rapidly stained dry and oven dried for 48 h at 80 °C for measuring dry weight (W_d). RWC was calculated from the following equation [49]:

$$RWC = \frac{\left(W_f - W_d\right)}{(W_t - W_d)} \times 100$$

Proline content was determined in the normal and stressed cane leaves using the procedure described by Bates et al. [50]. Moreover, leaf temperature was recorded by using a hand–held infrared thermometer [13].

4.4. Statistical Analysis

Two years of experimentation were executed using a randomized complete block under split plot arrangements, and collected data were analyzed using OriginPro 2022 software. Treatment means were presented using paired comparison plot techniques, and the LSD test was employed to differentiate means at 5% probability. However, Pearson correlation analysis was done by deploying a two–tailed *t*–test (df–2).

5. Conclusions

Regardless of the complex interactions of physiological parameters in cane genetics, drought, and environmental effects during year I and year II, it was concluded that S2006–US–658, S2007–AUS–384, and HSF–240 sugarcane genotypes had better resistance to water deficit and low water availability. However, there is still a need to assess the lowest water requirement that will help these clones grow sustainably. Therefore, the aforementioned cane genotypes are screened out for drought conditions, and S2006–US–658, S2007–AUS–384, and HSF–240 for irrigated conditions under the sandy loam texture.

Supplementary Materials: The following supporting information can be downloaded at: https://www.mdpi.com/article/10.3390/plants12152759/s1, Figure S1: Weather data during the course of study of 2020–2021 and 2021–2022; Table S1: Soil profile analysis of 2020–2021 and 2021–2022.

Author Contributions: Conceptualization, M.S. and H.M.; methodology, M.S. and M.A. (Muhammad Amjid); software, M.S.; validation, M.S., M.A. (Muhammad Amjid) and U.Z.; formal analysis, M.S.; investigation, M.A. (Muhammad Amjid); resources, H.M.; data curation, U.Z.; writing—original draft preparation, M.S.; writing—review and editing, M.A. (Muhammad Ahmad) and M.F.A.; visualization, M.F.A.; supervision, H.M.; project administration, H.M.; funding acquisition, M.A.F., M.A.A.A. and A.A. All authors have read and agreed to the published version of the manuscript.

Funding: This work was supported by Researchers Supporting Project number (RSPD2023R694), King Saud University, Riyadh, Saudi Arabia.

Data Availability Statement: All data generated or analyzed during this study are included in this published article.

Acknowledgments: The authors would like to extend their sincere appreciation to the Researchers Supporting Project number (RSPD2023R694), King Saud University, Riyadh, Saudi Arabia.

Conflicts of Interest: The authors declare no conflict of interest.

References

1. Inman-Bamber, N.; Smith, D. Water relations in sugarcane and response to water deficits. *Field Crops Res.* **2005**, *92*, 85–202. [CrossRef]
2. Alamgir, A.; Khan, M.A.; Manino, I.; Shaukat, S.S.; Shahab, S. Vulnerability to climate change of surface water resources of coastal areas of Sindh, Pakistan. *Desalination Water Treat.* **2016**, *57*, 18668–18678. [CrossRef]
3. Javed, T.; Shabbir, R.; Ali, A.; Afzal, I.; Zaheer, U.; Gao, S.-J. Transcription factors in plant stress responses: Challenges and potential for sugarcane improvement. *Plants* **2020**, *9*, 491. [CrossRef]
4. Javed, T.; Zhou, J.-R.; Li, J.; Hu, Z.-T.; Wang, Q.-N.; Gao, S.-J. Identification and expression profiling of WRKY family genes in sugarcane in response to bacterial pathogen infection and nitrogen implantation dosage. *Front. Plant Sci.* **2022**, *13*, 917953. [CrossRef] [PubMed]
5. Ramesh, P. Effect of different levels of drought during the formative phase on growth parameters and its relationship with dry matter accumulation in sugarcane. *J. Agron. Crop Sci.* **2000**, *185*, 83–89. [CrossRef]
6. Shabbir, R.; Javed, T.; Afzal, I.; Sabagh, A.E.; Ali, A.; Vicente, O.; Chen, P. Modern biotechnologies: Innovative and sustainable approaches for the improvement of sugarcane tolerance to environmental stresses. *Agronomy* **2021**, *11*, 1042. [CrossRef]
7. Ali, A.; Chu, N.; Ma, P.; Javed, T.; Zaheer, U.; Huang, M.; Fu, H.; Gao, S. Genome–wide analysis of mitogen-activated protein (MAP) kinase gene family expression in response to biotic and abiotic stresses in sugarcane. *Physiol. Plant.* **2021**, *171*, 86–107. [CrossRef]
8. Shabbir, R.; Singhal, R.K.; Mishra, U.N.; Chauhan, J.; Javed, T.; Hussain, S.; Kumar, S.; Anuragi, H.; Lal, D.; Chen, P. Combined abiotic stresses: Challenges and potential for crop improvement. *Agronomy* **2022**, *12*, 2795. [CrossRef]
9. Silva, M.d.A.; Jifon, J.L.; dos Santos, C.M.; Jadoski, C.J.; da Silva, J.A.G. Photosynthetic capacity and water use efficiency in sugarcane genotypes subject to water deficit during early growth phase. *Braz. Arch. Biol. Technol.* **2013**, *56*, 735–748. [CrossRef]
10. Lawlor, D.W.; Tezara, W. Causes of decreased photosynthetic rate and metabolic capacity in water–deficient leaf cells: A critical evaluation of mechanisms and integration of processes. *Ann. Bot.* **2009**, *103*, 561–579. [CrossRef]
11. Galmés, J.; Ribas-Carbó, M.; Medrano, H.; Flexas, J. Rubisco activity in Mediterranean species is regulated by the chloroplastic CO_2 concentration under water stress. *J. Exp. Bot.* **2011**, *62*, 653–665. [CrossRef] [PubMed]
12. Kusvuran, S. Effects of drought and salt stresses on growth, stomatal conductance, leaf water and osmotic potentials of melon genotypes (*Cucumis melo* L.). *Afr. J. Agric. Res.* **2012**, *7*, 775–781.
13. Silva, M.D.A.; Jifon, J.L.; Da Silva, J.A.G.; Sharma, V. Use of physiological parameters as fast tools to screen for drought tolerance in sugarcane. *Braz. J. Plant Physiol.* **2007**, *19*, 193–201. [CrossRef]
14. Altinkut, A.; Kazan, K.; Ipekçi, Z.; Gozukirmizi, N. Tolerance to paraquat is correlated with the traits associated with water stress tolerance in segregating F2 populations of barley and wheat. *Euphytica* **2001**, *121*, 81. [CrossRef]
15. Gonçalves, E.R.; Ferreira, V.M.; Silva, J.V.; Endres, L.; Barbosa, T.P.; Duarte, W.D.G. Gas exchange and chlorophyll a fluorescence of sugarcane varieties submitted to water stress. *Rev. Bras. Eng. Agrícola Ambient.* **2010**, *14*, 378–386. [CrossRef]
16. Medeiros, D.B.; da Silva, E.C.; Nogueira, R.J.M.C.; Teixeira, M.M.; Buckeridge, M.S. Physiological limitations in two sugarcane varieties under water suppression and after recovering. *Theor. Exp. Plant Physiol.* **2013**, *25*, 213–222. [CrossRef]
17. Ramzan, T.; Shahbaz, M.; Maqsood, M.F.; Zulfiqar, U.; Saman, R.U.; Lili, N.; Irshad, M.; Maqsood, S.; Haider, A.; Shahzad, B.; et al. Phenylalanine supply alleviates the drought stress in mustard (*Brassica campestris*) by modulating plant growth, photosynthesis and antioxidant defense system. *Plant Physiol. Biochem.* **2023**, *201*, 107828. [CrossRef] [PubMed]

18. Maqsood, M.F.; Shahbaz, M.; Kanwal, S.; Kaleem, M.; Shah, S.M.R.; Luqman, M.; Iftikhar, I.; Zulfiqar, U.; Tariq, A.; Naveed, S.A.; et al. Methionine promotes the growth and yield of wheat under water deficit conditions by regulating the antioxidant enzymes, reactive oxygen species, and ions. *Life* **2022**, *12*, 969. [CrossRef]

19. Randhawa, M.S.; Maqsood, M.; Shehzad, M.A.; Chattha, M.U.; Chattha, M.B.; Nawaz, F.; Yasin, S.; Abbas, T.; Nawaz, M.M.; Khan, R.D.; et al. Light interception, radiation use efficiency and biomass accumulation response of maize to integrated nutrient management under drought stress conditions. *Turk J. Field Crops* **2017**, *22*, 134–142. [CrossRef]

20. Smith, D.; Inman-Bamber, N.; Thorburn, P. Growth and function of the sugarcane root system. *Field Crops Res.* **2005**, *92*, 169–183. [CrossRef]

21. Hunsigi, G. *Production of Sugarcane: Theory and Practice*; Springer Science & Business Media: Berlin, Germany, 2012; Volume 21.

22. Javed, T.; Shabbir, R.; Hussain, S.; Naseer, M.A.; Ejaz, I.; Ali, M.M.; Ahmar, S.; Yousef, A.F. Nanotechnology for endorsing abiotic stresses: A review on the role of nanoparticles and nanocompositions. *Funct. Plant Biol.* **2022**. [CrossRef] [PubMed]

23. Shabbir, R.; Javed, T.; Hussain, S.; Ahmar, S.; Naz, M.; Zafar, H.; Pandey, S.; Chauhan, J.; Siddiqui, M.H.; Pinghua, C. Calcium homeostasis and potential roles in combatting environmental stresses in plants. *South Afr. J. Bot.* **2022**, *148*, 683–693. [CrossRef]

24. Quizenberry, J.E. Breeding plants for drought tolerance and plant water use efficiency. In *Breeding Plants for Less Favorable Environments*; Christiansen, M.N., Lewis, C.F., Eds.; Wiley–Inter–Science: New York, NY, USA, 1982; pp. 193–212.

25. Smit, M.; Singels, A. The response of sugarcane canopy development to water stress. *Field Crops Res.* **2006**, *98*, 91–97. [CrossRef]

26. Ali, A.; Javed, T.; Zaheer, U.; Zhou, J.-R.; Huang, M.-T.; Fu, H.-Y.; Gao, S.-J. Genome–wide identification and expression profiling of the bHLH transcription factor gene family in Saccharum spontaneum under bacterial pathogen stimuli. *Trop. Plant Biol.* **2021**, *14*, 283–294. [CrossRef]

27. Silva, M.D.A.; Jifon, J.L.; DA Silva, J.A.G.; DOS Santos, C.M.; Sharma, V. Relationships between physiological traits and productivity of sugarcane in response to water deficit. *J. Agric. Sci.* **2014**, *152*, 104–118. [CrossRef]

28. Silva, M.d.A.; da Silva, J.A.G.; Enciso, J.; Sharma, V.; Jifon, J. Yield components as indicators of drought tolerance of sugarcane. *Sci. Agric.* **2008**, *65*, 620–627. [CrossRef]

29. Silva, M.d.A.; Soares, R.A.B.; Landell, M.G.d.A.; Campana, M.P. Agronomic performance of sugarcane families in response to water stress. *Bragantia* **2008**, *67*, 655–661. [CrossRef]

30. Inman-Bamber, N.G. Sugarcane water stress criteria for irrigation and drying off. *Field Crops Res.* **2004**, *89*, 107–122. [CrossRef]

31. Filek, M.; Walas, S.; Mrowiec, H.; Rudolphy-Skórska, E.; Sieprawska, A.; Biesaga-Kościelniak, J. Membrane permeability and micro– and microelement accumulation in spring wheat cultivars during the short–term effect of salinity and PEG–induced water stress. *Acta Physiol. Plant.* **2012**, *34*, 985–995. [CrossRef]

32. Bouchemal, K.; Bouldjadj, R.; Belbekri, M.N.; Ykhlef, N.; Djekoun, A. Differences in antioxidant enzyme activities and oxidative markers in ten wheat (*Triticum durum*) genotypes in response to drought, heat and paraquat stress. *Arch. Agron. Soil Sci.* **2016**, *63*, 710–722. [CrossRef]

33. Fang, Y.; Xiong, L. General mechanisms to drought response and their application in drought resistance improvement in plants. *Cell. Mol. Life Sci.* **2015**, *72*, 673–689. [CrossRef]

34. Jain, R.; Chandra, A.; Venugopalan, V.K.; Solomon, S. Physiological changes and expression of SOD and P5CS genes in response to water deficit in sugarcane. *Sugar Technol.* **2015**, *17*, 276–282. [CrossRef]

35. Zhao, D.; Glaz, B.; Comstock, J.C. Sugarcane leaf photosynthesis and growth characters during development of water–deficit stress. *Crop Sci.* **2013**, *53*, 1066–1075. [CrossRef]

36. Natarajan, S.; Basnayake, J.; Lakshmanan, P.; Fukai, S. Genotypic variation in intrinsic transpiration efficiency correlates with sugarcane yield under rainfed and irrigated field conditions. *Physiol. Plant.* **2020**, *172*, 976–989. [CrossRef] [PubMed]

37. da Graça, J.P.; Rodrigues, F.A.; Farias, J.R.B.; de Oliveira, M.C.N.; Hoffmann-Campo, C.B.; Zingaretti, S.M. Physiological parameters in sugarcane cultivars submitted to water deficit. *Braz. J. Plant Physiol.* **2010**, *22*, 189–197. [CrossRef]

38. Endres, L.; dos Santos, C.M.; Silva, J.V.; Barbosa, G.V.D.S.; Silva, A.L.J.; Froehlich, A.; Teixeira, M.M. Inter–relationship between photosynthetic efficiency, 13C, antioxidant activity and sugarcane yield under drought stress in field conditions. *J. Agron. Crop Sci.* **2019**, *205*, 433–446. [CrossRef]

39. De Silva, A.; De Costa, W. Varietal variation in stomatal conductance, transpiration and photosynthesis of commercial sugarcane varieties under two contrasting water regimes. *Trop. Agric. Res. Ext.* **2009**, *12*, 97–102. [CrossRef]

40. Songsri, P.; Nata, J.; Bootprom, N.-A.; Jongrungkl, N. Performances of elite sugarcane genotypes for agro–physiological traits in relation to yield potential and ratooning ability under rain–fed conditions. *J. Agron.* **2020**, *19*, 1–13. [CrossRef]

41. Hemaprabha, G.; Swapna, S.; Lavanya, D.L.; Sajitha, B.; Venkataramana, S. Evaluation of drought tolerance potential of elite genotypes and progenies of sugarcane (*Saccharum* sp. hybrids). *Sugar Technol.* **2013**, *15*, 9–16. [CrossRef]

42. Nair, N.V.; Mohanraj, K.; Sunadaravelpandian, K.; Suganya, A.; Selvi, A.; Appunu, C. Characterization of an intergeneric hybrids of *Erianthus procerus* × *Saccharum officinarum* and its backcross progenies. *Euphytica* **2017**, *213*, 267–277. [CrossRef]

43. Zhou, J.-R.; Sun, H.-D.; Ali, A.; Rott, P.C.; Javed, T.; Fu, H.-Y.; Gao, S.-J. Quantitative proteomic analysis of the sugarcane defense responses incited by Acidovorax avenae subsp. avenae causing red stripe. *Ind. Crops Prod.* **2021**, *162*, 113275. [CrossRef]

44. Hong, D.-K.; Talha, J.; Yao, Y.; Zou, Z.-Y.; Fu, H.-Y.; Gao, S.-J.; Xie, Y.; Wang, J.-D. Silicon enhancement for endorsement of *Xanthomonas albilineans* infection in sugarcane. *Ecotoxicol. Environ. Saf.* **2021**, *220*, 112380. [CrossRef]

45. Cui, D.; Huang, M.T.; Hu, C.Y.; Su, J.B.; Lin, L.H.; Javed, T.; Deng, Z.H.; Gao, S.J. First report of *Pantoea stewartii* subsp. stewartii causing bacterial leaf wilt of sugarcane in China. *Plant Dis.* **2021**, *105*, 1190. [CrossRef] [PubMed]

46. Hu, Z.T.; Ntambo, M.S.; Zhao, J.Y.; Javed, T.; Shi, Y.; Fu, H.Y.; Huang, M.T.; Gao, S.J. Genetic Divergence and Population Structure of *Xanthomonas albilineans* Strains Infecting *Saccharum* spp. Hybrid and *Saccharum officinarum*. *Plants* **2023**, *12*, 1937. [CrossRef]
47. Shabbir, R.; Zhaoli, L.; Yueyu, X.; Zihao, S.; Pinghua, C. Transcriptome Analysis of Sugarcane Response to Sugarcane Yellow Leaf Virus Infection Transmitted by the Vector Melanaphis sacchari. *Front. Plant Sci.* **2022**, *13*, 921674. [CrossRef] [PubMed]
48. Nassif, D.S.P.; Marin, F.R.; Costa, L.G. Evapotranspiration and transpiration coupling to the atmosphere of sugarcane in southern Brazil: Scaling up from leaf to field. *Sugar Technol.* **2014**, *16*, 250–254. [CrossRef]
49. Jamaux, I.; Steinmetz, A.; Belhassen, E. Looking for molecular and physiological markers of osmotic adjustment in sunflower. *New Phytolo.* **1997**, *137*, 117–127. [CrossRef]
50. Bates, L.S.; Waldren, R.P.; Teare, I.D. Rapid determination of free proline for water–stress studies. *Plant Soil* **1973**, *39*, 205–207. [CrossRef]

Article

Climate Variations in the Low-Latitude Plateau Contribute to Different Sugarcane (*Saccharum* spp.) Yields and Sugar Contents in China

Yong Zhao [1,2], Ling-Xiang Yu [3], Jing Ai [1,2], Zhong-Fu Zhang [1,2], Jun Deng [1,2] and Yue-Bin Zhang [1,2,*]

[1] Sugarcane Research Institute, Yunnan Academy of Agricultural Sciences, Kaiyuan 661699, China; 18087395132@163.com (Y.Z.); zhongfu_zhang@yeah.net (Z.-F.Z.)
[2] National Key Laboratory for Biological Breeding of Tropical Crops, Kunming 650205, China
[3] Climate Center, Yunnan Meteorological Bureau, Kunming 650034, China; yulx@163.com
* Correspondence: ynzyb@sohu.com; Tel.: +86-136-0873-2528

Abstract: In China, the main sugarcane (*Saccharum* spp.) planting areas can be found in the low-latitude plateau (21° N–25° N, 97° E–106° E), which has most of the natural ecological types. However, there is limited information on the climate conditions of this region and their influence on sugarcane yield and sucrose content. Monthly variations in the main climate factors, namely, average air temperature (AAT), average relative humidity (ARH), average rainfall amount (ARA), and average sunshine duration (ASD), from 2000 to 2019 and sugarcane yield and sucrose content of 26 major sugarcane-producing areas from 2001/2002 to 2018/2019 were collected from the low-latitude plateau in Yunnan for studying the impact of climate variations on sugarcane yield and sucrose content. The results showed that AAT in the mid-growth season had a significant positive correlation with sucrose content ($p < 0.05$), and AAT in the late-growth season had a very significant positive correlation with sucrose content ($p < 0.01$). ARH in the mid-growth season had a significant positive correlation with sugarcane yield ($p < 0.05$). ARA in the early-growth season showed a significant positive correlation with sugarcane yield ($p < 0.05$). ASD in the late-growth season had a significant positive correlation with sugarcane yield ($p < 0.05$) and sucrose content ($p < 0.01$). The rainy and humid sugarcane areas were characterized by high ARA and ARH during the entire growth period, low AAT and ASD in the mid-growth season, and low AAT in the late-growth season, contributing to a high sugarcane yield, but not a high sucrose content. The low temperature and sunshine semi-humid sugarcane areas were characterized by the lowest AAT in the early and middle stages of sugarcane growth, less ASD in the early and middle stages, and less ARA in the early and late stages, which are unfavorable for sugarcane yield and sucrose content. The high temperature and humidity sugarcane areas were characterized by higher AAT and ARA, and moderate ASD during the entire growth period, resulting in good sugarcane growth potential and contributing to the sugarcane yield and sucrose content. The semi-humid and multi-sunshine sugarcane areas were characterized by the lowest ARH in the entire growth period, the lowest ARA in the middle and late seasons, and the longest ASD, contributing to an increase in sucrose content. The humid and sunny areas were characterized by the longest ASD and high ARH in the early and late seasons of sugarcane growth and moderate AAT and ARA during the entire growth season, which are beneficial for high sugarcane yield and sucrose content. Overall, these findings suggest that the sugarcane variety layout should be based on the climate type (of which there are five in the plateau), and corresponding cultivation practices should be used to compensate for the climatic conditions in various growth stages.

Keywords: sugarcane (*Saccharum* spp.); climate condition; average air temperature; average relative humidity; average rainfall amount; average sunshine duration

Citation: Zhao, Y.; Yu, L.-X.; Ai, J.; Zhang, Z.-F.; Deng, J.; Zhang, Y.-B. Climate Variations in the Low-Latitude Plateau Contribute to Different Sugarcane (*Saccharum* spp.) Yields and Sugar Contents in China. *Plants* **2023**, *12*, 2712. https://doi.org/10.3390/plants12142712

Academic Editor: Yasutomo Hoshika

Received: 5 June 2023
Revised: 7 July 2023
Accepted: 17 July 2023
Published: 21 July 2023

1. Introduction

Sugarcane (*Saccharum* spp.) is a perennial tropical and subtropical herb and a C4 crop, and it is an important raw material for sugar production worldwide. Because of its strong photosynthetic capacity, high biological yield, and total fermentable sugar, sugarcane has shown good developmental prospects for utilization as an energy material [1]. Currently, sugarcane is mainly used for sugar production in China, accounting for 90% of the total sugar yield. Increasing the sugar yield has always been an important aim of sugarcane planting; this is mainly affected by the biomass yield and sugar content of sugarcane, which depend on many factors such as climate and meteorological factors [2], varieties [3], cultivation measures [4], soil fertility [5], and pest control [6]. Meteorological factors directly affect growth and regional distribution of sugarcane planting [7]. The climate of Yunnan basically belongs to the subtropical plateau monsoon type, with remarkable three-dimensional climate characteristics. There are many climate types in Yunnan, with small annual temperature differences, large daily temperature differences, distinct dry and wet seasons, and obvious vertical variations in temperature with terrain height. Yunnan has cold, warm, and hot (including the subtropical zone) climates. In Yunnan, the low-latitude plateau (21° N–25° N, 97° E–106° E) is one of the most important sugarcane-planting areas and has most of the ecological types in China [8]. The sugarcane areas show diverse climate types, large climate differences, and significant differences in sugarcane yield and sucrose content. A comprehensive understanding of the climate characteristics of this region can provide basic data support for improving sugarcane cultivation measures, implementing an optimized variety layout, increasing the sucrose yield, and tapping the sugarcane-production potential.

Climate conditions have a significant impact on sugarcane yield and sucrose content, of which temperature, rainfall, humidity, and sunshine are important factors that limit sugarcane production [9]. Previous studies have shown that sunshine and accumulated temperature have a very significant positive correlation with sugarcane yield [10], and rainfall has a key impact on sugarcane elongation and maturity, and a positive correlation with yield [9]. Aboveground dry weight, cane yield, and cane sucrose yield are significantly related to temperature and rainfall [11]. Atmospheric relative humidity, temperature, and sunshine are important factors that affect sugarcane elongation, sucrose accumulation, and yield [12]. Fabio and Gustavo used multi-year data from 37 meteorological stations to simulate and evaluate the potential and achievable yield of sugarcane and found that climate factors could explain 43% variability in sugarcane yield efficiency [13]. The order of importance of meteorological factors that affect yield is solar radiation, water shortage, maximum temperature, rainfall, and minimum temperature. Araujo et al. found that rainfall was an important factor that affected the yield of sugarcane [14]. The sucrose content of sugarcane is more vulnerable than the yield to the impact of air temperature; the process of sucrose accumulation in stems is affected by many factors, especially climate conditions such as temperature and water [15].

Sugarcane has a prolonged growth cycle that is typically divided into five stages: germination, seedling, tillering, elongation, and maturity. The germination and seedling stages are the early stages of sugarcane production, which are characterized by low temperature and inadequate rainfall in the regions of the low-latitude plateau. The middle stage of sugarcane growth comprises tillering and elongation, and it is characterized by elevated temperature and abundant rainfall. The maturity stage is the final phase of sugarcane development; this period is crucial for sugarcane yield and sucrose content, with a large difference in temperature between morning and evening, and moderate rainfall. The growth characteristics of sugarcane are variable across different time periods and are significantly influenced by climatic factors [8]. Previous studies [12] have extensively investigated the impact of climatic factors on sugarcane production. These studies used various methodologies, including relevant models, to assess sugarcane production [10], conducted regression analyses to establish a correlation between climatic factors and sugarcane yield [11], and examined the impacts of climatic factors during specific planting periods on sugarcane growth [14].

However, few studies have examined the impacts of climatic factors on the overall yield and sucrose content of sugarcane at different growth stages on a large scale. Additionally, there is a lack of information on the distribution of sugarcane yield and sucrose content potential in relation to specific climatic characteristics. Moreover, the ecological characteristics of sugarcane-growing areas in the low-latitude plateau are complex, and the climate characteristics of each sugarcane-growing area are unique. There is limited comprehensive data regarding changes in the characteristics of the main climate-influencing factors. Furthermore, few studies have elucidated the impacts of climate change on sugarcane yield and sugar content in each growth period. Therefore, the appropriate distribution of sugarcane varieties and use of cultivation measures in different growing regions need to be considered for improving both yield and sucrose content, particularly in the context of climate change. The objectives of this study were to (1) understand the detailed characteristics of the ecological climate in the low-latitude plateau sugarcane region, (2) comprehend the characteristics of the main climate factors in different periods of sugarcane growth and their correlation with sugarcane yield and sucrose content, and (3) delicately divide the ecological types of sugarcane areas and understand the impact of current climate conditions on sugarcane production in the low-latitude plateau.

2. Results

2.1. Differences in Climate Factors at Different Sites

From 2000 to 2009, monthly AAT, ARH, ARA, and ARD values of the test sites in 26 counties in seven prefectures in the low-latitude plateau of Yunnan (classified and analyzed by administrative regions) were statistically analyzed (Figure 1). Differences were detected in monthly AAT, ARH, ARA, and ARD of 26 test sites, but the overall trend was consistent. Monthly AAT was the lowest in December and January (5–15 °C), highest in June, July, and August (20–25 °C), increased month by month from February to June, and declined significantly from August to January. Monthly ARH was the highest in June, July, and August (75% to 90%), lowest in March (50% to 70%), increased month by month from March to July, and decreased month by month from July to February. For monthly ARA, obvious differences in monthly ARA values were observed among the main producing counties. Generally, monthly ARA was the highest in July (between 150 mm and 500 mm), lowest in November to March (between 0 mm and 50 mm), mainly concentrated in May to October, increased month by month from March to July, and decreased month by month from July to February. For monthly ASD, obvious differences in monthly ASD values were observed among the test sites. Monthly ASD was the shortest in July (60–150 h), longest from November to March and from January to March (100–260 h), decreased month by month from March to July, and increased month by month from July to February. Monthly ARH, ARA, and ARD showed a consistent trend among the test sites (counties) in the same prefecture, but differed in different prefectures. Monthly AAT, ARH, ARA, and ARD of 26 test sites were obviously different in the same prefecture (Figure 2).

2.2. Climatic Differences in Different Growth Seasons of Sugarcane

Differential analysis of the main climatic factors in three growth seasons of sugarcane showed significant differences in AAT, ARH, ARA, and ASD in different growth seasons ($p < 0.01$). AAT was the lowest in the late-growth season and highest in the mid-growth season. ARH was the lowest in the early-growth season and highest in the mid-growth season. ARA was the highest in the mid-growth season and lower in the late-growth season. ASD was the longest in the late-growth season and shortest in the mid-growth season (Figure 3).

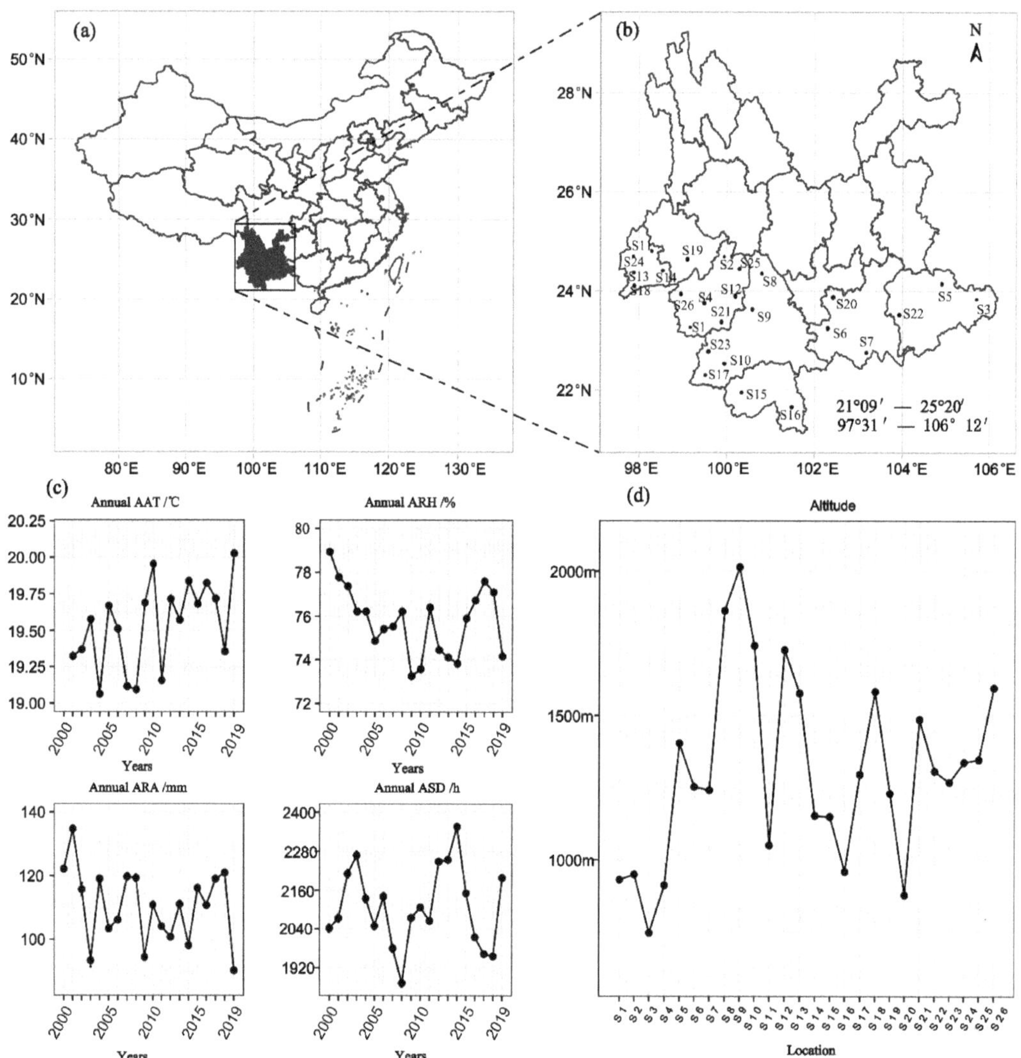

Figure 1. Study area. (**a**) Location of Yunnan in China. (**b**) Distribution of the study sites. (**c**) Annual average air temperature (AAT), average relative humidity (ARH), average rainfall amount (ARA), and average sunshine duration (ASD) across all sites over 20 years. (**d**) Average altitude of the test sites.

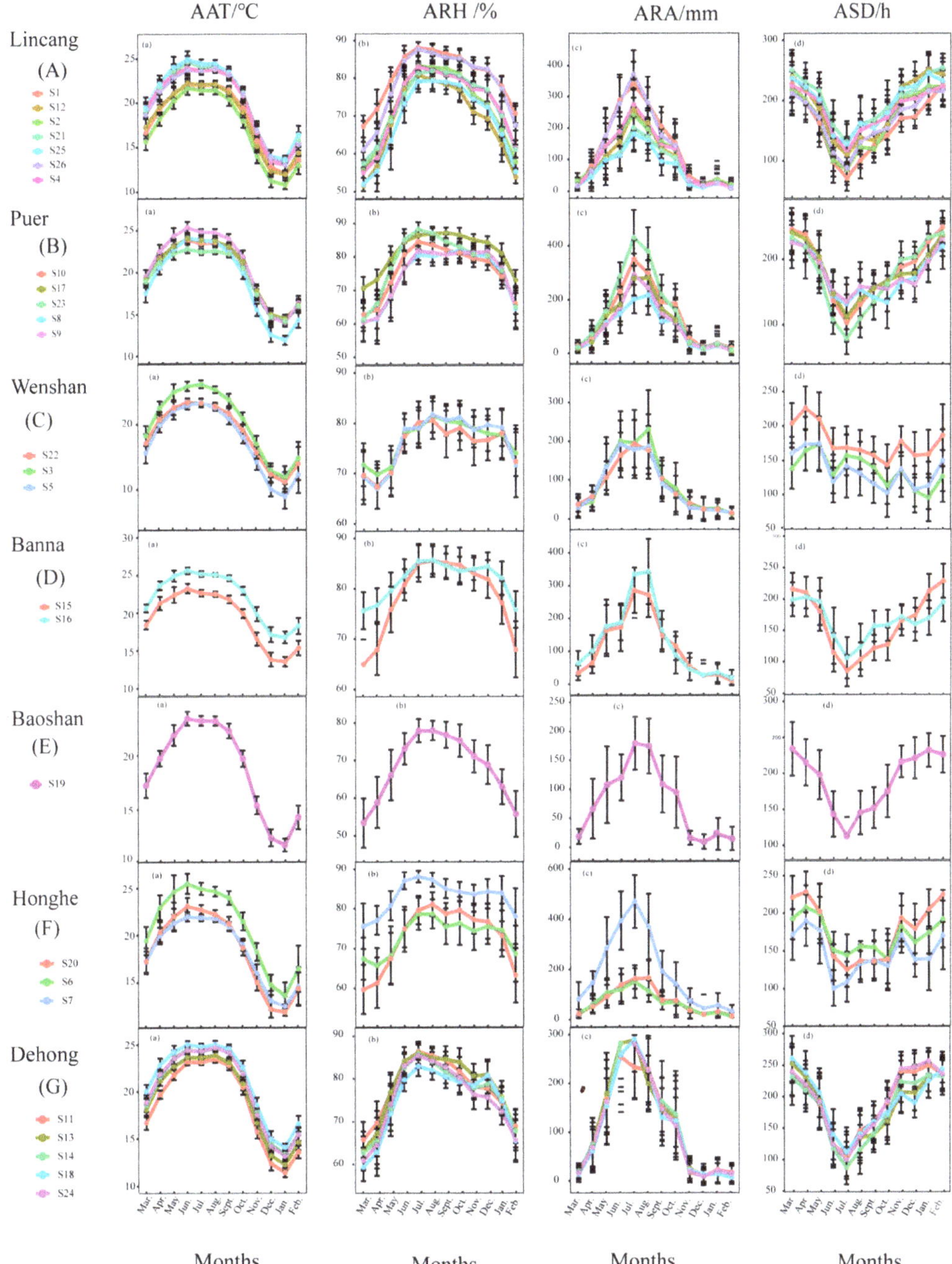

Figure 2. (a) Monthly average air temperature (AAT), (b) average relative humidity (ARH), (c) average rainfall amount (ARA), and (d) average sunshine duration (ASD) in 20 years (2000 to 2019) in 26 counties in seven prefectures in the low-latitude plateau of Yunnan. Error bars show standard error of the mean.

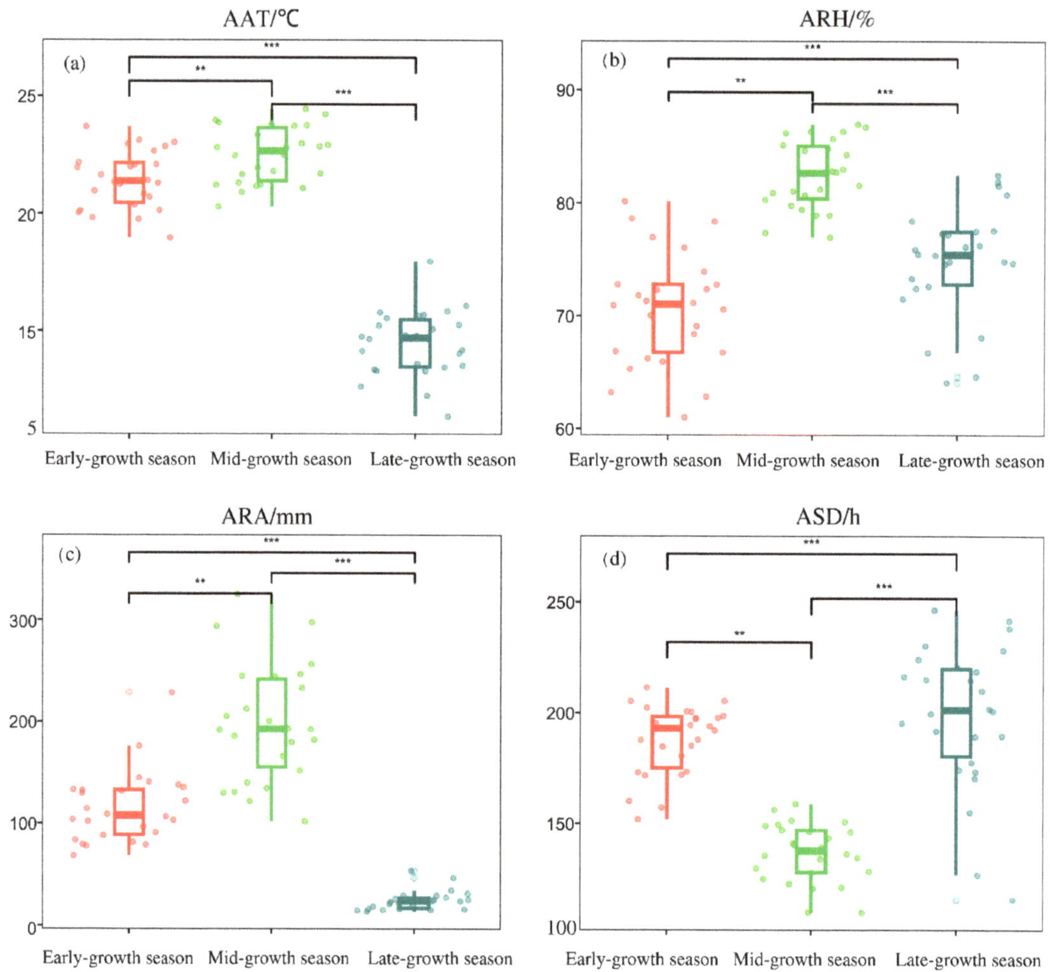

Figure 3. Differential analysis of average air temperature (AAT), average relative humidity (ARH), average rainfall amount (ARA), and average sunshine duration (ASD) in different sugarcane growth seasons. (**a**) average air temperature (AAT) in different seasons. (**b**) average relative humidity (ARH) in different seasons. (**c**) average rainfall amount (ARA) in different seasons. (**d**) average sunshine duration (ASD) in different seasons. ** and *** *F*-values are significant at $p < 0.01$ and $p < 0.001$ levels, respectively. Scatter plots show the mean of weather data from 2000 to 2019 at 26 sites.

2.3. Correlation between Climate Factors in Different Growth Periods and Sugarcane Production

2.3.1. Correlation between AAT and Sugarcane Production

Linear analysis of the sugarcane yield and sucrose content of 26 sites from 2001/2002 to 2015/2016 with the corresponding AAT of early-growth, mid-growth, and late-growth seasons showed that AAT in the mid-growth season had a significant positive correlation with sucrose content ($p < 0.05$) and AAT of the late-growth season had a significant positive correlation with sucrose content ($p < 0.01$; Figure 4).

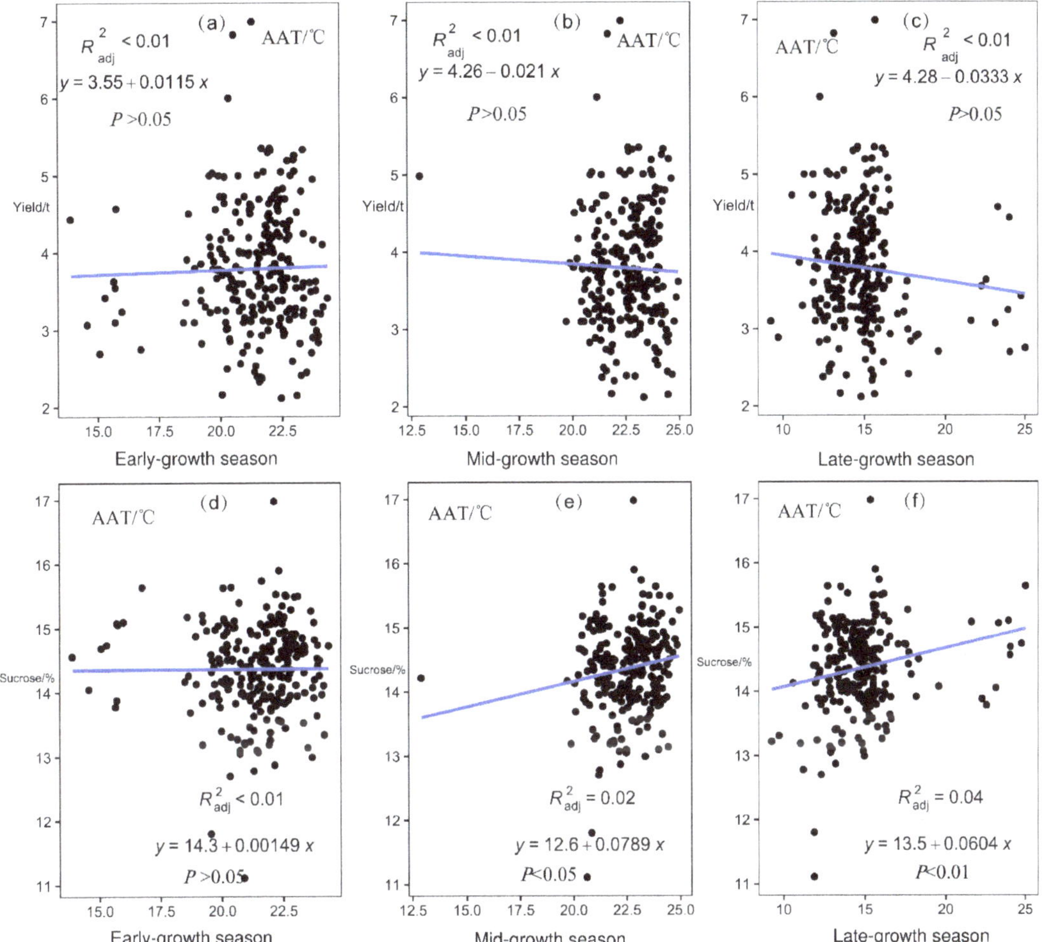

Figure 4. Linear analysis of average air temperature (AAT) and yield and sucrose content across all sites in different sugarcane growth seasons. (**a**) Linear analysis of average air temperature (AAT) and yield in early-growth season. (**b**) Linear analysis of average air temperature (AAT) and yield in mid-growth season. (**c**) Linear analysis of average air temperature (AAT) and yield in late-growth season. (**d**) Linear analysis of average air temperature (AAT) and sucrose in early-growth season. (**e**) Linear analysis of average air temperature (AAT) and sucrose in mid-growth season. (**f**) Linear analysis of average air temperature (AAT) and sucrose in late-growth season. Scatter plots show the mean of weather data from 2005/2006 to 2015/2016 at 26 sites.

2.3.2. Correlation between ARH and Sugarcane Production

Linear analysis of the sugarcane yield and sucrose content of 26 sites from 2001/2002 to 2015/2016 with the corresponding ARH of early-growth, mid-growth, and late-growth seasons showed that ARH in the mid-growth season had a significant positive correlation with sugarcane yield ($p < 0.05$; Figure 5).

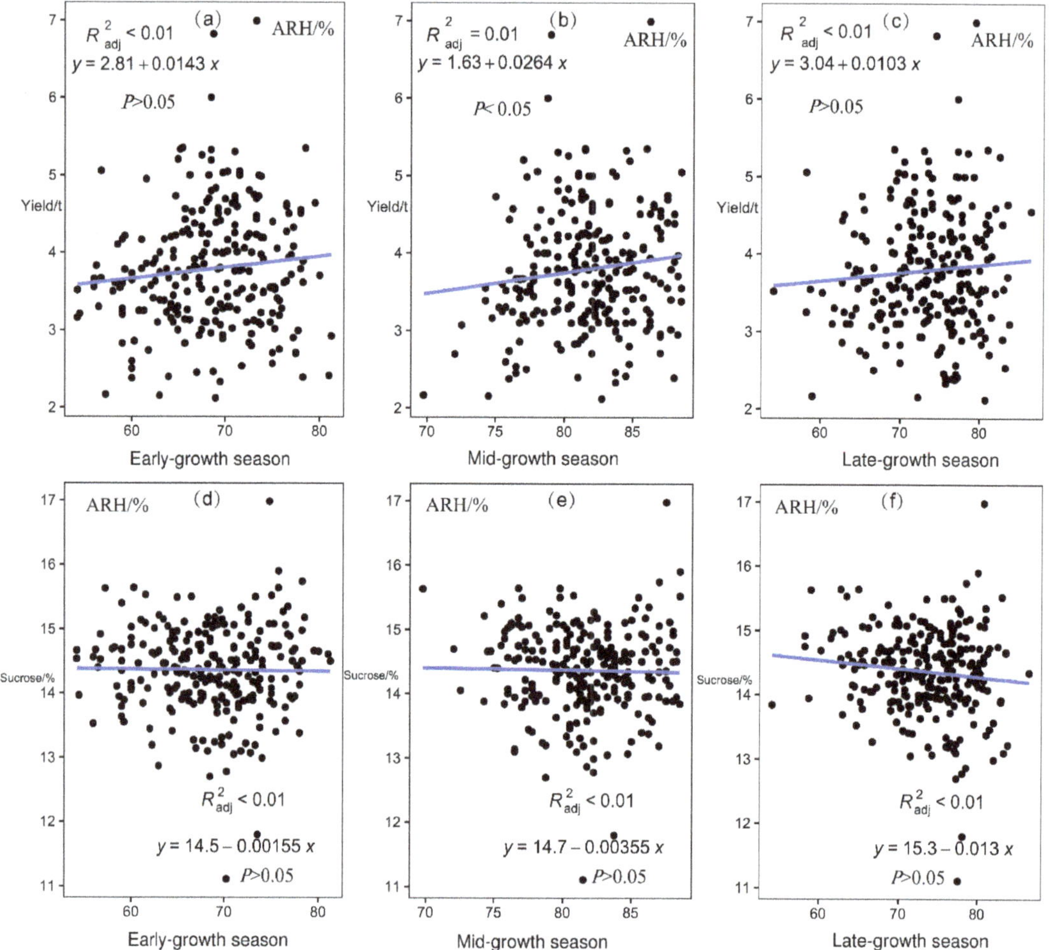

Figure 5. Linear analysis of average relative humidity (ARH) and yield and sucrose content across all sites in different sugarcane growth seasons. (**a**) Linear analysis of average relative humidity (ARH) and yield in early-growth season. (**b**) Linear analysis of average relative humidity (ARH) and yield in mid-growth season. (**c**) Linear analysis of average relative humidity (ARH) and yield in late-growth season. (**d**) Linear analysis of average relative humidity (ARH) and sucrose in early-growth season. (**e**) Linear analysis of average relative humidity (ARH) and sucrose in mid-growth season. (**f**) Linear analysis of average relative humidity (ARH) and sucrose in late-growth season. Scatter plots show the mean of weather data from 2005/2006 to 2015/2016 at 26 sites.

2.3.3. Correlation between ARA and Sugarcane Production

Linear analysis of the sugarcane yield and sucrose content of 26 sites from 2001/2002 to 2015/2016 with the corresponding ARA of early growth, mid-growth, and late-growth seasons showed that ARA in the early-growth season had a significant positive correlation with sugarcane yield ($p < 0.05$; Figure 6).

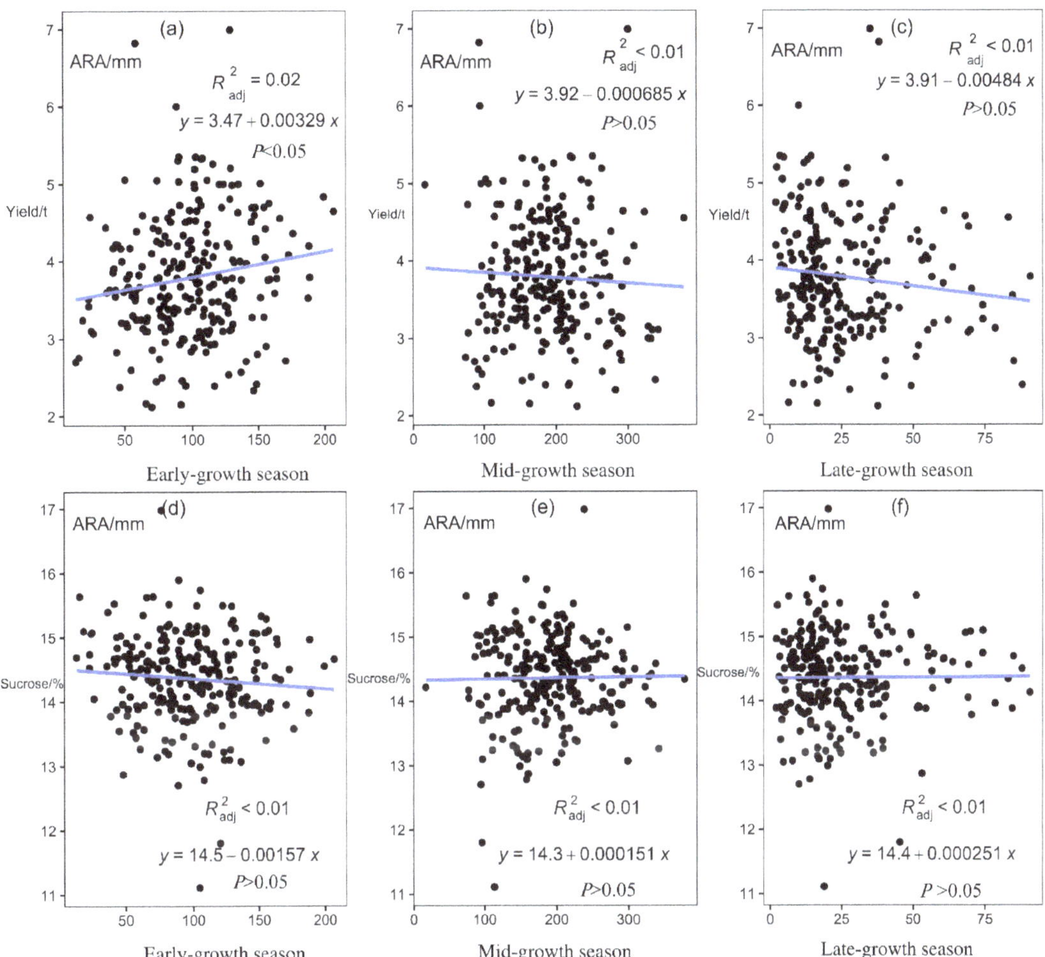

Figure 6. Linear analysis of average rainfall amount (ARA) and yield and sucrose content across all sites in different sugarcane growth seasons. (**a**) Linear analysis of average rainfall amount (ARA) and yield in early-growth season. (**b**) Linear analysis of average rainfall amount (ARA) and yield in mid-growth season. (**c**) Linear analysis of average rainfall amount (ARA) and yield in late-growth season. (**d**) Linear analysis of average rainfall amount (ARA) and sucrose in early-growth season. (**e**) Linear analysis of average rainfall amount (ARA) and sucrose in mid-growth season. (**f**) Linear analysis of average rainfall amount (ARA) and sucrose in late-growth season. Scatter plots show the mean of weather data from 2005/2006 to 2015/2016 at 26 sites.

2.3.4. Correlation between ASD and Sugarcane Production

Linear analysis of the sugarcane yield and sucrose content of 26 sites from 2001/2002 to 2015/2016 with the corresponding ASD of early-growth, mid-growth, and late-growth seasons showed that ASD in the late-growth season had a significant positive correlation with sugarcane yield ($p < 0.05$) and sucrose content ($p < 0.01$; Figure 7).

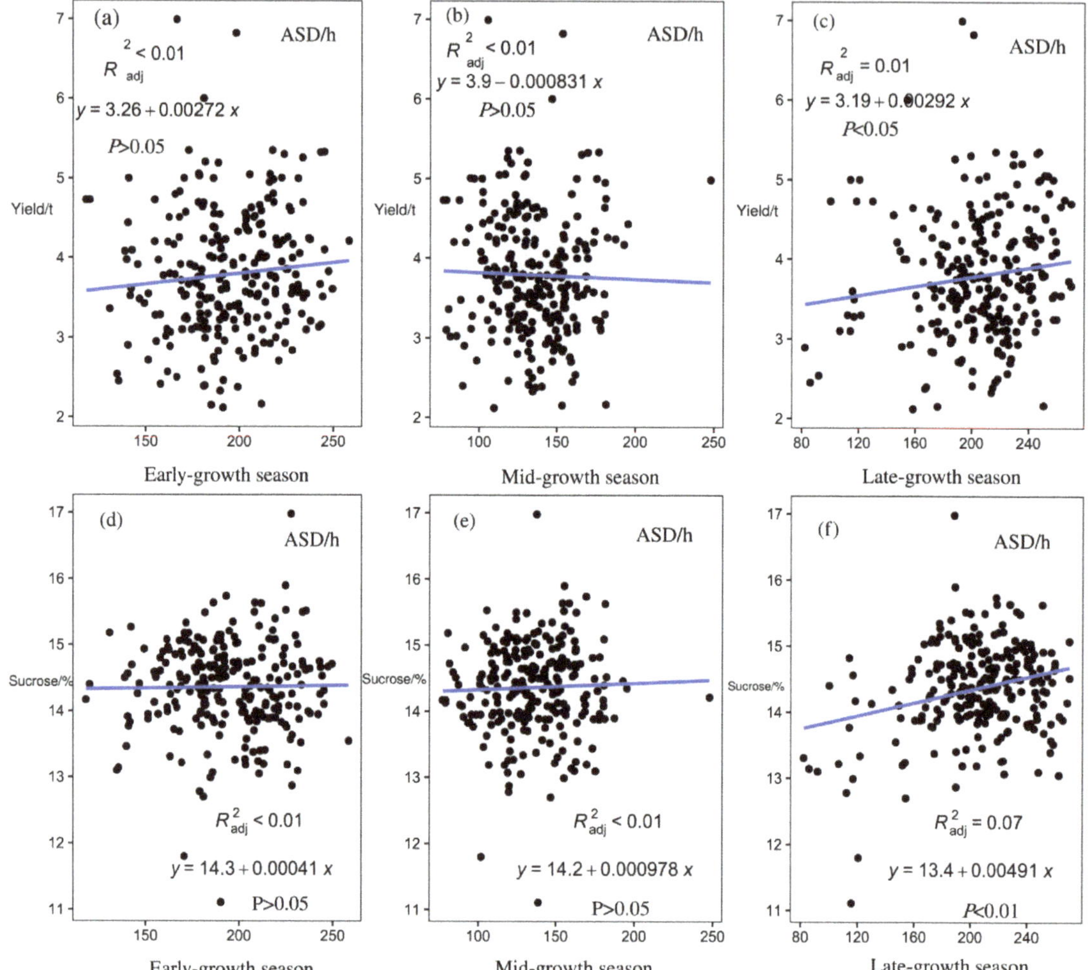

Figure 7. Linear analysis of average sunshine duration (ASD) and yield and sucrose content across all sites in different sugarcane growth seasons. (**a**) Linear analysis of average sunshine duration (ASD) and yield in early-growth season. (**b**) Linear analysis of average sunshine duration (ASD) and yield in mid-growth season. (**c**) Linear analysis of average sunshine duration (ASD) and yield in late-growth season. (**d**) Linear analysis of average sunshine duration (ASD) and sucrose in early-growth season. (**e**) Linear analysis of average sunshine duration (ASD) and sucrose in mid-growth season. (**f**) Linear analysis of average sunshine duration (ASD) and sucrose in late-growth season. Scatter plots show the mean of weather data from 2005/2006 to 2015/2016 at 26 sites.

2.4. Division of Sugarcane Growth Climate Types

2.4.1. Cluster Analysis of 26 Sites

According to the differences in AAT, ARH, ARA, and ASD in the early-growth, middle-growth, and late-growth seasons, 26 major sugarcane-producing regions were divided into five groups at a European distance of 70, where s7 was a separate group; s3 and s5 were the second group; s23, s10, s14, s16, s1, and s26 were the third group; and s21, s19, s25, s22, s6, and s20 were the fourth group. The rest of the sites formed the fifth group (Figure 8).

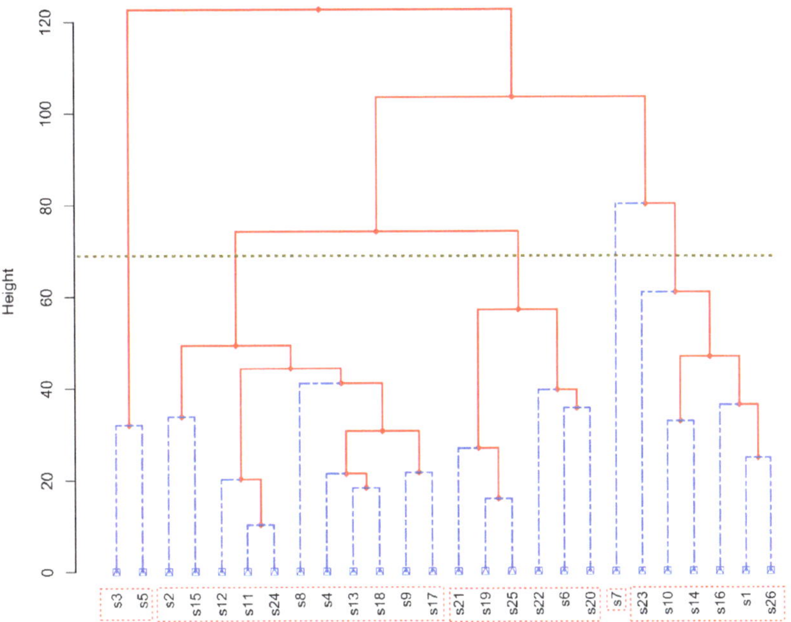

Figure 8. Cluster analysis of 26 sites on the basis of average air temperature (AAT), average relative humidity (ARH), average rainfall amount (ARA), and average sunshine duration (ASD) from 2000 to 2019 (20 years) in different sugarcane growth seasons.

2.4.2. Climate Differences among Different Cluster Groups

According to the differential analysis of climate factors of the five groups, AAT in the early-growth season of the second group was significantly lower than that of other groups and AAT in the early-growth season of the third group was significantly higher than that of other groups. ARA of the first group was significantly higher than that of other groups. The fifth group showed the longest ASD in the early-growth season, and the second group had the shortest ASD. The first group showed the highest ARH, and the fourth group had the lowest humidity in the early-growth season (Figure 9a). In the mid-growth season, the AAT of the first group was significantly lower than that of the other groups. The ARA of the first group was significantly higher than that of the other groups, and the ARA of the fourth group was significantly lower than that of the other groups. ASD values of the first and second groups were the lowest, and ASD values of the fourth and fifth groups were the highest. The ARH of the first group was the highest, and the ARH of the fourth group was the lowest (Figure 9b). In the late-growth season, the AAT of the first group was significantly lower than that of the other groups, and the ARA of the first group was significantly higher than that of the other groups, followed by the third group. The ATA values of the second and fourth groups were the lowest. The ASD of the fourth group was the longest and significantly higher than that of the other groups. The ARH was the highest in the first group and lowest in the fourth group (Figure 9c). The climate first type can be found in rainy and humid sugarcane areas characterized by high ARA and ARH during the entire growth period, low AAT and ASD in the mid-growth season, and low AAT in the late-growth season. The second type can be found in low-temperature and sunshine semi-humid sugarcane areas characterized by the lowest AAT in the early and middle stages of sugarcane growth, less ASD in the early and middle stages, and less ARA in the early and late stages. The third type can be found in high temperature and humidity sugarcane areas characterized by higher AAT and ARA, and moderate ASD during the entire growth period. The fourth type can be found in semi-humid and multi-sunshine sugarcane areas characterized by the lowest ARH in the entire growth period, lowest ARA in the middle and late seasons, and longest ASD. The fifth type can be found in humid and sunny areas characterized by the longest ASD and high ARH in the early and late seasons of sugarcane growth and moderate AAT and ARA during the entire growth season.

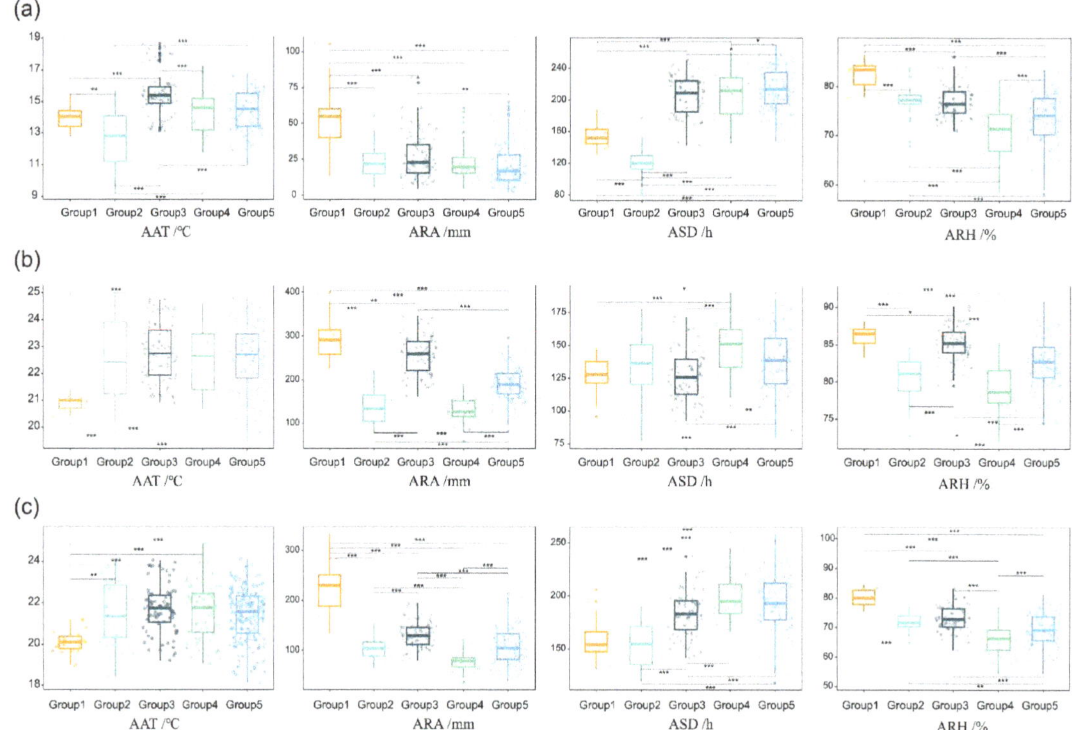

Figure 9. Differences in the average air temperature (AAT), average relative humidity (ARH), average rainfall amount (ARA), and average sunshine duration (ASD) of five cluster groups in three growth seasons. *, **, and *** F-values significant at $p < 0.05$, $p < 0.01$, and $p < 0.001$ levels, respectively. (**a**) Early-growth season; (**b**) mid-growth season; (**c**) later-growth season.

3. Discussion

Sugarcane is a crop that requires high temperature and humidity and strong light. Sugarcane planting is mainly concentrated between southern latitude 25° and northern latitude 25°. The growth of sugarcane has an important relationship with climate factors. Some researchers have reported that the growth of sugarcane is closely related to climate factors such as sunshine duration, temperature, rainfall, and humidity, and the factors are required in different amounts in different periods of sugarcane growth [16]. The climate characteristics of the low-latitude plateau of Yunnan are unique and cover the main climate types in China's sugarcane regions, which are generally characterized by high temperature, humidity, and sunshine [17]. The low-latitude plateau of Yunnan is mountainous and hilly; sugarcane is mainly planted on dry and sloping land, which is barren and characterized by low soil moisture. Therefore, rainfall has become the main factor for the early growth of sugarcane in Yunnan. In this study, the climate of the sugarcane region in the low-latitude plateau of Yunnan was found to be generally dry in winter and early spring, with very little rainfall from November to March (0–50 mm; Figure 2). This season is also the key period for sugarcane planting. Therefore, early rainfall has become the main limiting factor for sugarcane emergence and tillering, which directly affects sugarcane yield. The accumulated rainfall of 10–20 days after sugarcane planting is the key meteorological factor that determines the emergence of sugarcane seedlings [18]. Water deficits can even reduce sugarcane yield by up to 60% [19], and the development of sugarcane is directed to areas with high water availability for growth and production [20]. In this study, ARA values in the mid-growth and late-growth seasons were not related to the sugarcane yield and sucrose content. Some studies have found that, in the field, sugarcane height, stem diameter, and yield are not easily affected by rainfall, which has little impact on sucrose content in the early harvest period [21]. Sugarcane ripening has been associated with both incident sunlight and temperature, but not with rainfall [11], and sugarcane production was reduced in only the dry and rainy years [22]. This is consistent with the results of this study; we found that rainfall in the mid-growth season was sufficient and rainfall in the later-growth season was low (Figure 3c), which are favorable conditions for sugarcane growth on the basis of the analysis of rainfall in the last 20 years.

Temperature is the main factor that affects the germination of sugarcane buds. Within a certain range, with an increase in temperature, the activity of enzymes in the seedlings increases as respiratory metabolism increases, and the germination speed of seed buds accelerates. The minimum temperature for bud germination in most varieties is 13 °C, and the optimum temperature is 20–30 °C; the minimum temperature for rooting is 10 °C, and the optimum temperature is 25 °C [8]. The analysis of 26 major sugarcane-producing areas showed that the lowest AAT in the low-latitude plateau area was mainly in January and February (10–13 °C), and the temperature began to increase in March (at about 15 °C) and then increased month by month. The highest average temperature was in June and July (about 25 °C), and the temperature characteristics in this area were more suitable for sugarcane growth. We found that the AAT in the middle and late stages affected the sucrose content of sugarcane and detected a significant positive correlation between temperature and sucrose content. A number of other modelling studies have reported the positive effect of increased temperatures on sugarcane yield [23]. Lower maximum and higher minimum temperatures and light wind have been found to be favorable for the growth of sugarcane in the active and elongation stages [24]. Luo et al. reported that temperature has a lower impact on sugarcane growth; temperature mainly affects sucrose content [21], but it has little impact on other aspects. The emergence rate, plant emergence rate, and tillering rate of sugarcane are significantly affected by rainfall, whereas plant height, stem diameter, and yield are not easily affected by rainfall. Cold and freezing injury of sugarcane is one of the main meteorological disasters in the low-latitude plateau region of Yunnan [25]. The low-temperature and high-humidity conditions during the start of the crushing season affect sugar accumulation in sugarcane [23]. In this study, sugarcane growth was divided into three periods. In the mid-growth season, the rainfall was sufficient and temperature and radiation became the main limiting factors for sugarcane growth and sucrose accumulation. Since 2008, severe low temperatures and drought were recorded almost every year, which caused a continuous decrease in cane and sucrose productivity in Guangxi and Yunnan; the decreasing trend reached the bottom in the milling year 2010/2011 [26]. In the late-growth season, low temperature is the main threat to sugarcane production, and an increase in average accumulated temperature is beneficial to the increase in sucrose content in sugarcane.

Rupa found that maximum and minimum temperatures and relative humidity during the first three months of crop production (germination and tillering phases) have a profound influence on the yield [27]. In this study, a significant correlation was found between relative humidity and sugarcane yield in the mid-growth season, which is consistent with the results of a previous study. The yields of sugarcane are strongly correlated across irrigated and rainfed environments [28]. A previous study revealed that maximum temperature, morning humidity, and sunshine hours play the most important roles during the germination stage Generally, variations in relative humidity result in flowering in certain varieties of sugarcane [29] and occurrence of the sugarcane leaf disease [30,31]. In the low-latitude plateau area, ASD was the longest in the late-growth season of sugarcane and shortest in the mid-growth season. In this study, ASD was found to be average (>200 h/month) in the late-growth stage, with a significant positive correlation between the yield and sucrose content of sugarcane. Sugarcane plant height and stem diameter have been significantly or extremely significantly correlated with average temperature, precipitation, and sunshine duration [18]. Huang also found that, in the early stage of sugarcane growth (from April to June) [32], the three factors of sunshine duration, temperature, and water were relatively well-matched and the seedling condition was good, which laid a good foundation for the later high yield of sugarcane. In sugarcane, sucrose accumulation and ASD showed a significant positive correlation ($p < 0.01$) [33]. In fact, the yield and sucrose content of sugarcane are not affected by a single climatic factor, but by multiple factors. The growth and yield of sugarcane are profoundly influenced by climate elements. This may be due to a large growing season for good phenological development, high germination, more tillers, and maximally millable canes. Appropriate temperature, sunshine duration, humidity, rainfall, etc., are favorable for germination, tillering, and cane formation.

Sugarcane planting can be subdivided into four growth stages according to physiological and growth characteristics: seed germination, tillering, stalk elongation to maturity, and flowering. In this study, sugarcane planting was divided into three important growth periods according to seasonal changes: early-growth season, mid-growth season, and late-growth season. Meteorological factors such as AAT, ARH, ARA, and ASD were found to significantly affect the growth of sugarcane in different seasons ($p < 0.05$, $p < 0.01$; Figures 4–7). Moreover, in different growth seasons, these meteorological factors have different effects on sugarcane yield and sucrose content. Therefore, according to the differences in meteorological factors in different growth seasons, the low-latitude plateau of Yunnan was divided into five different ecological types (Figure 8). This will help us understand the growth potential of different ecological types of sugarcane, which is of great significance in improving productivity. To identify areas suitable for expansion, the Brazilian government performed Sugarcane Agroecological Zoning (AEZ-Sugarcane); Silva et al. modeled sugarcane expansion in São Paulo in 2041 and 2060 to map low-risk areas on the basis of climate change scenarios, in addition to comparing the expansion areas from the model with suitable AEZ-Sugarcane areas [34]. According to the ecological adaptability of sugarcane, 69 sugarcane production areas in Guangxi, Guangdong, and Yunnan in China were divided into six groups of ecological sugarcane areas on the basis of natural climate characteristics of the main sugarcane production areas, namely, subtropical humid low-light sugarcane areas, subtropical humid low-evaporation sugarcane areas, tropical humid sugarcane areas, subtropical semi-humid sugarcane areas, subtropical high-humidity sugarcane areas, and subtropical humid multi-light sugarcane areas [17]. The climatic types of the main sugarcane-producing areas in Yunnan were classified using the systematic clustering method. The six sugarcane climate types have good regional distribution; the climate characteristics of each type have obvious differences, which are mainly reflected in the average sunshine hours from April to October, average precipitation from May to October, average sunshine hours from November to February, and accumulated temperature differences from September to December [35]. The division of climate types in the sugarcane-growing areas provided a reference for our research. However, it is essentially different from previous studies; the division of climate types is more detailed and a significant guide for production. In this study, 26 major sugarcane-producing counties in the low-latitude plateau region could be divided into five climatic ecological types for sugarcane growth. In the first ecological type, both ARA and ARH are high throughout the growth period, AAT and ASD are low in the mid-growth season, and AAT is low in the late-growth season of sugarcane. Using Jinping County (s2) as the representative model, this type is conducive to an increase in sugarcane yield and not conducive to an increase in sucrose content. In the second ecological type, AAT is low in the early and middle seasons of sugarcane growth, the ASD is short in the early season, and ARA is low in the late season, with Funing (s3) and Guangnan County (s5) as the representative models. This type of sugarcane has a medium yield, which is unfavorable for an increase in sucrose content. In the third ecological type, the AAT of sugarcane is significantly higher than that of other sugarcane areas in the early-growth season, rainfall and humidity are relatively high in the late-growth season, and climatic characteristics in the mid-growth season are not obvious, with Cangyuan (s1), Lancang (s10), Mangshi (s14), Mengla (s16), Ximeng (s23), and Zhenkang County (s26) as the representative models. This type of sugarcane has a moderate yield and sucrose content, which is not particularly conducive or detrimental to an increase in sugarcane and sucrose yield. In the fourth ecological type, air humidity is the lowest in the early, middle, and late seasons of sugarcane growth, rainfall is the lowest in the middle and late seasons, and light duration is the longest, with Honghe (s6), Shidian (s19), Shiping (s20), Shuangjiang (s21), Wenshan (s22), and Yunxian (s25) as the representative models. This type is conducive to an increase in sucrose content, and the yield is medium. In the fifth ecological type, light duration is the longest in the early and late seasons of sugarcane growth and rainfall is relatively low, with Fengqing (s2), Menghai (s15), Linxiang District (s12), Lianghe (s11), Yingjiang (s24), Jingdong (s8), Gengma (s4), Longchuan (s13), Ruili (s18), Jinggu (s9), and Menglian County (s17) as the representative models. This type is beneficial to sugarcane yield and sucrose content. Riajaya found that the rainy season in sugarcane regions in Indonesia ranges from 20–25 ten-day periods in 59% of the seasonal zone in Sumatra [20], 15–20 ten-day periods in 73% of the regions in Java and Madura, and 20–25 and 10–15 ten-day periods in 33% and 34% of the regions in Sulawesi. Areas with 25–30 ten-day rainy periods are more suitable for the development of sugarcane as the raw material for the brown sucrose industry, and areas with 10–15 ten-day rainy periods require additional irrigation to meet the water needs of sugarcane plants. Areas with 15–20 ten-day rainy periods are ideal for sugarcane growth. In this study, the climate types were precisely divided, which will help guide sugarcane production in the low-latitude plateau region.

4. Materials and Methods

4.1. Data Collection and Test Sites

4.1.1. Data Collection

The data collected in this study were mainly divided into two categories obtained from the sugarcane planting (test) region. The above-mentioned data were provided by the archives of the Climate Center of Yunnan Meteorological Bureau. The second category was composed of data of the annual sugarcane-planting area, sugarcane harvest amount, and average sucrose content of sugarcane in each major sugarcane-producing county (test sites) from 2001/2002 to 2018/2019. The average yield of sugarcane was calculated from the sugarcane harvest amount divided by the sugarcane planting area. The above-mentioned data were provided by the sugarcane research institute of the Yunnan Academy of Agricultural Sciences. The yield and sucrose content were the average values of New plant, Ratoon1, and Ratoon2 planting periods collected from varieties Roc22 and YT93-159. The test soil type was clay; organic matter content, 12.5–18.6 g/kg; pH value, 5.1–5.9; available phosphorus, 32.3–52.1 mg/kg; available potassium, 42.1–56.0 mg/kg; and alkali hydrolyzed nitrogen, 57.9–72.3 mg/kg. Roc22 (planting areas, 60%) and YT93-159 (planting areas, 40%) were planted at an altitude between 826 m and 1100 m by using one-time fertilization and full-film coverage technology with a row spacing of 1.1 m. The planting time of each variety was between January and February.

4.1.2. Overview of Test Sites

In this study, the data of climatic factors, yield, and sucrose content of sugarcane from 26 counties (test sites) in seven prefecture-level cities that represented the main sugarcane-producing areas were collected from the low-latitude plateau of Yunnan (21° N–25° N, 97° E–106° E) between 2000 and 2019. The locations of the test sites were as follows: Lianghe County (s11), Longchuan County (s13), Ruili City (s18), Mangshi (s14) and Yingjiang County (s24) in Dehong prefecture-level city, Cangyuan County (s1), Fengqing County (s2), Gengma County (s4), Linxiang District (s12), Shuangjiang County (s21),

Zhenkang County (s26) and Yun County (s25) in Lincang prefecture-level city, Shidian County (s19) in Baoshan prefecture-level city, Shiping County (s20), Jinping County (s7) and Honghe County (s6) in Honghe prefecture-level city, Jingdong County (s8), Jinggu County (s9), Lancang County (s10), Menglian County (s17) and Ximeng County (s23) in Pu'er prefecture-level city, Menghai County (s15) and Mengla County (s16) in Xishuangbanna prefecture-level city, Funing County (s3), and Guangnan County (s5) and Wenshan City (s22) in Wenshan prefecture-level city. They represent more than 85% of the sugarcane-planting area in Yunnan, with an altitude of 500 to 2000 m. The annual AAT in the last 20 years was mainly between 19.0 and 20.1 °C; annual ARH, between 73% and 79%; annual ARA, between 50 mm and 140 mm; and annual ASD, between 1800 and 2400 h (Figure 1).

4.2. Study Factors

4.2.1. Division of Sugarcane Growth Season

The sugarcane growth season was divided into three periods, namely, early-growth season (seedling stage and tillering stage), mid-growth season (jointing stage and big growth stage), and late-growth season (formation and stability stage for sugarcane yield and sucrose content). On the basis of the planting habit and growth characteristics of spring sugarcane in the low-latitude plateau of Yunnan, the early-growth season was from March to June, the mid-growth season was from July to October, and the late-growth season was from November to February.

4.2.2. Climatic Characteristics of Growth Seasons and Sugarcane Production

The average sugarcane yield and sucrose content data were collected from 26 sites between 2005/2006 and 2015/2016, and the data for AAT, ARH, ARA, and ASD in the corresponding years were also obtained. Correlation analysis of the main climatic factors in different growth seasons with sugarcane yield and sucrose content was performed to study the relationship between climate and sugarcane production. A summary of average annual AAT, ARH, AAR, AAS, yield, and sucrose content of the corresponding sites and years has been provided in Tables S1 and S2.

4.2.3. Classification of Climate Types on the Basis of Sugarcane Growth

According to the climate variation characteristics of the 26 sites in different growth seasons, the climate types of the sugarcane-planting region in the low-latitude plateau of Yunnan were classified and differences among the different climate types were analyzed.

4.3. Statistical Analysis

Microsoft Excel 2019 (Microsoft, Redmond, WA, USA) was used for sorting the data of the climate, sugarcane yield, and sucrose content. DPS v14.10 software (Zhejiang University, Hangzhou, China) was used for the statistical analysis, including the calculation of mean value, standard deviation, and coefficient of variation. R software (version 4.1.0, R core team, 2021) was used for creating the figures. The package "ggplot2" of R software was used to create the line charts, box diagrams, scatter charts, and fitting curve and perform the cluster analysis. The package "ggsignif" was used for the LSD test (Detailed programming codes can be obtained from us).

5. Conclusions

The climate in different seasons in the low-latitude plateau region was significantly different, which has varied effects on sugarcane yield and sucrose content. Winter and early spring (from November to March) are dry and rainless, which is not conducive to the germination and growth of sugarcane buds. The rainfall is mainly concentrated in June to September, which is conducive to an increase in sugarcane yield. The low-temperature period is short, and the lowest temperature is mainly concentrated in December and January. The difference in sunshine duration in the main sugarcane-producing region was significant, which was the main factor that affected the sugarcane yield and sucrose content.

The relative humidity of sugarcane was relatively high throughout the growth season. Average air temperature in the mid-growth season and late-growth season significantly affected sucrose content, and low temperature was not conducive to sucrose formation. Average relative humidity in the mid-growth season affected sugarcane yield, and high humidity was beneficial to an increase in yield. Average rainfall amount in the early-growth season affected the sugarcane yield, mainly affecting the emergence and tillering of sugarcane. Average sunshine duration in the late-growth season was the main factor that affected the yield and sucrose content of sugarcane, and it was beneficial for high yield and sucrose content. The low-latitude plateau region has five climate types that contribute to different sugarcane yields and sucrose contents. The rainy and humid type contribute to a high sugarcane yield, but not to a high sucrose content. The low temperature and sunshine semi-humid type is unfavorable for a high sugarcane yield and sucrose content. The high temperature and humidity type has good sugarcane growth potential and contributes to sugarcane yield and sucrose content. The semi-humid and multi-sunshine type contributes to an increase in sucrose content. The humid and sunny type is beneficial for a high sugarcane yield and sucrose content.

Supplementary Materials: The following supporting information can be downloaded at https://www.mdpi.com/article/10.3390/plants12142712/s1, Table S1: Summary of average (±standard error) annual AAT, ARH, AAR, AAS, yield, and sucrose content of the sites across all years. Table S2: Summary of average (±standard error) AAT, ARH, AAR, AAS, yield, and sucrose content of the sites across all years.

Author Contributions: Methodology, Y.Z. and L.-X.Y.; software, Y.Z. and Z.-F.Z.; formal analysis, Y.Z., J.A., L.-X.Y. and J.D.; writing—original draft, Y.Z.; writing—review and editing, Y.Z. and Y.-B.Z.; funding acquisition, Y.Z. and Y.-B.Z.; conceptualization, L.-X.Y. and Y.-B.Z.; investigation, L.-X.Y. and J.D.; resources, J.A. and J.D. All authors have read and agreed to the published version of the manuscript.

Funding: This study was supported by the National Natural Science Foundation of China [grant number (31860341); the Earmarked Fund for China Agriculture Research System (grant number CARS-17); the National Key R&D Program of China (grant number 2022YFD2301100); and the Yunnan Joint Laboratory of Seed Science and Industry (grant number 202205AR070001-09).

Data Availability Statement: Raw data are available upon request.

Conflicts of Interest: The authors declare that they have no conflict of interest.

References

1. Tew, T.; Cobill, R.M. Genetic improvement of sugarcane (*Saccharum* spp.) as an energy crop. In *Genetic Improvement of Bioenergy Crops*; Springer: New York, NY, USA, 2008; pp. 249–272. [CrossRef]
2. Binbol, N.L.; Adebayo, A.A.; Kwon-Ndung, E.H. Influence of climatic factors on the growth and yield of sugar cane at Numan, Nigeria. *Clim. Res.* **2006**, *32*, 247–252. [CrossRef]
3. Zhao, Y.; Liu, J.; Huang, H.; Zan, F.; Zhao, P.; Zhao, J.; Deng, J.; Wu, C. Genetic Improvement of Sugarcane (*Saccharum* spp.) Contributed to High Sucrose Content in China Based on an Analysis of Newly Developed Varieties. *Agriculture* **2022**, *12*, 1789. [CrossRef]
4. Rakkiyappan, P.; Thangavelu, S.; Malathi, R.; Radhamani, R. Effect of biocompost and enriched pressmud on sugarcane yield and quality. *Sugar Tech* **2001**, *3*, 92–96. [CrossRef]
5. Boschiero, B.N.; Mariano, E.; Torres-Dorante, L.O.; Sattolo, T.M.S.; Otto, R.; Garcia, P.L.; Dias, C.T.S.; Trivelin, P.C.O. Nitrogen fertilizer effects on sugarcane growth, nutritional status, and productivity in tropical acid soils. *Nutr. Cycl. Agroecosyst.* **2020**, *117*, 367–382. [CrossRef]
6. Snyman, S.; Meyer, G.; Koch, A.; Banasiak, M.; Watt, M.P. Applications of in vitro culture systems for commercial sugarcane production and improvement. *In Vitro Cell. Dev. Biol. Plant* **2011**, *47*, 234–249. [CrossRef]
7. Liang, Y.; Yu, S.; Zhen, Z.; Zhao, Y.; Deng, J.; Jiang, W. Climatic change impacts on Chinese sugarcane planting: Benefits and risks. *Phys. Chem. Earth* **2020**, *116*, 102856. [CrossRef]
8. Zhang, Y. Sugarcane biology in low latitude plateau. In *Theory and Practice of High Yield, High Sugar and High Efficiency of Sugarcane in Low Latitude Plateau*; China Agriculture Press: Beijing, China, 2015.
9. Liao, Z.; He, Y.; Mo, S. The effect of meteorological factors on sugarcane yield and research progress of environmental interaction genes. *Chin. Agric. Sci. Bull.* **2022**, *38*, 82–87. [CrossRef]

10. Verma, A.K.; Garg, P.K.; Prasad, K.S.H.; Dadhwal, V.K. Variety-specific sugarcane yield simulations and climate change impacts on sugarcane yield using DSSAT-CSM-CANEGRO model. *Agric. Water Manag.* **2023**, *275*, 108034. [CrossRef]
11. Legendre, B.L. Ripening of Sugarcane: Effects of Sunlight, Temperature, and Rainfall1. *Crop Sci.* **1975**, *15*, 349–352. [CrossRef]
12. Mbhamali, T.; Jury, M. Climate-sensitivity of sugarcane yield in the southeastern Africa lowlands. *Int. J. Biometeorol.* **2021**, *41*, 4187–4200. [CrossRef]
13. Fabio, R.M.; Gustavo, L.C. Spatio-temporal variability of sugarcane yield efficiency in the state of São Paulo, Brazil. *Pesqui. Agropecu. Bras.* **2012**, *47*, 149–156. [CrossRef]
14. Araujo, R.; Júnior, J.; Casaroli, D.; Pêgo Evangelista, A. Variation in the sugar yield in response to drying-off of sugarcane before harvest and the occurrence of low air temperatures. *Bragantia* **2016**, *75*, 118–127. [CrossRef]
15. Cardozo, N.; Sentelhas, P. Climatic effects on sugarcane ripening under the influence of cultivars and crop age. *Sci. Agric.* **2013**, *70*, 449–456. [CrossRef]
16. Flack-Prain, S.; Shi, L.; Zhu, P.; da Rocha, H.R.; Cabral, O.; Hu, S.; Williams, M. The impact of climate change and climate extremes on sugarcane production. *GCB Bioenergy* **2021**, *13*, 408–424. [CrossRef]
17. Zhang, Y.; Fan, X.; Mao, J.; LI, R.d.; Yu, X. Ecological region division for China's main sugarcane producing area through meteorological and ecological characters. *Southwest Chin. J. Agric. Sci.* **2021**, *34*, 2281–2288. [CrossRef]
18. Wu, X.; Ma, D.; Huang, W.; Liu, Y.; Wei, J.; Yao, Y. Effects of meteorological factors on the emergence rate and seedling growth of sugarcane. *Chin. J. Trop. Crops* **2022**, *43*, 1411–1416. [CrossRef]
19. Gentile, A.; Dias, L.I.; Mattos, R.S.; Ferreira, T.H.; Menossi, M. MicroRNAs and drought responses in sugarcane. *Front. Plant Sci.* **2015**, *6*, 58. [CrossRef]
20. Riajaya, P.D. Rainy season period and climate classification in sugarcane plantation regions in indonesia. *IOP Conf. Ser. Earth Environ. Sci.* **2020**, *418*, 012058. [CrossRef]
21. Luo, S.; He, H.; Tang, L.; Liu, L.; Lu, Z. Effects of temperature and rainfall on sugarcane growth. *Sugar Crops Chin.* **2018**, *40*, 13–15. [CrossRef]
22. Fu, J.; Sun, D.; Chen, L.; Lu, Y.; Zhao, H.; Dai, S.; Gong, H.; An, Y. Relationship Analysis of the Temperature and Precipitation with the Yield of Sugarcane. *Chin. J. Trop. Agric.* **2018**, *38*, 13–17.
23. Pathak, S.K.; Singh, P.; Singh, M.M.; Sharma, B.L. Impact of temperature and humidity on sugar recovery in Uttar Pradesh. *Sugar Tech* **2019**, *21*, 176–181. [CrossRef]
24. Samui, R.P.; John, G.; Kulkarni, M.B. Impact of climate on yield of sugarcane at different growth stages. *Agric. Phys.* **2003**, *3*, 119–125. [CrossRef]
25. Tan, Z.; Huang, C.; Meng, C.; Pan, Y. Study on the Grade Indexes of Sugar-cane Chilly Injury or Freezing Injury and Loss. *Chin. Agric. Sci. Bull.* **2014**, *30*, 169–181. [CrossRef]
26. Li, Y.; Yang, L. Sugarcane Agriculture and Sugar Industry in China. *Sugar Tech* **2015**, *17*, 1–8. [CrossRef]
27. Rupa, K.K. Yield response of sugarcane to climate variations in Northeast Andhra Pradesh, India. *Arch. Meteorol. Geophys. Bioklimatol. Ser. B* **1984**, *35*, 265–276. [CrossRef]
28. Liu, J.; Basnayake, J.; Jackson, P.A.; Chen, X.; Zhao, J.; Zhao, P.; Yang, L.; Bai, Y.; Xia, H.; Zan, F.; et al. Growth and yield of sugarcane genotypes are strongly correlated across irrigated and rainfed environments. *Field Crops Res.* **2016**, *196*, 418–425. [CrossRef]
29. Abu-Ellail, F.F.B.; McCord, P.H. Temperature and Relative Humidity Effects on Sugarcane Flowering Ability and Pollen Viability Under Natural and Seminatural Conditions. *Sugar Tech* **2019**, *21*, 83–92. [CrossRef]
30. Mansoor, S.; Khan, M.; Khan, N. Screening of Sugarcane Varieties/Lines against Whip Smut Disease in Relation to Epidemiological Factors. *J. Plant Pathol. Microbiol.* **2016**, *7*, 2. [CrossRef]
31. Chinnaraja, C.; Viswanathan, R. Variability in yellow leaf symptom expression caused by the Sugarcane yellow leaf virus and its seasonal influence in sugarcane. *Phytoparasitica* **2015**, *43*, 339–353. [CrossRef]
32. Huang, Z. Analysis and prediction of climatic characteristics and sugar yield trend in Yunnan sugarcane region in 2007. *Yunnan Agric. Sci. Technol.* **2008**, *3*, 5–12. [CrossRef]
33. Kuang, Z.; Qu, Z.; Li, L.; Chen, D.; Li, Z.; Liu, Z. Evaluation of the influence of daily air temperature range on sucrose content of sugarcane. *Sugarcane Canesugar* **2022**, *51*, 24–29. [CrossRef]
34. Silva, G.J.d.; Berg, E.C.; Calijuri, M.L.; Santos, V.J.d.; Lorentz, J.F.; Alves, S.d.C. Aptitude of areas planned for sugarcane cultivation expansion in the state of São Paulo, Brazil: A study based on climate change effects. *Agric. Ecosyst. Environ.* **2021**, *305*, 107164. [CrossRef]
35. Lu, W.; Zhou, Y.; He, Y. Climatic types and characteristics of sugarcane planting in Yunnan. *Meteorol. Sci. Technol.* **2016**, *44*, 848–853. [CrossRef]

Article

Comparative Analysis of Sucrose-Regulatory Genes in High- and Low-Sucrose Sister Clones of Sugarcane

Qaisar Khan [1], Ying Qin [1], Dao-Jun Guo [1], Yu-Yan Huang [1], Li-Tao Yang [1], Qiang Liang [2], Xiu-Peng Song [2,*], Yong-Xiu Xing [1,*] and Yang-Rui Li [2,*]

[1] Guangxi Key Laboratory of Sugarcane, College of Agriculture, Guangxi University, Nanning 530004, China; qaisar.khan@yahoo.com (Q.K.); yingqinemail@163.com (Y.Q.); gdj0506@163.com (D.-J.G.); yygxu1028@163.com (Y.-Y.H.); liyr@gxu.edu.cn (L.-T.Y.)

[2] Guangxi Key Laboratory of Sugarcane Genetic Improvement, Key Laboratory of Sugarcane Biotechnology and Genetic Improvement (Guangxi), Ministry of Agriculture and Rural Affairs, Sugarcane Research Institute of Guangxi Academy of Agricultural Sciences, Nanning 530003, China; liangqiangde@163.com

* Correspondence: xiupengsong@gxaas.net (X.-P.S.); document126@126.com (Y.-X.X.); liyr@gxaas.net (Y.-R.L.)

Abstract: Sugarcane is a significant primitive source of sugar and energy worldwide. The progress in enhancing the sugar content in sugarcane cultivars remains limited due to an insufficient understanding of specific genes related to sucrose production. The present investigation examined the enzyme activities, levels of reducing and non-reducing sugars, and transcript expression using RT-qPCR to assess the gene expression associated with sucrose metabolism in a high-sucrose sugarcane clone (GXB9) in comparison to a low-sucrose sister clone (B9). Sucrose phosphate synthase (SPS), sucrose phosphate phosphatase (SPP), sucrose synthase (SuSy), cell wall invertase (CWI), soluble acid invertase (SAI), and neutral invertase (NI) are essential enzymes involved in sucrose metabolism in sugarcane. The activities of these enzymes were comparatively quantified and analyzed in immature and maturing internodes of the high- and low-sucrose clones. The results showed that the higher-sucrose-accumulating clone had greater sucrose concentrations than the low-sucrose-accumulating clone; however, maturing internodes had higher sucrose levels than immature internodes in both clones. Hexose concentrations were higher in immature internodes than in maturing internodes for both clones. The SPS and SPP enzymes activities were higher in the high-sucrose-storing clone than in the low-sucrose clone. SuSy activity was higher in the low-sucrose clone than in the high-sucrose clone; further, the degree of SuSy activity was higher in immature internodes than in maturing internodes for both clones. The *SPS* gene expression was considerably higher in mature internodes of the high-sucrose clones than the low-sucrose clone. Conversely, the *SuSy* gene exhibited up-regulated expression in the low-sucrose clone. The enhanced expression of *SPS* in the high-sucrose clone compared to the low-sucrose clone suggests that *SPS* plays a major role in the increased accumulation of sucrose. These findings provide the opportunity to improve sugarcane cultivars by regulating the activity of genes related to sucrose metabolism using transgenic techniques.

Keywords: sucrose phosphate synthase; sucrose phosphate phosphatase; sucrose synthase; cell wall invertase; neutral invertase; reducing sugar; non-reducing sugar; RT-qPCR

Citation: Khan, Q.; Qin, Y.; Guo, D.-J.; Huang, Y.-Y.; Yang, L.-T.; Liang, Q.; Song, X.-P.; Xing, Y.-X.; Li, Y.-R. Comparative Analysis of Sucrose-Regulatory Genes in High- and Low-Sucrose Sister Clones of Sugarcane. *Plants* **2024**, *13*, 707. https://doi.org/10.3390/plants13050707

Academic Editors: Alex Troitsky and San-Ji Gao

Received: 23 December 2023
Revised: 21 February 2024
Accepted: 28 February 2024
Published: 1 March 2024

1. Introduction

Sugarcane (*Saccharum officinarum* L.) is a very valuable crop used for the manufacture of sugar and bioethanol [1]. The amount and quality of the soluble sugar, mainly sucrose, directly affect the sugar yield in sugarcane [2]. Sugarcane is a major source of sucrose, which is used as the principal source of sugar in human consumption, animal feed, and biofuel generation [3,4]. In addition, sugarcane is a crucial crop for the worldwide economy due to its production of many byproducts, such as bagasse, bagasse ash, press mud, molasses, vinasse, alcohol, yeast, wasted wash, and other derivatives [5,6]. Physiological changes occur sequentially in different internodes of the sugarcane stalk, including immature, maturing, and mature internodes located at the top, middle, and bottom. Immature internodes of sugarcane stalks are composed of fibrous tissue containing a significant amount of sugar that can be reduced while having a low quantity of sucrose. As the internodes mature, the growth rate gradually decreases until maturity is achieved [7–9].

The escalating emissions of greenhouse gases (GHGs) and subsequent global warming because of climate change have led to a rise in the incidence and harshness of immoderate weather phenomena [10]. The extent of the influence of climate change on sugarcane is determined by the geographical position and the ability to adapt to it. Climate unevenness and variation are expected to cause alterations in sea levels, precipitation patterns, and the occurrence of exceptionally hot and cold events, floods, droughts, and other non-living environmental pressures, including tornadoes and hurricanes [11–13]. These variables greatly impede the growth and development of sugarcane plants [14]. Sugarcane plants respond to varied environmental factors by employing multiple mechanisms and generating different metabolites, including antioxidants, osmoprotectants, and heat shock proteins, along with activating distinct metabolic processes. One area of scientific inquiry that lacks a comprehensive understanding is the process of sugarcane maturity and the impact of diverse cultivars, crop growth stage, and climate on this process. Sugarcane maturity refers to the process of sucrose accumulation in the stalks, which is significantly impacted by several circumstances [15]. To a certain extent, research has been carried out on the process of sugarcane maturation [16]; however, the consequence of an intricate array of environmental variables, the genetic capacity of varieties and crop management, soil moisture, ambient temperature, and other ecological factors interacting with sugarcane are the primary areas involved in sugarcane ripening [17], which still needs further investigation. The amalgamation of these factors induces the pace of growth, yield, ripening process, and sucrose accumulations, so additional scrutiny is required to determine the pinpoint role of each factor [18].

Generally, plants are split into "source" where carbon dioxide fixation takes place, and "sink" where excess assimilates are stored [19]. In sugarcane, leaves are classified as "source" due to their role in fixing carbon dioxide through photosynthesis [20]. In contrast, stalks are regarded as "sink" where excess sugar is stored [21]. The quantity of sucrose originating from the source is contingent upon the photosynthetic activity and the allocation of carbon among different activities, such as the synthesis of starch in the chloroplast [22] and the transfer of triose-phosphates from the chloroplast. These processes control the production and buildup of sucrose in the vacuole. An alteration in these regulators affects the connection between sources and sinks in sugarcane. The source–sink relationship is an intricate phenomenon that requires thorough clarification. Nevertheless, it is widely recognized that the ultimate location for storing sucrose is the parenchyma cells found in the stalk of sugarcane, rather than the sink tissues [23–25].

The main goal of commercial sugarcane cultivation is the sucrose content. Sugar transporters help the phloem carry sucrose from the leaves, where photosynthesis produces it, to the internodes [26–28]. After that, it can either be broken down into glucose and fructose to supply the energy needed for development and growth or stored in the internodes [29]. While growth and development decrease the sucrose content in internodes, they also allow the plant factory to enhance sucrose production and store more sucrose in the storage sink [30]. The grand growth phase is of utmost importance for the accumulation of sucrose, as it is during this period that a balance is achieved for sucrose production, growth use, and storage in the sink [31]. During the maturation phase of the cane, any extra fructose, glucose, and other soluble sugars are transformed into sucrose and deposited in sinks for accumulation [32].

Multiple genes are involved in the production, breakdown, and storage of sucrose in sugarcane; for example, *sucrose phosphate synthase (SPS)*, *sucrose-phosphate phosphatase (SPP)*, *sucrose synthase (SuSy)*, *β-fructofuranosidase*, *soluble acid invertase (SAI)*, *acid invertase (AI)*, *neutral invertase (NI)*, and *cell wall invertase (CWI)* have been identified in the apoplast and cell walls of sugarcane [33]. Vacuoles contain soluble *acid invertase (SAI)* [34], which primarily operates in rapidly developing tissues such as immature internodal tissue, root apices, and cell and tissue culture. Upon entering parenchyma cells, hexoses undergo a transformation into sucrose, facilitated by the enzymes SPS and SPP [35]. The sucrose level varies between high- and low-sugar sugarcane genotypes and between immature and mature internodes. Recent advancements in sugarcane breeding have resulted in notable enhancements in agronomic features [36,37]. However, further research is required to fully understand the intricate mechanism of sucrose metabolism in sugarcane [9,38,39]. An essential target for future studies is to investigate the genes linked to sugar synthesis, sucrose metabolism, and accumulation in internodes of sugarcane with variation in age.

SuSy is associated with sucrose metabolism in plant sinks, where it catalyzes the production of uridine diphosphate glucose or adenosine diphosphate glucose and fructose from the decomposition of sucrose [40]. SuSy also contributes to plant growth and development, particularly in the shoot apical meristem [41–43]. SuSy activities positively influence glucose and fructose concentrations, while negatively affecting sucrose accumulation in sugarcane. The genotypes with high SuSy expression had low sucrose levels but greater hexose levels, and vice versa. SuSy may provide hexose sugars for glycolysis in fast-growing cells and be the precursor to other growth progressions [44]. The breakdown of sucrose into hexoses is required for the extraction of carbon and energy [45,46]. SuSy participates in physiological activities in plants, including the production or breakdown of sucrose in different sinks. However, SuSy appears to play a function in sucrose breakdown rather than in production, primarily in young developing tissues [40,47].

The present work aimed to examine the regulation mechanism of sucrose metabolism by comparing the immature and maturing internodes of a high-sucrose sugarcane clone "GXB9" with those of a low-sucrose clone "B9" after 250 days of planting (DAP), using RT-qPCR. Additionally, tests were conducted to measure the activities of enzymes and the concentration of sugars. The objective of the study was to compare the expression of genes associated with sucrose accumulation and metabolism regulation, including SPS, SPP, SuSy, SAI, CWI, and NI, by RT-qPCR analysis in the immature and maturing internodes of both clones. To the best of our knowledge, this is the first study to analyze the GXB9 and B9 sugarcane clones comparatively for sucrose metabolism.

2. Results

In this study, the activities of SPS, SPP, SAI, NI, CWI, and SuSy in the high-sucrose clones GXB9 and the low-sucrose clones B9 were compared, and the transcript expression in immature and maturing internodes was analyzed by RT-qPCR.

2.1. Reducing and Non-Reducing Sugar Concentrations

Disparities in sucrose and sugar concentrations were observed between the sister clones that accumulate high and low amounts of sucrose. Furthermore, there were variations in the sucrose and hexose levels between the immature and maturing internodes of the same clone. The sucrose concentration in immature internodes was lower than in the maturing internodes, while the contents of glucose and fructose were higher in immature internodes compared to the maturing internodes for the high-sucrose clone. Immature internodes have lower sucrose concentrations, and maturing internodes have high sucrose concentrations in the low-sucrose clones. However, glucose and fructose concentrations were more significant in immature internodes than in maturing internodes in the low-sucrose-accumulating clone. For both clones, sucrose concentration was higher in maturing internodes, while glucose and fructose concentrations were higher in immature internodes. Overall, the high-sucrose-accumulating clone had higher sucrose concentrations than the low-sucrose-accumulating sister clone (Figure 1).

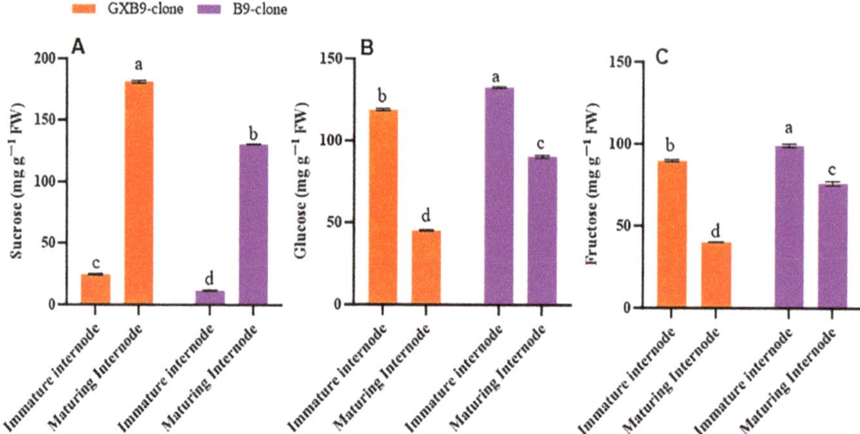

Figure 1. Figure 1 illustrates differences in (**A**) sucrose, (**B**) glucose, and (**C**) fructose concentrations between high- (GXB9) and low-sucrose (B9) sister clones of sugarcane and immature and maturing internodes of identical clones. The figure legends GXB9 and B9 represent sugarcane sister clones. The *x*-axis represents the different types of internodes, while the *y*-axis represents the concentration of sugars. Different lowercase letters on bars show a significant difference, at $p < 0.05$, among both internode types and sister clones using the. The data have been presented as mean ± SE (standard error). The unit "mg g⁻¹ FW" is milligram per gram of fresh weight.

2.2. Enzymes Activities Level

The activities of SPS, SPP, SuSy, SAI, CWI, and NI in the two sister sugarcane clones were assessed and subjected to comparative analysis. The activities of the SPS and SPP were greater in the clone with greater sucrose accumulation compared to that with lower sucrose accumulation. The SPS and SPP activities were shown to be higher in mature internodes of the high-sucrose clone compared to immature internodes of the same clone. The activity of SuSy was higher in the clone with lower sucrose accumulation compared to that with higher sucrose accumulation. The low-sucrose clone exhibited higher SuSy activity in immature internodes compared to matured internodes. The participation of SuSy was shown to be higher in immature internodes, regardless of the sucrose concentration in both clones. Conversely, the activity of SPS was seen to be higher in the maturing internodes of both clones (Figure 2). SAI, CWI, and NI activities were greater in immature internodes of the low-sucrose clone than in the same internodes of the high-sucrose clone. In contrast, the activity levels of CWI and NI were higher in maturing internodes of the high-sucrose clone compared to the low-sucrose clone. The SAI activity was higher in maturing internodes of the low-sucrose clone in contrast to those of the high-sucrose clone (Figure 3).

2.3. RT-qPCR Data Analysis

The analysis of RT-qPCR data derived from the pools of immature and ripening internode samples of high- and low-sucrose sugarcane clones revealed varying levels of gene expression in different clones and internodes (Figure 4). The *SPS*, *SPP*, *SuSy*, *CWI*, *SAI*, and *NI* genes were selected for assessing the relative expression in the high-sucrose clone "GXB9" and low-sucrose clone "B9". Both clones exhibited expression of the selected genes, and the *SPS* gene was up-regulated in the clone with a higher sucrose level and down-regulated in the clone with a lower sucrose level. The *SPS* gene exhibited notable differences in expression level between immature and maturing internodes of the high-sucrose clone. Specifically, it was up-regulated in maturing internodes and down-regulated in immature internodes. The *SPP* gene was down-regulated in immature internodes of both clones; however, it was up-regulated in maturing internodes of the high-sucrose clone but down-regulated in the low-sucrose clone. The *SuSy* gene exhibited a higher expression level in immature internodes compared to maturing internodes in both clones. The higher expression of the *SPS* gene in the maturing internode of the high-sucrose clone compared to the low-sucrose clone suggests its significant contribution to the increased sucrose accumulation. *CWI*, *SAI*, and *NI* genes were up-regulated in immature and maturing internodes of both clones; however, immature internodes showed higher expression levels compared to maturing internodes.

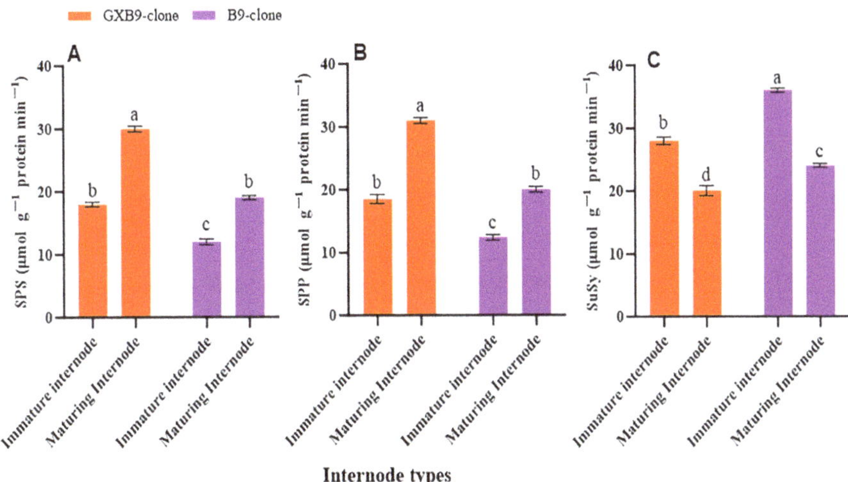

Figure 2. Shows variation in activities of (**A**) sucrose phosphate synthase (SPS), (**B**) sucrose phosphate phosphatase (SPP), and (**C**) sucrose synthase (SuSy) between high (GXB9) and low (B9) sucrose sister clones of sugarcane and immature and maturing internodes of identical clones. The figure legends GXB9 and B9 represent sugarcane sister clones. The *x*-axis denotes the different types of internodes, whereas the *y*-axis represents the magnitude of enzyme activities. Different lowercase letters on bars show a significant difference, at $p < 0.05$, among both internode types and sister clones using the least significant difference (LSD) test. The data have been presented as mean $\pm$ SE (Standard error). The unit "μmol g^{-1} protein min^{-1}" is micromole per gram of protein per minute.

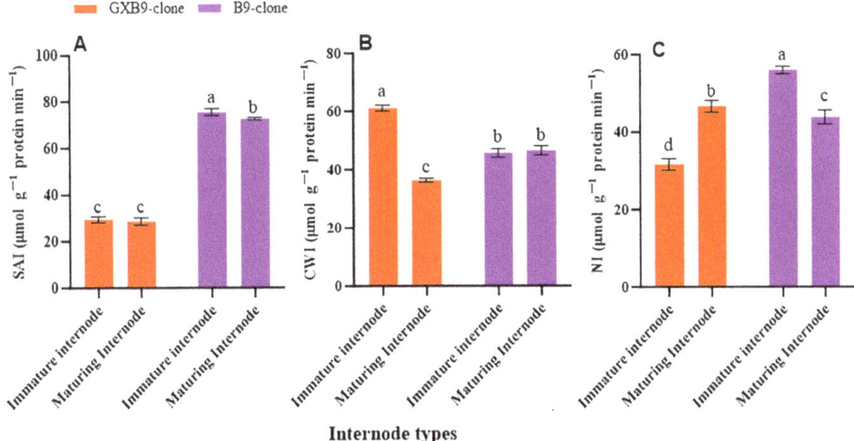

Figure 3. Figure 3 displays disparity in activities of (**A**) soluble acid invertase (SAI), (**B**) cell wall invertase (CWI), and (**C**) neutral invertase (NI) between high- (GXB9) and low-sucrose (B9) sister

clones of sugarcane and immature and maturing internodes of identical clones. The figure legends GXB9 and B9 represent sugarcane sister clones. The *x*-axis denotes the different types of internodes, whereas the *y*-axis represents the magnitude of enzyme activities. Different lowercase letters on bars show a significant difference, at $p < 0.05$, among both internode types and sister clones using the least significant difference (LSD) test. The data have been presented as mean $\pm$ SE (standard error). The unit "μmol g^{-1} protein min^{-1}" is micromole per gram of protein per minute.

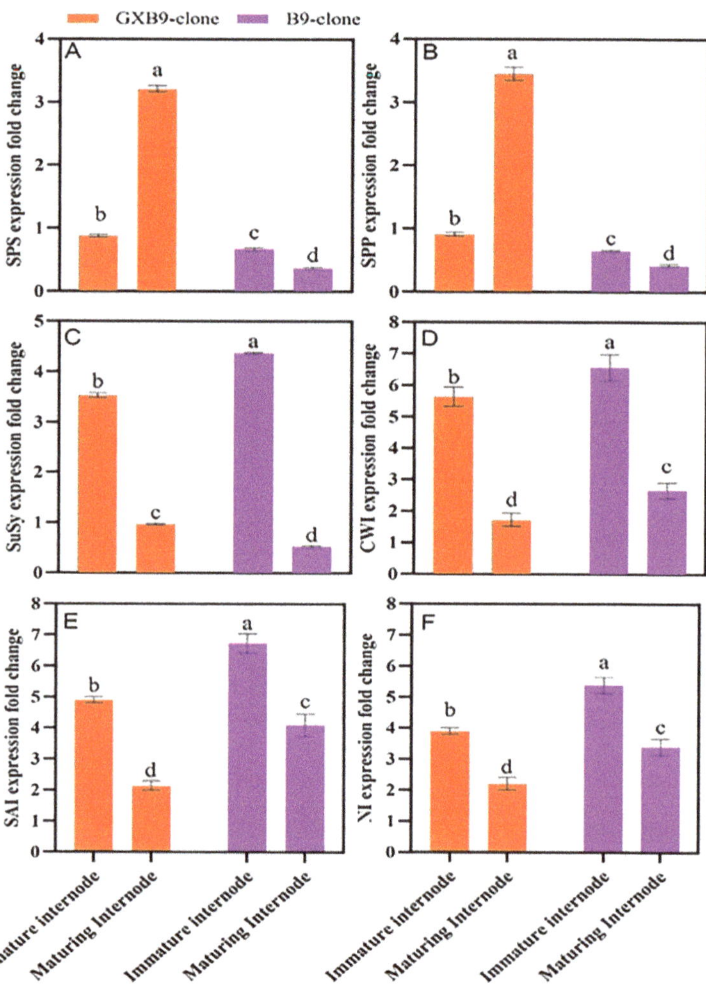

Figure 4. Figure 4 indicates differential expression regulation of sucrose-metabolism-related genes (**A**) *sucrose phosphate synthase* (*SPS*), (**B**) *sucrose phosphate phosphatase* (*SPP*), (**C**) *sucrose synthase* (*SuSy*), (**D**) *cell wall invertase* (*CWI*), (**E**) *soluble acid invertase* (*SAI*), and (**F**) *neutral invertase* (*NI*), between high- (GXB9) and low-sucrose (B9) sister clones of sugarcane and immature and maturing internodes of identical clones. The figure legends GXB9 and B9 represent sugarcane sister clones. The *x*-axis denotes the different types of internodes, while the *y*-axis shows the fold change in values of gene expression. Different lowercase letters on bars show a significant difference, at $p < 0.05$, among both internode types and sister clones using the least significant difference (LSD) test. The data have been presented as mean $\pm$ SE (standard error). Genes with a fold change (FC) value ≥ 1 were up-regulated, whereas genes with a fold change value ≤ 1 were down-regulated.

3. Discussion

Sugarcane is a significant agricultural commodity responsible for around 80 percent of the world's sugar production. China ranks fourth among the major sugar-producing nations [48]. RT-qPCR is a crucial tool for determining the expression of target genes in samples of interest [49]. The primary metabolic pathways for sucrose in the internodes of sugarcane involve SPS, SPP, SuSy [50], SAI, NI, and CWI [51]. Sucrose synthesis in green leaves and transport to storage and consuming tissues are co-ordinated appropriately in sugarcane. Within the parenchyma cells of internodal tissues, the sugars undergo metabolic processes and are, subsequently, resynthesized through the actions of SPS and SuSy [52,53].

3.1. Environmental Interaction and Sucrose Improvement

Climate change poses a threat to sugarcane both directly, through variations in temperature and precipitation, and indirectly, through shifting the intensity of other biotic and abiotic factors, an abundance of pollutants, and the functioning of other ecosystem actors that impact sugarcane productivity [11,54]. A complex interplay between climate factors, cultivar genetic potential, and agricultural management determines when sugarcane ripens. Ripening is a natural process of sucrose accumulation acting as a shield against unfavorable environmental conditions, which is not the same as the maturity of the stalk [55]. Sugarcane crops have numerous difficulties in terms of adjusting of varieties to various production environments, with an emphasis on choosing desired traits in accordance with environmental circumstances [56]. Technological criteria, which include the percentages of moisture, polarizable sugar, fiber, and apparent sucrose and soluble solids in the juice, as well as other intrinsic features of the plant, can determine the quality of sugarcane used to produce both sugar and ethanol [57]. Therefore, it is suggested that future research based on characteristics of importance to sugarcane agro-industries for high sugarcane yields to gain sustainability should concentrate on identifying potential sugarcane varieties for increased sucrose production in variable environmental conditions.

3.2. Free Sugars

The current study discovered the differences in the sucrose and sugar storage between immature and maturing internodes of identical clones and between the low- and high-sucrose-accumulating sister clones. Compared to the low-sucrose clone, the high-sucrose-accumulating clone had a greater sucrose concentration. Higher glucose and fructose concentrations were found in immature internodal tissues compared to maturing internodes, whereas lower sucrose contents were found in the latter. Our results are in line with the findings of other studies that were published [16,58,59]; that is, immature internodes had high hexoses and little sucrose, while maturing internodes had the opposite pattern. The current study also observed that the activity of the SPS enzyme was greater in the clone with a higher sucrose accumulation than in that with a lower sucrose level. Maturing internodes exhibited superior SPS performance compared to immature ones. The presence of SuSy activity was observed to be higher in immature internodes compared to maturing internodes. The observed disparities in the levels of reducing and non-reducing sugars and enzyme activity suggest that the GXB9 clone may have undergone genetic mutations, leading to the observed variations despite both clones having identical genetic backgrounds. However, further experimental verification is still required.

3.3. Invertase Participation in Sucrose Metabolism

Invertase is the primary enzyme responsible for the breakdown of sucrose into glucose and fructose, which serves as an energy source for cellular development, elongation, and various metabolic activities [60,61]. SAI is thought to play a significant role in controlling hexose levels in specific tissues and is concerned with the rate at which sugar returning from storage. There is a noticeable seasonal variance in SAI activity, which is strong during a period of rapid growth and less in mature tissues [62]. The SAI enzyme potentially contributes to the release of stored sucrose from the vacuole, and its activity level was particularly elevated during the elongation phase [63]. In our finding, the SAI activity was higher in the immature internode of both clones as compared to the maturing internode, which is consistent with the findings of the following researchers. The increased SAI activity observed in the immature internode, as compared to the maturing internode, for both clones aligns with the findings of [64], who reported a higher SAI activity in the immature internode compared to the maturing internode. The activity of SAI reduces as internodes mature, suggesting a correlation between the decrease in SAI activity and the maturation of internodes [65].

It is commonly believed that cell wall invertase allows sucrose to enter cells in developing tissues via phloem loading [66]. Sugarcane top internodes had increased cell wall invertase expression, which decreased significantly with maturity and was very low in mature internodes [67]. In our results, the activity of CWI was more significant in immature internodes than in maturing internodes of both clones; however, maturing internodes of the low-sucrose clone have higher CWI activity than the high-sucrose clone, which is in alignment with the above-reported results.

Ebrahim et al. [68] and Vorster et al. [69] found that the activity of NI increases in developing internodes and decreases in maturing internodes. Additionally, it is shown that NI is the sole enzyme responsible for breaking down sucrose and is directly associated with sugar levels in fully developed sugarcane internodal tissue, where a positive correlation between NI activity levels and hexose concentrations in the respective tissues is found reported by Gayler et al. [70] and Bosch et al. [71], and a negative correlation between NI activity levels and sucrose quantity has been noticed by Rose and Botha [72]. In the current study, it was found that NI activity was higher in immature internodes than maturing internodes of both low- and high-sucrose sugarcane clones. Overall immature and maturing internodes of the low-sucrose clone showed higher NI activity than the high-sucrose one. These findings agree with earlier reports cited here stating that NI activities are higher in maturing internodes than mature internodes of sugarcane.

3.4. Predominant Role of SPS Gene in Sucrose Accretion

SPS is the primary gene required for plant sucrose production and control [73–75]. Previous research has found that the *SPS* gene is expressed more intensely in maturing internodes of sugarcane than in immature internodes [76–78]. However, an earlier report on *SPS* expression showed that *SPS* activity is higher in immature internodes than in mature internodes [79]. *SPS* gene activity has been found to be higher in high-sucrose sugarcane growing tissues than in low-sucrose tissues [77]. The SPS activity declines with the sugarcane plant's maturity [67,80]. It transforms fructose-6-phosphate and uridine diphosphate-glucose into sucrose-6-phosphate, an important precursor of sucrose synthesis [81]. The current study revealed a significant up-regulation of the *SPS* and *SPP* genes in high-sucrose sugarcane clones compared to low-sucrose clones. Moreover, *SPS* expression was higher in maturing internodes than immature ones of the high-sucrose sugarcane clone. Thus, the up-regulation of the *SPS* gene in the high-sucrose clone and maturing internodes shows that it plays a dominating role in synthesizing and accumulating increased sucrose content in high-sucrose sugarcane clones.

3.5. Role of SuSy Gene in Sucrose Metabolism

The current investigation showed that the *SuSy* gene was more highly expressed in the lower-sucrose clone than in the higher-sucrose clone, and it was more abundant in immature internodes than maturing internodes, demonstrating its importance in sugarcane-growing tissues via sucrose lysis. As a result, internodes with higher *SuSy* activity have lower sucrose concentrations. The current study's findings about the *SuSy* gene are consistent with prior findings cited here [40,44,47]. The expression of *CWI*, *SAI*, and *NI* genes were higher in immature internodes than in maturing internodes of both high- and low-sucrose sister clones. Finally, the expression level of invertase genes was comparatively greater in the low-sucrose clone compared to the high-sucrose clone, which support the results of relevant enzyme activities.

4. Materials and Methods

4.1. Experimental Material and Sample Collection

The Guangxi Academy of Agriculture Science (GAAS) in Nanning, China, provided the sugarcane clone GXB9, which contains high sucrose, and B9, which has low-sucrose content, for experimental use. The "B9" clone is known for its low sugar content and great yield, as well as its remarkable development and growth [82]. B9 and GXB9 clones share the same genetic lineage, but GXB9 has a higher sugar content than B9. In October 2013, the high-sugar-content clone was identified among the population of the low-sugar clone B9 (parent) during the regular sugar-testing procedure [83]. The recently identified clone with elevated sugar levels was designated as Guixuan B9 (GXB9). Over the course of several years, both clones underwent testing at various field stations to determine any disparities in sugar concentration. The findings continually revealed a notable difference in sugar content between B9 and GXB9 [84].

On 7 March 2022, sugarcane setts were planted at the experimental field of the College of Agriculture, Guangxi University in Nanning, China. The cultivation of sugarcane followed all normal standards, including regular irrigation, weed removal, disease inspection, fertilization (N-P$_2$O$_5$-K$_2$O: 22-8-12) (Kingenta Ecological Engineering Group Co., Ltd., Binzhou, China), metsulfuron herbicides (Shandong Qiaochang Chemical Co., Ltd., Binzhou, China), and carbendazim fungicide (Jiangyin Fuda Agrochemical Co., Ltd., Jiangyin, China) [85].

A total of six sugarcane plants were selected at random from each plot. Samples from the upper immature (1–4) and middle maturing (12–15) internodes of both clones were obtained in mid-November, 250 DAP, for the purpose of analyzing enzymes activities and transcript expression, as well as reducing and non-reducing sugars. The samples were obtained during the early morning hours (8:00–9:00 a.m.) and promptly immersed in liquid nitrogen to halt the metabolic processes related to sucrose and sugar levels. The samples were preserved at a temperature of $-80\ °C$ for subsequent study.

4.2. Reducing and Non-Reducing Sugar Extraction and Analysis

Samples were prepared by removing the rind of the fresh immature and maturing internodes collected from high- and low-sucrose clones GXB9 and B9, and soft tissue were subsequently diced into minute fragments and combined. These prepared basic samples were stored at $-80\ °C$ and used in the upcoming analysis according to the purpose.

Sucrose, glucose, and fructose were obtained from a 15 g sample of immature and maturing internodes of low- and high-sucrose clones, respectively. The weighed samples were subjected to boiling in 85% ethanol and, subsequently, in 75% ethanol for durations of 18 and 20 min, respectively. The extract was transformed into syrup by submitting it to a process of reduced pressure at a temperature of 40 °C. The presence of sucrose was identified using the approach described by the method described by Roe [86] with modification. The sucrose concentration was determined using a 1 mL mixture consisting of 500 µL of extract and 500 µL of a 6% KOH solution. The tubes containing the reaction mixture were immersed in an 80 °C water bath for 22 min, and then cooled to ambient temperature. The cooled reaction mixture was supplemented with 12 mL of resorcinol (0.2%) and 5 mL of HCl (32%), and thereafter subjected to incubation at 90 °C for 15 min. The sucrose concentration was determined at a wavelength of 540 nm using Shimadzu UV 1600 Double Beam Spectrophotometer Kyoto, Japan, and then compared to a standard curve. The hexose concentrations in the samples were measured using the methodology described by Somogyi [87] with some amendments. In summary, 120 µL of samples were collected in a test tube, and the volume was adjusted to 2.5 mL by adding distilled water. Next, 1.5 mL of alkaline tartrate reagent was added, and the mixture was subjected to incubation in a boiling bath for 12 min. The samples were chilled to ambient temperature, and then 1.2 mL of arsenomolybdic reagent was added. After incubating at room temperature for 15 min, the reaction volume was adjusted to 12 mL using distilled water, and the absorbance was measured at 620 nm using Shimadzu UV 1600 Double Beam Spectrophotometer, Kyoto, Japan. The quantity of reducing sugars was determined with the help of a glucose standard curve.

4.3. Enzyme Isolation

The internodal tissue samples that were prepared in advance and stored at $-80\ ^\circ\text{C}$ were used for the extraction of enzymes, including SPS, SPP, SuSy, CWI, SAI, and NI, according to the related methodologies.

The enzymes were extracted from 350 g of diced internodal tissues by homogenizing them at 5 $^\circ$C. The homogenization was carried out using a pH 7.4 Tris-HCl (0.12 M) buffer containing TritonX-100 (1.5%), cysteine-HCl (0.03 M), $MgCl_2$ (0.02 M), EDTA (0.03 M), DIECA (0.03 M), and mannitol (0.35 M). Prior to usage, the buffer was supplemented with complete protease inhibitor cocktail tablets (Takara Biomedical Technology Co., Ltd., Beijing, China) according to the manufacturer's instructions. The homogenate was cleaned from debris using a double nylon filter, resulting in the collection of the filtrate. The extraction process was iterated for the residue using the same buffer. The residue was then filtered through a nylon filter in a similar manner and, subsequently, washed with more buffer. Afterward, the liquid that passed through the filter was centrifuged at a speed of 14000 rpm for 12 min. The pellet was resuspended in buffer, and centrifugation was repeated. Three different concentrations of $(NH4)_2SO_4$, that is, 0–30%, 30–60%, and 60–80%, were added to the mutual supernatant of the enzymes. This resulted in the formation of three different fractions of precipitate. It is worth noting that no decrease in enzyme activity was observed. The fraction of precipitate obtained at 30–60% $(NH4)_2SO_4$ was selected for enzyme analysis due to the optimal activity of both enzymes seen within this concentration of $(NH4)_2SO_4$. The chosen precipitate was fragmented and, subsequently, dissolved in a small volume of extraction buffer. The solubilized precipitate was subjected to dialysis against the solvent buffer by placing it for 24 h at a temperature of 5 $^\circ$C. Subsequently, a solution containing 22% glycerol was added to the dialysate. The mixture was then dipped in liquid nitrogen and stored at a temperature of $-80\ ^\circ\text{C}$ for subsequent analysis.

4.4. Enzymes Assay

The functional level of SPS and SPP were measured in a total volume of 1 mL containing 200 μL Tris-HCl (1.2 M, pH 8) composed of $MgCl_2$ (1.1 M), sodium fluoride (NaF, 2.2 mM), 200 μL UDPG (4.2 mM), and 200 μL fructose-6-phosphate (8.2 mM), and 400 μL dialyzed preparation having 22–24 μg protein μL^{-1}. The reaction mixture was incubated at a temperature of 37 $^\circ$C for 30 min. The reaction was terminated by adding 200 μL NaOH (1.0 N). The reaction mixture was immersed in the water bath for 15 min. Subsequently, 300 μL of resorcinol solution (0.15%, dissolved in 90% ethanol) and 800 μL 28% hydrochloric acid (HCl) were added to the reaction mixture. The combination underwent incubation at 80 $^\circ$C for 12 min. Then, the reaction mixture was allowed to cool to ambient temperature, and the optical density of the reaction mixture was measured at a wavelength of 520 nm in a UV–visible spectrophotometer (Shimadzu UV 1600 Double Beam Spectrophotometer, Kyoto, Japan). The OD output was compared to the standard sucrose absorption curve. The sucrose lysis activity of SuSy was evaluated using 1 mL of reaction mixture consisting of 200 μL Tris-HCl (pH 7.4) including $MgCl_2$ 1.1 M, NAD 2.2 mM, ATP 1.2 mM, 150 μL sucrose (325 mM), and 400 μL dialyzed preparation containing 22–24 μg protein μL^{-1}. Then, 100 μL of UDP (2 M) was mixed in the mixture to onset the reaction, and the mixture was put in an incubator for 30 min at 37 $^\circ$C. The procedure for protein concentration detection [88] was followed. For CWI, SAI, and neutral invertase (NI) activities, measure and analysis of the procedure described by [89,90] were used per requirement. The activity of CWI, SAI, and NI was measured in a reaction mixture with a final volume of 120 μL. The reaction mixture consisted of 30 μL of 50 mM Na-citrate buffer at pH 3.7 for CWI activity assay,

30 µL of 50 mM Na-citrate buffer at pH 5.2 for SAI activity assay, and 30 µL of 50 mM potassium phosphate buffer at pH 7.6 for NI activity assay. Additionally, the reaction mixture contained 60 µL of 125 mM sucrose and 30 µL of the sample. The incubation temperature was set at 37 °C. The reactions were halted at 20 min by adding 120 µL of alkaline reagent, followed by boiling for 18 min, and then cooling on ice for 5 min. Absorbance was detected at 540 nm using UV–visible spectrophotometer (Shimadzu UV 1600 Double Beam Spectrophotometer, Kyoto, Japan).

4.5. RNA Extraction and RT-qPCR Expression

RNA was extracted from the immature and maturing internodal tissues of the high- and low-sucrose sugarcane clones using TRIzolR Reagent (Plant RNA Purification Reagent for plant tissue) according to the manufacturer's instructions (Invitrogen, Shanghai, China), and the genomic DNA was removed using DNase I (TaKaRa). The isolated RNA was quantified using ND-2000 (NanoDrop Thermo Scientific, Wilmington, NC, USA), and high-quality RNA (OD260/280 = 1.9–2.4, OD260/230 $\geq$ 2.5, RIN $\geq$ 8.5, 29S: 19S $\geq$ 1.2, >2 µg) was cast off to construct cDNA. RT-qPCR was performed to examine the expression of sucrose metabolic genes, including SPS, SPP, SuSy, CWI, SAI, and NI, using TSINGKE biological technology (www.tsingke.net: 28-10-2022) primers (Table 1) in Light CyclerR480 II (Roche, Basel, Switzerland). Then, 1 µg RNA of each clone was used to create cDNA using the first-strand cDNA synthesis kit (Vazyme Biotech Co., Ltd., Nanjing, China) following the manufacturer's instructions. The RT-qPCR reaction was conducted in a total volume of 20 µL, including 2 µL of cDNA (template), 0.5 µL primer mix (10 µm each of forward and reverse primers), 10.5 µL 2× ChamQ Universal SYBR qPCR master mix (Vazyme Biotech Co., Ltd.), and 6.5 µL sterile water. The amplification parameters were as follows: 1 cycle of 30 s at 95 °C, followed by 40 cycles of 5 s at 95 °C and 15 s at 60 °C, and 1 cycle of 15 s at 95 °C, 1 min at 60 °C, and 15 s at 95 °C. The sugarcane housekeeping gene, glyceraldehyde-3-phosphate dehydrogenase (GAPDH), was used as the internal reference gene to normalize the expression level. Triplicate (n = 3) biological and technical replicates were used for each sample. The data analysis was performed using Roche's Light Cycler R 480 version 1.5.1. The relative fold change in the expression of the gene was calculated using the $2^{-\triangle\triangle Ct}$ algorithm [91].

Table 1. Primer sequences used in RT-qPCR amplification.

Genes Name	Forward/Reverse	5′-3′ Sequence	Product Size (bp)
Glyceraldehyde 3-phosphate	F	CTCTGCCCCAAGCAAAGATG	100
dehydrogenase (GAPDH)	R	TGTTGTGCAGCTAGCATTGGA	
Sucrose phosphate synthase (SPS)	F	CCATCTGTATGTTGCTGTGTGC	99
	R	GTCGGTGTCGCCCTTGTC	
Sucrose synthase (SuSy)	F	TGAAAATGGGATACTTAAGAAATGG	92
	R	ATAACGAACCAATGATGATATTCACCTC	
Cell wall invertase (CWI)	F	TCTGTACAAGCCAACCTTCG	104
	R	CCGCTTGAAATGTCAATGTC	
Sucrose phosphate phosphatase (SPP)	F	GGCTTTGTGCTAACCCACAT	98
	R	TTACGCACCAAATCCTCTCC	
Soluble acid invertase (SAI)	F	TCCTTGCTTGCCTCTCAAAT	97
	R	ACAAATGTAGCCCTGCCTTG	
Neutral invertase (NI)	F	ATAAACAGCCGCACCAATTC	112
	R	GCCTCTGAGGTGGAGTCTTG	

4.6. Statistical Analysis

The data acquired from examined variables were statistically analyzed using Statistix 8.1 software. The data were subjected to a two-way ANOVA, with clones as factor 1 and internodes as factor 2. The LSD test was applied to compare the significant difference, at $p < 0.05$, among both internode types and sister clones. Furthermore, figures were created by using GraphPad Prism (Version 10).

5. Conclusions

The main product of photosynthesis is sucrose, which is produced in sugarcane leaves and accumulates in the internodes. Its metabolism is a constant hydrolysis and resynthesis process to support respiration and store excess sucrose in the stalks.

The present investigation discovered variations in sucrose and sugar concentrations between two sister clones with high and low sucrose accumulation, GXB9 and B9. The sucrose and hexose levels varied between the immature and maturing internodes of the same clone. The clone GXB9 with high sucrose storage showed reduced sucrose content in immature internodes compared to maturing ones. In immature internodes, both clones showed higher fructose and glucose levels than in maturing internodes. Immature internodes exhibited minimal sucrose content, whereas maturing internodes demonstrated elevated sucrose levels in the low-sucrose-accumulating clone B9. The concentrations of glucose and fructose were higher in immature internodes compared to maturing internodes in B9. The sucrose concentration was higher in maturing internodes, while glucose and fructose were more abundant in immature internodes in both clones. The high-sucrose-accumulating clone exhibited a higher total sucrose content than the low-sucrose-accumulating clone.

A comparison of SPS, SPP, and SuSy activities revealed that the high-sucrose sugarcane clone exhibited elevated SPS and SPP activities. The SPS and SPP activities in maturing internodes were greater than those in immature internodes for the high-sucrose sugarcane clone. The activity of SuSy was greater in the low-sucrose clone compared to the high-sucrose clone. Additionally, the SuSy activity was higher in immature internodes compared to maturing internodes for the low-sucrose-storing clone. Susy exhibited increased activity in immature internodes of both high- and low-sucrose clones, while SPS and SPP had higher activities in maturing internodes.

The up-regulation of *SPS* and *SPP* was only found in the high-sucrose clone but not in the low-sucrose clone. *SPS* was significantly up-regulated in the maturing internodes compared with immature internodes of the high-sucrose clone. *SuSy* was significantly up-regulated in immature internodes of both clones. Hence, the significant up-regulation of *SPS* in maturing internodes presents the predominant role in the high sucrose accumulation in the high-sucrose clone. Both clones have a similar genetic background, but considerable variations in sucrose content make them appealing material for further study.

The regulatory network governing sucrose metabolism in sugarcane internodes (Figure 5) still needs comprehensive elucidation. Further spatiotemporal analysis of internodes would lead to a better understanding of the roles of SPS, SPP, and SuSy in sugarcane, despite the existing knowledge about these enzymes and their involvement in sucrose metabolism.

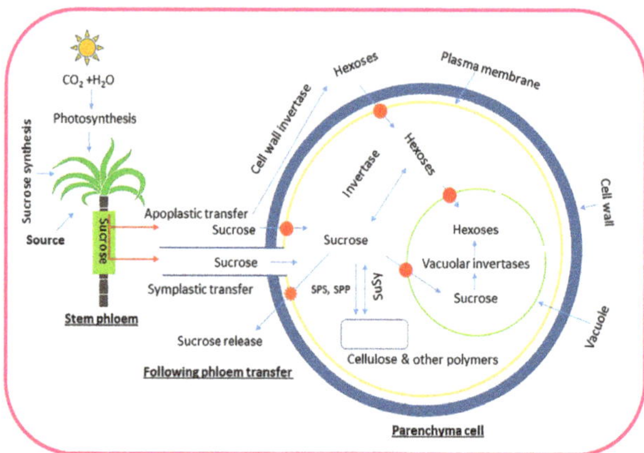

Figure 5. Schematic illustration depicting the process of sucrose transportation and metabolism in sugarcane, starting with its production in source leaves and ending with its storage in stems. Red

arrows show sucrose transport, whereas its subsequent breakdown and storage are represented by thin blue arrows. Red ovals with blue arrows show transporters, a large dark blue circle indicates the cell wall, and the vacuole is depicted in light green. The plasma membrane is gold in color. Sucrose produced in photosynthetic leaves is transported through the phloem to stem parenchyma cells. From there, it can be transferred in two ways: symplastic (via plasmodesmata, believed to be the main route in mature sugarcane internodes) and/or apoplastic (through the cell wall space, which was considered a potential factor in earlier stages of sugarcane stem development). Sucrose may be transported to storage parenchyma by either pathway; however, apoplastic transfer may need sucrose to be broken down into hexoses by cell wall invertase. Hexoses and sucrose both enter parenchyma cells via transporters. Neutral invertases in the cytoplasm or vacuolar acid invertases may produce hexoses from sucrose inside cells. Sucrose is stored in vacuoles and cell wall space, with the equilibrium between them regulated by transporters and sucrose release into the apoplast.

Therefore, an integrated transcriptomic and proteomic spatiotemporal comprehensive elucidation of the sucrose metabolism mechanism in sugarcane stalks of the two sister clones should be considered in order to evaluate possible mutation in the GXB9 clone. The finding could pave the way for genetic modifications of sugarcane to increase sucrose content.

Author Contributions: Conceptualization, Y.-R.L., Y.-X.X. and Q.K.; methodology monitoring, Y.-Y.H., X.-P.S., L.-T.Y., D.-J.G. and Q.K.; formal analysis, Q.K., Y.Q., Y.-R.L., D.-J.G. and Q.L.; data curation, X.-P.S., L.-T.Y., Y.-Y.H., Q.L. and Y.Q.; writing—original draft preparation, Y.-X.X., Q.K., Y.Q., Y.-Y.H., X.-P.S. and D.-J.G.; writing—review and editing, Y.-R.L., Y.-X.X., Q.L., Q.K. and L.-T.Y.; funding acquisition, Y.-R.L. All authors have read and agreed to the published version of the manuscript.

Funding: This work was funded by the Guangxi R & D Program (GKZY23055011, GKAA22117009-7), Fund for Guangxi Innovation Teams of Modern Agriculture Technology (gjnytxgxcxtd-03-01), Fund of Guangxi Academy of Agricultural Sciences (2021YT011), and Fund of Guangxi Key Laboratory of Sugarcane Genetic Improvement (21-238-16-K-03-03).

Institutional Review Board Statement: No experiments with humans or animals were performed in this study, so ethical clearance was not required.

Informed Consent Statement: All authors agreed to the contributions to the manuscript.

Data Availability Statement: All the necessary data are included in the manuscript.

Acknowledgments: We are highly thankful to the Guangxi Academy of Agriculture Science (GXAAS) for providing the research material of high- and low-sugar sugarcane clones GXB9 and B9.

Conflicts of Interest: The authors declare no conflict of interest.

References

1. Nair, P.N.; Sachan, H. The Improvement of sugarcane (*Saccharum officinarum* L.) for sugar, ethanol and biofuel production through innovative biotechnology: A perspective view on its scope, importance & challenges. In *Omics Approaches for Sugarcane Crop Improvement*; CRC Press: Boca Raton, FL, USA, 2022; pp. 233–249.
2. Den Besten, N.; Steele-Dunne, S.; Aouizerats, B.; Zajdband, A.; De Jeu, R.; Van Der Zaag, P. Observing sucrose accumulation with sentinel-1 backscatter. *Front. Remote Sens.* **2021**, *2*, 778691. [CrossRef]
3. Slewinski, T.L.; Baker, R.F.; Stubert, A.; Braun, D.M. Tie-dyed2 encodes a callose synthase that functions in vein development and affects symplastic trafficking within the phloem of maize leaves. *Plant Physiol.* **2012**, *160*, 1540–1550. [CrossRef] [PubMed]
4. Qin, C.-X.; Chen, Z.-L.; Wang, M.; Li, A.-M.; Liao, F.; Li, Y.-R.; Wang, M.-Q.; Long, M.-H.; Lakshmanan, P.; Huang, D.-L. Identification of proteins and metabolic networks associated with sucrose accumulation in sugarcane (*Saccharum* spp. interspecific hybrids). *J. Plant Interact.* **2021**, *16*, 166–178. [CrossRef]
5. Caieiro, J.T.; Panobianco, M.; Bespalhok Filho, J.C.; Ohlson, O.d.C. Physical purity and germination of sugarcane seeds (caryopses) (*Saccharum* spp.). *Rev. Bras. De Sementes* **2010**, *32*, 140–145. [CrossRef]
6. Raza, Q.-U.-A.; Bashir, M.A.; Rehim, A.; Sial, M.U.; Ali Raza, H.M.; Atif, H.M.; Brito, A.F.; Geng, Y. Sugarcane industrial byproducts as challenges to environmental safety and their remedies: A review. *Water* **2021**, *13*, 3495. [CrossRef]
7. PerezLopez, J.F.; Alvarez, J. (Eds.) *Reinventing the Cuban Sugar Agroindustry*; Lexington Books, c2005: Cloth Pbk; Rowman & Littlefield Publishers, Inc.: New York, NY, USA, 2005; p. 209.
8. Solomon, S. Sugarcane production and development of sugar industry in India. *Sugar Tech* **2016**, *18*, 588–602. [CrossRef]

9.   Wang, M.; Li, A.M.; Liao, F.; Qin, C.X.; Chen, Z.L.; Zhou, L.; Li, Y.R.; Li, X.F.; Lakshmanan, P.; Huang, D.L. Control of sucrose accumulation in sugarcane (*Saccharum* spp. hybrids) involves miRNA-mediated regulation of genes and transcription factors associated with sugar metabolism. *GCB Bioenergy* **2022**, *14*, 173–191. [CrossRef]

10. Zhao, D.; Li, Y.-R. Climate change and sugarcane production: Potential impact and mitigation strategies. *Int. J. Agron.* **2015**, *2015*, 547386. [CrossRef]

11. Trenberth, K.E.; Jones, P.D.; Ambenje, P.; Bojariu, R.; Easterling, D.; Klein Tank, A.; Parker, D.; Rahimzadeh, F.; Renwick, J.A.; Rusticucci, M. Observations. Surface and atmospheric climate change. In *Climate Change 2007: The Physical Science Basis*; Cambridge University Press: Cambridge, UK; New York, NY, USA, 2007; Chapter 3; pp. 235–336.

12. Dhillon, R.; von Wuehlisch, G. Mitigation of global warming through renewable biomass. *Biomass Bioenergy* **2013**, *48*, 75–89. [CrossRef]

13. Gawander, J. Impact of climate change on sugar-cane production in Fiji. *World Meteorol. Organ. Bull.* **2007**, *56*, 34–39.

14. Hussain, S.; Khaliq, A.; Mehmood, U.; Qadir, T.; Saqib, M.; Iqbal, M.A.; Hussain, S. Sugarcane production under changing climate: Effects of environmental vulnerabilities on sugarcane diseases, insects and weeds. *Clim. Change Agric.* **2018**, 1–17.

15. Cardozo, N.P.; Sentelhas, P.C. Climatic effects on sugarcane ripening under the influence of cultivars and crop age. *Sci. Agric.* **2013**, *70*, 449–456. [CrossRef]

16. Pereira, L.F.; Ferreira, V.M.; OLIVEIRA, N.G.; Sarmento, P.L.; Endres, L.; Teodoro, I. Sugars levels of four sugarcane genotypes in different stem portions during the maturation phase. *An. Da Acad. Bras. De Ciências* **2017**, *89*, 1231–1242. [CrossRef] [PubMed]

17. Ftwi, M.; Mekibib, F.; Tesfa, M. Maturity classification of sugarcane (*Saccharum officinarum* L) genotypes grown under different production environments of Ethiopia. *Adv. Crop Sci. Technol.* **2017**, *5*, 304.

18. Zhao, Y.; Yu, L.-X.; Ai, J.; Zhang, Z.-F.; Deng, J.; Zhang, Y.-B. Climate variations in the low-latitude plateau contribute to different sugarcane (*Saccharum* spp.) yields and sugar contents in China. *Plants* **2023**, *12*, 2712. [CrossRef] [PubMed]

19. Zepeda, A.C.; Heuvelink, E.; Marcelis, L.F. Carbon storage in plants: A buffer for temporal light and temperature fluctuations. *Silico Plants* **2023**, *5*, diac020. [CrossRef]

20. Takaragawa, H.; Matsuda, H. Rapid evaluation of leaf photosynthesis using a closed-chamber system in a C4 plant, sugarcane. *Plant Prod. Sci.* **2023**, *26*, 174–186. [CrossRef]

21. Khan, Q.; Qin, Y.; Guo, D.-J.; Yang, L.-T.; Song, X.-P.; Xing, Y.-X.; Li, Y.-R. A Review of the diverse genes and molecules involved in sucrose metabolism and innovative approaches to improve sucrose content in sugarcane. *Agronomy* **2023**, *13*, 2957. [CrossRef]

22. Lemoine, R.; Camera, S.L.; Atanassova, R.; Dédaldéchamp, F.; Allario, T.; Pourtau, N.; Bonnemain, J.-L.; Laloi, M.; Coutos-Thévenot, P.; Maurousset, L. Source-to-sink transport of sugar and regulation by environmental factors. *Front. Plant Sci.* **2013**, *4*, 272. [CrossRef]

23. Moore, P.H.; Cosgrove, D.J. Developmental changes in cell and tissue water relations parameters in storage parenchyma of sugarcane. *Plant Physiol.* **1991**, *96*, 794–801. [CrossRef]

24. Bihmidine, S.; Baker, R.F.; Hoffner, C.; Braun, D.M. Sucrose accumulation in sweet sorghum stems occurs by apoplasmic phloem unloading and does not involve differential sucrose transporter expression. *BMC Plant Biol.* **2015**, *15*, 186. [CrossRef]

25. Dhungana, S.R.; Braun, D.M. Sugar transporters in grasses: Function and modulation in source and storage tissues. *J. Plant Physiol.* **2021**, *266*, 153541. [CrossRef]

26. Sage, R.F.; Peixoto, M.M.; Sage, T.L. Photosynthesis in sugarcane. In *Sugarcane: Physiology, Biochemistry, and Functional Biology*; John Wiley & Sons, Inc.: Hoboken, NJ, USA, 2013; pp. 121–154.

27. Chen, L.-Q. Improved understanding of sugar transport in various plants. *Int. J. Mol. Sci.* **2022**, *23*, 10260. [CrossRef]

28. Zhang, Q.; Hua, X.; Liu, H.; Yuan, Y.; Shi, Y.; Wang, Z.; Zhang, M.; Ming, R.; Zhang, J. Evolutionary expansion and functional divergence of sugar transporters in *Saccharum* (*S. spontaneum* and *S. officinarum*). *Plant J.* **2021**, *105*, 884–906. [CrossRef]

29. Glassop, D.; Roessner, U.; Bacic, A.; Bonnett, G.D. Changes in the sugarcane metabolome with stem development. Are they related to sucrose accumulation? *Plant Cell Physiol.* **2007**, *48*, 573–584. [CrossRef]

30. Uys, L.; Hofmeyr, J.-H.S.; Rohwer, J.M. Coupling kinetic models and advection–diffusion equations. 1. Framework development and application to sucrose translocation and metabolism in sugarcane. *Silico Plants* **2021**, *3*, diab013. [CrossRef]

31. Khan, I.A.; Bibi, S.; Yasmin, S.; Khatri, A.; Seema, N.; Abro, S.A. Correlation studies of agronomic traits for higher sugar yield in sugarcane. *Pak. J. Bot* **2012**, *44*, 969–971.

32. Whittaker, A.; Botha, F.C. Carbon partitioning during sucrose accumulation in sugarcane internodal tissue. *Plant Physiol.* **1997**, *115*, 1651–1659. [CrossRef] [PubMed]

33. Albertson, P.L.; Peters, K.F.; Grof, C.P. An improved method for the measurement of cell wall invertase activity in sugarcane tissue. *Funct. Plant Biol.* **2001**, *28*, 323–328. [CrossRef]

34. Grof, C.P.; Campbell, J.A. Sugarcane sucrose metabolism: Scope for molecular manipulation. *Funct. Plant Biol.* **2001**, *28*, 1–12. [CrossRef]

35. Gutiérrez-Miceli, F.A.; Rodríguez-Mendiola, M.A.; Ochoa-Alejo, N.; Méndez-Salas, R.; Dendooven, L.; Arias-Castro, C. Relationship between sucrose accumulation and activities of sucrose-phosphatase, sucrose synthase, neutral invertase and soluble acid invertase in micropropagated sugarcane plants. *Acta Physiol. Plant.* **2002**, *24*, 441–446. [CrossRef]

36. Meena, D.M.R.; Reddy, G.; Kumar, R.; Pandey, S.; Hemaprabha, G. Recent advances in sugarcane genomics, physiology, and phenomics for superior agronomic traits. *Front. Genet.* **2022**, *13*, 854936. [CrossRef]

37. Sica, P. Sugarcane breeding for enhanced fiber and its impacts on industrial processes. In *Sugarcane-Biotechnology for Biofuels*; IntechOpen: London, UK, 2021.

38. Partida, V.G.S.; Dias, H.M.; Corcino, D.S.M.; Van Sluys, M.-A. Sucrose-phosphate phosphatase from sugarcane reveals an ancestral tandem duplication. *BMC Plant Biol.* **2021**, *21*, 23. [CrossRef] [PubMed]

39. Wang, J.; Zhao, T.; Yang, B.; Zhang, S. Sucrose metabolism and regulation in sugarcane. *J. Plant Physiol. Pathol.* **2017**, *5*, 2. [CrossRef]

40. Stein, O.; Granot, D. An overview of sucrose synthases in plants. *Front. Plant Sci.* **2019**, *10*, 95. [CrossRef] [PubMed]

41. Ogawa, A.; Ando, F.; Toyofuku, K.; Kawashima, C. Sucrose metabolism for the development of seminal root in maize seedlings. *Plant Prod. Sci.* **2009**, *12*, 9–16. [CrossRef]

42. Winter, H.; Huber, S.C. Regulation of sucrose metabolism in higher plants: Localization and regulation of activity of key enzymes. *Crit. Rev. Plant Sci.* **2000**, *19*, 31–67. [CrossRef]

43. Bansal, R. Cell Wall Invertase and Sucrose Synthase Regulate Sugar Metabolism During Seed Development in Isabgol (*Plantago ovata* Forsk.). *Proc. Natl. Acad. Sci. India Sect. B Biol. Sci.* **2018**, *88*, 73–78. [CrossRef]

44. Fujii, S.; Hayashi, T.; Mizuno, K. Sucrose synthase is an integral component of the cellulose synthesis machinery. *Plant Cell Physiol.* **2010**, *51*, 294–301. [CrossRef]

45. Tang, G.-Q.; Sturm, A. Antisense repression of sucrose synthase in carrot (*Daucus carota* L.) affects growth rather than sucrose partitioning. *Plant Mol. Biol.* **1999**, *41*, 465–479. [CrossRef]

46. Ahmad, S.; Ali, M.A.; Aita, G.M.; Khan, M.T.; Khan, I.A. Source-sink relationship of sugarcane energy production at the sugar mills. In *Sugarcane Biofuels: Status, Potential, and Prospects of the Sweet Crop to Fuel the World*; Khan, M.T., Khan, I.A., Eds.; Springer: Berlin/Heidelberg, Germany, 2019; pp. 349–388.

47. Geigenberger, P.; Stitt, M. Sucrose synthase catalyses a readily reversible reaction in vivo in developing potato tubers and other plant tissues. *Planta* **1993**, *189*, 329–339. [CrossRef]

48. Verma, K.K.; Song, X.-P.; Yadav, G.; Degu, H.D.; Parvaiz, A.; Singh, M.; Huang, H.-R.; Mustafa, G.; Xu, L.; Li, Y.-R. Impact of agroclimatic variables on proteogenomics in sugar cane (*Saccharum* spp.) plant productivity. *ACS Omega* **2022**, *7*, 22997–23008. [CrossRef] [PubMed]

49. da Silva Santos, P.H.; Manechini, J.R.V.; Brito, M.S.; Romanel, E.; Vicentini, R.; Scarpari, M.; Jackson, S.; Pinto, L.R. Selection and validation of reference genes by RT-qPCR under photoperiodic induction of flowering in sugarcane (*Saccharum* spp.). *Sci. Rep.* **2021**, *11*, 4589. [CrossRef] [PubMed]

50. Glasziou, K. Accumulation and transformation of sugars in sugar cane stalks. *Plant Physiol.* **1960**, *35*, 895. [CrossRef] [PubMed]

51. Tana, B.; Chanprame, S.; Tienseree, N.; Tadakittisarn, S. Relationship between invertase enzyme activities and sucrose accumulation in sugarcane (*Saccharum* spp.). *Agric. Nat. Resour.* **2014**, *48*, 869–879.

52. Yamaki, S. Metabolism and accumulation of sugars translocated to fruit and their regulation. *J. Jpn. Soc. Hortic. Sci.* **2010**, *79*, 1–15. [CrossRef]

53. Zhang, L.; Sun, S.; Liang, Y.; Li, B.; Ma, S.; Wang, Z.; Ma, B.; Li, M. Nitrogen levels regulate sugar metabolism and transport in the shoot tips of crabapple plants. *Front. Plant Sci.* **2021**, *12*, 626149. [CrossRef] [PubMed]

54. Chandiposha, M. Potential impact of climate change in sugarcane and mitigation strategies in Zimbabwe. *Afican J. Agric. Res.* **2013**, *8*, 2814–2818.

55. de Almeida Silva, M.; Caputo, M.M. Ripening and the use of ripeners for better sugarcane management. In *Crop Management–Cases and Tools for Higher Yield and Sustainability*; Marin, F.R., Ed.; IntechOpen: London, UK, 2012; pp. 2–24.

56. de Oliveira, A.R.; Braga, M.B.; Santos, B.L.S.; Walker, A.M. Análise biométrica de cultivares de cana-de-açúcar cultivadas sob estresse hídrico no vale do submédio São Francisco. *Energ. Na Agric.* **2016**, *31*, 48–58. [CrossRef]

57. Fernandes, A.C. *Cálculos na Agroindústria da Cana-de-Açúcar*, 3rd ed.; STAB: Piracicaba, Brasil, 2011. (In Portuguese)

58. Lingle, S.E.; Thomson, J.L. Sugarcane internode composition during crop development. *BioEnergy Res.* **2012**, *5*, 168–178. [CrossRef]

59. Botha, F.C.; Scalia, G.; Marquardt, A.; Wathen-Dunn, K. Sink strength during sugarcane culm growth: Size matters. *Sugar Tech.* **2023**, *25*, 1047–1060. [CrossRef]

60. Sturm, A. Invertases. Primary structures, functions, and roles in plant development and sucrose partitioning. *Plant Physiol.* **1999**, *121*, 1–8. [CrossRef] [PubMed]

61. Roitsch, T.; González, M.-C. Function and regulation of plant invertases: Sweet sensations. *Trends Plant Sci.* **2004**, *9*, 606–613. [CrossRef]

62. Lontom, W.; Kosittrakun, M.; Lingle, S. Relationship of acid invertase activities to sugar content in sugarcane internodes during ripening and after harvest. *Thai J. Agric. Sci.* **2008**, *41*, 143–151.

63. Sacher, J.; Hatch, M.; Glasziou, K. Sugar accumulation cycle in sugar cane. III. Physical & metabolic aspects of cycle in immature storage tissues. *Plant Physiol.* **1963**, *38*, 348. [PubMed]

64. Verma, A.K.; Upadhyay, S.K.; Srivastava, M.K.; Verma, P.C.; Solomon, S.; Singh, S. Transcript expression and soluble acid invertase activity during sucrose accumulation in sugarcane. *Acta Physiol. Plant.* **2011**, *33*, 1749–1757. [CrossRef]

65. Kubo, T.; Hohjo, I.; Hiratsuka, S. Sucrose accumulation and its related enzyme activities in the juice sacs of satsuma mandarin fruit from trees with different crop loads. *Sci. Hortic.* **2001**, *91*, 215–225. [CrossRef]

66. Moore, P.H. Temporal and spatial regulation of sucrose accumulation in the sugarcane stem. *Funct. Plant Biol.* **1995**, *22*, 661–679. [CrossRef]

67. Chandra, A.; Verma, P.K.; Islam, M.; Grisham, M.; Jain, R.; Sharma, A.; Roopendra, K.; Singh, K.; Singh, P.; Verma, I. Expression analysis of genes associated with sucrose accumulation in sugarcane (*Saccharum* spp. hybrids) varieties differing in content and time of peak sucrose storage. *Plant Biol.* **2015**, *17*, 608–617. [CrossRef]

68. Ebrahim, M.K.; Zingsheim, O.; El-Shourbagy, M.N.; Moore, P.H.; Komor, E. Growth and sugar storage in sugarcane grown at temperatures below and above optimum. *J. Plant Physiol.* **1998**, *153*, 593–602. [CrossRef]

69. Vorster, D.J.; Botha, F.C. Sugarcane internodal invertases and tissue maturity. *J. Plant Physiol.* **1999**, *155*, 470–476. [CrossRef]

70. Gayler, K.; Glasziou, K. Physiological functions of acid and neutral invertases in growth and sugar storage in sugar cane. *Physiol. Plant.* **1972**, *27*, 25–31. [CrossRef]

71. Bosch, S.; Grof, C.; Botha, F. Expression of neutral invertase in sugarcane. *Plant Sci.* **2004**, *166*, 1125–1133. [CrossRef]

72. Rose, S.; Botha, F.C. Distribution patterns of neutral invertase and sugar contentin sugarcane internodal tissues. *Plant Physiol. Biochem.* **2000**, *38*, 819–824. [CrossRef]

73. Anur, R.M.; Mufithah, N.; Sawitri, W.D.; Sakakibara, H.; Sugiharto, B. Overexpression of sucrose phosphate synthase enhanced sucrose content and biomass production in transgenic sugarcane. *Plants* **2020**, *9*, 200. [CrossRef] [PubMed]

74. Sawitri, W.D.; Sugiharto, B. Revealing the important role of allosteric property in sucrose phosphate synthase from sugarcane with N-terminal domain deletion. *Proc. IOP Conf. Ser. Earth Environ. Sci.* **2018**, *217*, 012043. [CrossRef]

75. Li, Y.; Yao, Y.; Yang, G.; Tang, J.; Ayala, G.J.; Li, X.; Zhang, W.; Han, Q.; Yang, T.; Wang, H. Co-crystal Structure of Thermosyne-chococcus elongatus Sucrose Phosphate Synthase With UDP and Sucrose-6-Phosphate Provides Insight Into Its Mechanism of Action Involving an Oxocarbenium Ion and the Glycosidic Bond. *Front. Microbiol.* **2020**, *11*, 1050. [CrossRef] [PubMed]

76. Zhang, J.; Zhou, M.; Walsh, J.; Zhu, L.; Chen, Y.; Ming, R. Sugarcane genetics and genomics. In *Sugarcane: Physiology, Biochemistry, and Functional Biology*; John Wiley & Sons, Inc.: Hoboken, NJ, USA, 2013; pp. 623–643.

77. Verma, A.K.; Upadhyay, S.; Verma, P.C.; Solomon, S.; Singh, S.B. Functional analysis of sucrose phosphate synthase (SPS) and sucrose synthase (SS) in sugarcane (*Saccharum*) cultivars. *Plant Biol.* **2011**, *13*, 325–332. [CrossRef]

78. Botha, F.C.; Black, K.G. Sucrose phosphate synthase and sucrose synthase activity during maturation of internodal tissue in sugarcane. *Funct. Plant Biol.* **2000**, *27*, 81–85. [CrossRef]

79. Grof, C.P.; Albertson, P.L.; Bursle, J.; Perroux, J.M.; Bonnett, G.D.; Manners, J.M. Sucrose-phosphate synthase, a biochemical marker of high sucrose accumulation in sugarcane. *Crop Sci.* **2007**, *47*, 1530–1539. [CrossRef]

80. McCormick, A.; Watt, D.; Cramer, M. Supply and demand: Sink regulation of sugar accumulation in sugarcane. *J. Exp. Bot.* **2009**, *60*, 357–364. [CrossRef]

81. Bilska-Kos, A.; Mytych, J.; Suski, S.; Magoń, J.; Ochodzki, P.; Zebrowski, J. Sucrose phosphate synthase (SPS), sucrose synthase (SUS) and their products in the leaves of Miscanthus× giganteus and *Zea mays* at low temperature. *Planta* **2020**, *252*, 23. [CrossRef] [PubMed]

82. Su, Y.; Wang, Z.; Xu, L.; Peng, Q.; Liu, F.; Li, Z.; Que, Y. Early selection for smut resistance in sugarcane using pathogen proliferation and changes in physiological and biochemical indices. *Front. Plant Sci.* **2016**, *7*, 1133. [CrossRef] [PubMed]

83. Khan, Q.; Qin, Y.; Guo, D.-J.; Zeng, X.-P.; Chen, J.-Y.; Huang, Y.-Y.; Ta, Q.-K.; Yang, L.-T.; Liang, Q.; Song, X.-P. Morphological, agronomical, physiological and molecular characterization of a high sugar mutant of sugarcane in comparison to mother variety. *PLoS ONE* **2022**, *17*, e0264990. [CrossRef] [PubMed]

84. Khan, Q.; Chen, J.Y.; Zeng, X.P.; Qin, Y.; Guo, D.J.; Mahmood, A.; Yang, L.T.; Liang, Q.; Song, X.P.; Xing, Y.X. Transcriptomic explo-ration of a high sucrose mutant in comparison with the low sucrose mother genotype in sugarcane during sugar accumulating stage. *GCB Bioenergy* **2021**, *13*, 1448–1465. [CrossRef]

85. Shukla, S.; Sharma, L.; Jaiswal, V.; Pathak, A.; Tiwari, R.; Awasthi, S.; Gaur, A. Soil quality parameters vis-a-vis growth and yield attributes of sugarcane as influenced by integration of microbial consortium with NPK fertilizers. *Sci. Rep.* **2020**, *10*, 19180. [CrossRef] [PubMed]

86. Roe, J.H. A colorimetric method for the determination of fructose in blood and urine. *J. Biol. Chem.* **1934**, *107*, 15–22. [CrossRef]

87. Somogyi, M. Notes on sugar determination. *J. Biol. Chem.* **1952**, *195*, 19–23. [CrossRef]

88. Classics Lowry, O.; Rosebrough, N.; Farr, A.; Randall, R. Protein measurement with the Folin phenol reagent. *J. Biol. Chem.* **1951**, *193*, 265–275. [CrossRef]

89. Leite, G.H.P.; Alexandre, C.; Crusciol, C.; Siqueira, G.F.d.; Silva, M.d.A. Plant regulators and invertase activity in sugarcane at the beginning of the harvest season. *Cienc. Rural* **2015**, *45*, 1788–1794. [CrossRef]

90. Siswoyo, T.A.; Oktavianawati, I.; Djenal, D.; Sugiharto, B.; Murdiyanto, U.; XI, P.P.N. Changes of sucrose content and invertase activity during sugarcane stem storage. *Indones. J. Agric. Sci.* **2007**, *8*, 75–81. [CrossRef]

91. De Andrade, L.M.; dos Santos Brito, M.; Fávero Peixoto Junior, R.; Marchiori, P.E.R.; Nóbile, P.M.; Martins, A.P.B.; Ribeiro, R.V.; Creste, S. Reference genes for normalization of qPCR assays in sugarcane plants under water deficit. *Plant Methods* **2017**, *13*, 28. [CrossRef] [PubMed]

Article

Biochar Mitigates the Negative Effects of Microplastics on Sugarcane Growth by Altering Soil Nutrients and Microbial Community Structure and Function

Qihua Wu [1], Wenling Zhou [1], Diwen Chen [1], Jiang Tian [2] and Junhua Ao [1,*]

[1] Institute of Nanfan & Seed Industry, Guangdong Academy of Sciences, Guangzhou 510316, China; wqh5859@126.com (Q.W.); zwl2018@vip.163.com (W.Z.); chendiwen@126.com (D.C.)

[2] Root Biology Center, South China Agricultural University, Guangzhou 510642, China; jtian@scau.edu.cn

* Correspondence: junhuaao@163.com

Abstract: Microplastic pollution in sugarcane areas of China is severe, and reducing the ecological risks is critical. Biochar has been widely used in soil remediation. This study aims to explore the effects and mechanisms of microplastics combined with or without biochar on sugarcane biomass, soil biochemical properties in red soil through a potted experiment. The results show that, compared with control (CK), treatments with microplastics alone reduced the dry biomass of sugarcane, soil pH, and nitrogen (N) and phosphorus (P) contents by an average of 8.8%, 2.1%, 1.1%, and 2.0%, respectively. Interestingly, microplastics combined with biochar could alleviate the negative effects of microplastic accumulation on sugarcane growth and soil quality. There were significant differences in the bacterial community alpha diversity indices and compositions among different treatments. Compared with CK, treatments with microplastics alone obviously decreased the observed operational taxonomic units (OTUs) and the Chao1 and Shannon indices of soil total bacteria (16S rRNA gene-based bacteria) while increasing them in *phoD-harboring* bacteria. Microplastics combined with biochar treatments significantly increased the abundance of *Subgroup_10* for the 16S rRNA gene and treatments with microplastics alone significantly increased the relative abundance of *Streptomyces* for the *phoD* gene compared to CK. Moreover, compared with microplastics alone, the treatments with microplastics combined with biochar increased the relative abundance of *Subgroup_10*, *Bacillus*, *Pseudomonas* in soil total bacteria, and *Amycolatopsis* and *Bradyrhizobium* in *phoD-harboring* bacteria, most of which can inhibit harmful bacteria and promote plant growth. Additionally, different treatments also changed the abundance of potential microbial functional genes. Compared to CK, other treatments increased the abundance of aerobic ammonia oxidation and denitrification but decreased the abundance of nitrate respiration and nitrogen respiration; meanwhile, these four functional genes involved in N cycling processes were obviously higher in treatments with microplastics combined with biochar than in treatments with microplastics alone. In conclusion, microplastics combined with biochar could alleviate the negative effects of microplastic accumulation on sugarcane biomass by altering soil nutrients and microbial community structure and function.

Keywords: microplastics; biochar; sugarcane biomass; microbial community; red soil

Citation: Wu, Q.; Zhou, W.; Chen, D.; Tian, J.; Ao, J. Biochar Mitigates the Negative Effects of Microplastics on Sugarcane Growth by Altering Soil Nutrients and Microbial Community Structure and Function. *Plants* **2024**, *13*, 83. https://doi.org/10.3390/plants13010083

Academic Editor: San-Ji Gao

Received: 1 December 2023
Revised: 21 December 2023
Accepted: 22 December 2023
Published: 27 December 2023

1. Introduction

The prevalence of microplastic (plastic particles < 5 mm) pollution in the world and its potential animal, plant, and human health risks have attracted great attention in recent years [1,2]. Terrestrial ecosystems are an important gathering place of microplastics, so the impact of microplastics on soil ecosystems and plant growth and development has become a research hotspot [3]. The possible sources of microplastics in the soil environment mainly include plastic film mulching, sludge landfill, compost application, and irrigation [4]. Plastic film mulching is an effective measure used to increase crop yield. Sugarcane mulching has been applied in China for more than 30 years, and there is basically no recovery of plastic film in sugarcane fields [5]. Our investigation in the sugarcane area showed that the amount of plastic film used in sugarcane fields was 75–150 kg/ha/year. Assuming that all the plastic film is converted into microplastics and remains in the soil, the cumulative concentration of microplastics in the soil of the sugarcane area would reach 0.1–0.2% in 30 years, which indicates that the pollution of microplastics in sugarcane fields is serious. At the same time, sugarcane has the characteristics of a long growth cycle, high yield, and high nutrient demand [5,6]. Therefore, it is particularly necessary to explore the effects of microplastic pollution on sugarcane growth and soil nutrient content.

When microplastic particles enter the soil, they can combine with soil particles, thereby affecting the physical and chemical properties of the soil, the distribution of soil microorganisms, and the growth and development of plants [7,8]. Several researchers have found that microplastics affect soil physical and chemical properties, including soil porosity and moisture, pH, organic matter, and N, P, and potassium (K) nutrient content [4,9]. Polyethylene (PE) microplastics have been found to have increased soil pH and the content of soluble organic matter and ammonium N by microcosm incubation [10]. Microplastics have also been reported by many studies to affect soil microbial community structure and function [11,12]. Microplastics can directly or indirectly change soil properties, thereby affecting the composition and diversity of microbial communities [13,14]. At the same time, microplastics themselves can provide specific substrates or adsorption sites for microorganisms [15]. The microbial community structure in the biofilm of microplastics is clearly different from the microbial composition in the soil, thus changing the functional properties of the soil [12]. In addition, microplastics can also cause changes in rhizosphere microbial communities, which may affect plant growth and development to a certain extent [9]. However, few studies have focused on the effects of microplastic accumulation on soil microorganisms, especially those related to P cycling, and the relationship between microbial changes and sugarcane growth.

Biochar is formed by the pyrolysis of biomass under anaerobic conditions. Its carbon (C) content can reach 40–75%, and it is rich in mineral nutrients, functional groups, and pore structure. Biochar is stable and not easy to degrade, so it has great application prospects in promoting crop growth, soil improvement, and soil pollution remediation [16]. Biochar can alleviate the impact of microplastic pollution on soil microbial diversity [17,18]. It has been found that polyvinyl chloride (PVC) microplastic pollution has adverse effects on crop yield, soil enzyme activity, and microorganisms, whereas biochar mitigates the adverse effects [18]. However, to date, the effects and mechanisms of biochar application on sugarcane growth, soil nutrients, and microbial communities in soils contaminated with different concentrations of microplastics are still unclear.

In this study, sugarcane planting soil was collected to explore the effects of microplastics combined with or without biochar on sugarcane biomass soil biochemical properties through pot experiments. The study aimed to (1) identify the differences in the biomass of sugarcane soil physiochemical properties and microbial community characteristics under different treatments of PE plastics combined with or without corn straw biochar and (2) explore the relationship among sugarcane biomass soil physiochemical properties and microbial community characteristics and their mechanism. Based on previous research results [19] and sugarcane field survey data, we hypothesized that PE microplastics added alone will decrease the biomass of sugarcane and have a negative impact on soil physicochemical and microbial communities, whereas the addition of biochar may alleviate the negative effects of soil microplastic pollution on plant growth and microbial community structure and function, as biochar has rich nutrients and a large specific surface area. The purpose of this study is to provide a reference for the control of microplastic pollution in sugarcane fields and promote the green development of the sugarcane industry.

2. Materials and Methods

2.1. Experimental Soil, Biochar, and Microplastics

The tested red soil was collected from the sugarcane experimental base in Wengyuan, Guangdong (113.94° E, 24.28° N), and the soil physicochemical properties were as follows: pH 4.14, organic matter 14.2 g/kg, available N 158.67 mg/kg, available P 17.47 mg/kg, and available K 89.50 mg/kg. Corn stalk biochar (BC) was purchased from Henan Lize Environmental Protection Technology Co., Ltd. (Zhengzhou, China), and was obtained by pyrolysis at 600 °C. Its main properties were pH 9.0, organic carbon 51.1%, total N 0.9%, total phosphorus 0.2%, and total potassium 1.6%. High-density polyethylene microplastics (PEs) were purchased from Shanghai Aladdin Company. According to the data provided by the manufacturer, the polyethylene particles have a random spherical structure, a purity of 99.99%, a density of 0.96 g/cm^3, and a melting point of 132°. In this study, polyethylene spherical particles with a particle size of approximately 180 μm were selected for the experiment. PE microplastics were first placed in a solution to remove possible heavy metals, then washed with deionized water and air-dried, sterilized to eliminate microbial contamination, and stored in a refrigerator at 4 °C for later use.

2.2. Experimental Setup and Design

The pot experiment was conducted in a greenhouse, and six treatments were set up: adding 0.1% PE microplastic (w/w, soil dry weight, low PE), adding 1% PE microplastic (high PE), adding 1% BC and 0.1% PE microplastic combined with 1% biochar (low PE + BC), 1% PE microplastic combined with 1% biochar (high PE + BC), and control (no addition of PE microplastic or biochar, CK). Each treatment was repeated three times, with a total of 18 pots arranged randomly. The pot was 30 cm high, with a top diameter of 32 cm and a bottom diameter of 26 cm. Urea was used as the main source of N at a rate of 300 mg N/kg soil, and potassium dihydrogen phosphate was used as the source of P and K at a rate of 150 mg K/kg soil. Soil micronutrient additions were (mg/kg soil) CaCl$_2$ 125.67, MgSO$_4$·7H$_2$O 43.34, EDTA-FeNa 5.80, and MnSO$_4$·4H$_2$O 6.67. Before potting, all of the above microplastics, biochar, and nutrients were fully mixed with the soil. On 1 March 2022, one healthy sugarcane seedling (ROC22) of similar growth was transplanted into each pot. During the experiment, the weighing method was used to keep the water content at 70–80% of the field capacity to meet the water demand of sugarcane. It was harvested on 30 December 2022.

2.3. Sample Collection and Analysis

2.3.1. Plant and Soil Collection

Plant samples were collected, including aboveground (cane and leaves) and underground (root) parts. After cleaning, the sugarcane plants were quenched at 105 °C for 30 min and dried at 70 °C to constant weight, and the dry biomass was weighed [20]. For the collection and treatment of soil samples, the rhizosphere soil was obtained by shaking the soil, and the test plants were gently shaken to remove the larger soil clumps from the root system. Then, the soil attached to the root system was shaken off and put into a sterile self-sealing bag. Soil samples were quickly brought back to the laboratory, and one part of the fresh sample of rhizosphere soil was stored in a refrigerator at −40 °C for analysis of soil microorganisms. The other part of the rhizosphere soil was naturally air-dried and then used to determine the physicochemical properties of the soil after impurities were removed.

2.3.2. Measurement of Soil Physicochemical Properties

Soil pH was measured with a pH meter (1:2.5 soil/water). The electrical conductivity (EC) was determined by an electrical conductivity meter. Soil organic carbon (SOC) was measured by the digestion method. Soil total N and total K were determined using the micro-Kjeldahl digestion and flame photometry methods, respectively. Soil available N was determined using the alkaline diffusion method. Available K was extracted with 1 M NH$_4$OAc and then measured with a flame photometer. Total P was extracted with an acid solution (H_2SO_4-$HClO_4$), and the available P was extracted with 0.5 $NaHCO_3$ (Olsen P) and then measured through molybdenum blue colorimetry [21].

2.3.3. Soil DNA Extraction and Microbial Community Analysis

Soil DNA was extracted from 0.5 g of fresh soil samples using a Power Soil DNA isolation kit (MoBio Laboratories, Carlsbad, CA, USA) following the instructions in the accompanying manual, and each treatment contained 3 biological replicates. The quality and concentration of the extracted DNA were determined with a Qubit™ 3.0 fluorometer (Thermo Fisher Scientific Inc., Waltham, MA, USA). The extracted DNA samples were stored at $-40\,°C$.

The 16S rRNA gene was amplified with the primer F515/R907 (GTGCCAGCMGC-CGCGG/CCGTCAATTCMTTTRAGTTT) using the qualified DNA of the assay sample as the template. PCR was carried out on a MasterCycler gradient (Eppendorf, Germany) using 25 μL reaction volumes containing 5 μL 5 × reaction buffer, 5 μL 5 × GC buffer, 2 μL dNTP (2.5 mM), 1 μL forward primer (10 μM), 1 μL reverse primer (10 μM), 2 μL DNA template, 8.75 μL ddH_2O, and 0.25 μL Q5 DNA polymerase. The cycling parameters were 95 °C for 5 min, followed by 28 cycles of 95 °C for 45 s, 55 °C for 50 s, and 72 °C for 45 s, with a final extension at 72 °C for 10 min. Each sample was mixed after three replicates at the same time. Agarose gel electrophoresis (1%) was used to detect the quality of PCR amplification products, an Omega gel recovery and purification kit was used to purify them, and Qubit 2.0 was used to determine the concentration of purified PCR products. The PCR amplification products of different samples were mixed based on equimolar amounts, and the quality of the mixed PCR products was detected at the same time. Then, sequencing was completed by the Illumina MiSeq-PE300 sequencing platform.

In addition, primers ALPS-F730/ALPS-R1101 (CAGTGGGACGACCACGAGGT/GAGGCCGATCGGCATGTCG) were used for *phoD* gene amplification [22]. PCR was carried out on a Mastercycler gradient thermal cycler (Eppendorf, Germany) using 25 μL reaction volumes containing 5 μL 5 × reaction buffer, 5 μL 5 × GC buffer, 2 μL dNTP (2.5 mM), 1 μL forward primer (10 μM), 1 μL reverse primer (10 μM), 2 μL DNA template, 8.75 μL ddH_2O, and 0.25 μL Q5 DNA polymerase. The cycling parameters were 94 °C for 5 min, followed by 35 cycles of 94 °C for 30 s, 50 °C for 30 s, and 72 °C for 60 s, with a final extension at 72 °C for 7 min. Sequencing was completed by the Illumina MiSeq-PE250 sequencing platform after they were qualified.

The raw data were first screened and qualified and then separated using barcode sequences and trimmed with Trimmomatic (version 0.36) after the PCRs were carried out. Then, the dataset was analyzed using QIIME (v1.9). The sequences were clustered into OTUs by VSEARCH (v2.13.4) at a similarity level of 97%. The Basic Local Alignment Search Tool (BLAST) (v2.6.0) was used to classify all the sequences into taxonomic groups based on the Silva database. Then, alpha diversity indices (observed OTUs, Chao1, and Shannon indices) were calculated with Mothur software (version 1.30.1).

2.4. Data Analysis

All data were collated using Excel; SASV9 software was used for one-way analysis of variance, and the Duncan (SSR) method was used to test the significance of the mean value (3 replicates) of each index among different treatments ($p < 0.05$). The data of the species were analyzed by detrended correspondence analysis (DCA), and the length of the first DCA axis was 1.4 for 16S rRNA gene-based bacteria and 3.3 for *phoD-harboring* bacteria. Then, the relationships between soil microbial community structures and soil environmental factors were analyzed by redundancy analysis (RDA). Spearman's correlation coefficients were employed to test the relationships between the soil bacteria, functional genes, and soil properties.

3. Results

3.1. Biomass of Sugarcane

The aboveground biomass (AGB, the sum of cane and leaves) and below-ground biomass (BGB, the root) of sugarcane in different treatments showed some differences (Figure 1). Compared with the CK treatment, the AGB of the BC treatment significantly increased by 7.1%; the AGB and BGB of the low PE and high PE treatments significantly decreased by 4.2% and 8.6% and 22.6% and 37.9%, respectively; and the BGB of the low PE + BC and high PE + BC treatments significantly decreased by 8.8% and 21.9% ($p < 0.05$), respectively. In addition, the total biomass of sugarcane (the sum of BGB and BGB) was in the order of BC > low PE + BC > CK > high PE + BC > low PE > high PE. The AGB and BGB in the treatments with microplastics combined with biochar (low PE + BC and high PE + BC) were significantly higher than those in the treatments with microplastics alone (low PE and high PE).

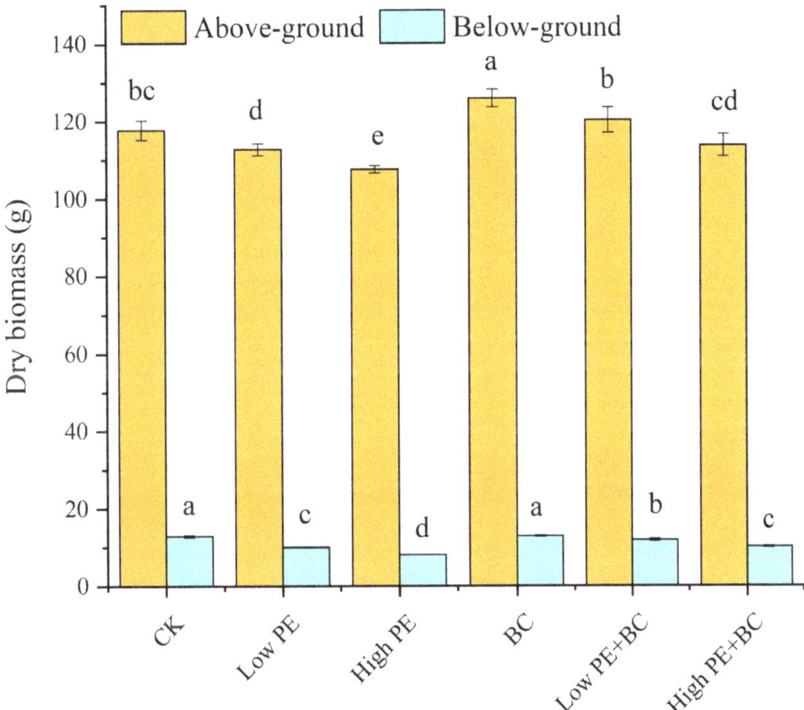

Figure 1. Biomass of sugarcane under different treatments. Different lowercase letters indicate significant difference among the six treatments ($p < 0.05$).

3.2. Physical and Chemical Properties of Soil

Compared with CK, treatments with microplastics alone decreased the pH values averagelly by 2.1%, whereas treatments with BC amendment of low PE + BC and high PE + BC and BC increased the pH values by 2.5%, 1.0% and 3.0%, respectively (Table 1). The other five treatments all increased the SOC contents compare to CK, but only treatments with BC amendment were significantly higher than CK ($p < 0.05$). Compared with CK, the contents of total N, P, and K and available N and P decreased on average by 1.1%, 2.0%, 0.6%, 1.2%, and 13.5%, respectively, whereas the available K content increased by 1.8% in treatments with microplastics alone; the corresponding indices in treatments with BC amendment significantly increased. In addition, the contents of pH, EC, SOC, total N, total

K, and available N, P, and K in the treatments of microplastics combined with biochar were significantly higher than those in treatments with microplastics alone.

Table 1. Changes in physical and chemical properties of soil under different treatments.

	pH (H_2O)	EC (ms/cm)	SOC (g/kg)	TN (g/kg)	TP (g/kg)	TK (g/kg)	AN (mg/kg)	AP (mg/kg)	AK (mg/kg)
CK	3.96 (0.04) bc	0.82 (0.01) b	12.71 (0.69) b	1.46 (0.03) b	0.54 (0.01) b	14.34 (0.03) c	196.02 (4.06) b	15.55 (2.79) cd	84.67 (3.06) b
Low PE	3.89 (0.04) cd	0.81 (0.02) b	13.32 (0.73) b	1.45 (0.02) b	0.53 (0.02) b	14.31 (0.20) c	194.73 (7.07) b	14.08 (1.41) cd	86.33 (3.51) b
High PE	3.86 (0.05) d	0.83 (0.02) b	13.37 (0.53) b	1.44 (0.03) b	0.53 (0.03) b	14.21 (0.12) c	193.05 (5.41) b	12.84 (0.33) d	86.01 (4.00) b
BC	4.08 (0.04) a	1.06 (0.07) a	19.14 (1.16) a	1.66 (0.02) a	0.60 (0.04) a	15.05 (0.32) b	215.69 (5.81) a	24.44 (1.46) a	95.33 (4.51) a
Low PE + BC	4.06 (0.04) a	1.07 (0.04) a	19.83 (0.49) a	1.66 (0.02) a	0.56 (0.02) ab	15.45 (0.10) a	213.92 (4.17) a	19.31 (1.66) b	97.67 (7.09) a
High PE + BC	4.00 (0.06) ab	1.11 (0.01) a	19.65 (0.37) a	1.66 (0.05) a	0.55 (0.04) ab	15.41 (0.15) a	211.63 (1.6) a	16.38 (1.99) bc	98.67 (4.51) a

Different lowercase letters in the same column indicate significant difference among the treatments ($p < 0.05$).

3.3. Soil Microorganisms

3.3.1. Diversity of Soil Microbial Communities

For the total bacteria (16S rRNA gene-based bacteria), compared with CK, the observed_OTUs and the Chao1 and Shannon indices of low PE and high PE were decreased by 5.0%, 5.1%, and 1.4% and by 7.5%, 7.3%, and 2.1%, respectively, whereas these three indices were increased for low PE + BC, high PE + BC, and BC treatments (Table 2). Moreover, compared with the treatment with microplastics alone, the treatments with microplastics combined with biochar significantly increased the observed_OTUs and the Chao1 and Shannon indices. For the *phoD-harboring* bacteria, compared with CK, the observed_OTUs and the Chao1 and Shannon indices of the other five treatments were significantly increased, and these indices were the highest in the high PE and lowest in the low PE treatments. In addition, the treatments of microplastics combined with biochar reduced the observed OTUs and the Chao1 and Shannon indices compared with the treatment with microplastics alone.

Table 2. Bacterial diversity index of soils under different treatments.

	Treatment	Observed OTUs	Chao1	Shannon
16S rRNA gene-based bacteria	CK	1539 (67) ab	1541 (66) ab	9.34 (0.07) b
	Low PE	1462 (73) bc	1463 (37) bc	9.21 (0.05) c
	High PE	1424 (48) c	1428 (48) c	9.14 (0.03) c
	BC	1590 (31) a	1626 (56) a	9.63 (0.08) a
	Low PE + BC	1554 (43) ab	1557 (42) ab	9.56 (0.01) a
	High PE + BC	1538 (66) ab	1542 (67) ab	9.55 (0.03) a
phoD-harboring bacteria	CK	934 (68) c	1282 (36) c	3.59 (0.18) c
	Low PE	1207 (69) b	1504 (82) b	4.58 (0.20) ab
	High PE	1361 (47) a	1749 (75) a	4.84 (0.24) a
	BC	1197 (39) b	1557 (73) b	4.63 (0.09) ab
	Low PE + BC	1179 (44) b	1465 (38) b	4.51 (0.11) b
	High PE + BC	1311 (39) a	1722 (63) a	4.75 (0.08) ab

Different lowercase letters in the same column indicate significant difference among the treatments ($p < 0.05$).

3.3.2. Soil Microbial Community Composition

There were significant differences in the composition and proportion of dominant bacterial species in soil samples from different treatments (Figure 2). For total bacteria (16S rRNA gene-based bacteria), at the phylum level, the dominant bacteria were Proteobacteria, Chloroflexi, Actinobacteria, Acidobacteria, and Gemmatimonadota in all treatments, with average abundances of 41.0%, 9.1%, 11.8%, 9.0%, and 8.0%, respectively. There were significant differences in the relative abundances of two of the top ten bacterial phyla

($p < 0.05$) among the six treatments (Table S1). Compared with CK, the other five treatments significantly decreased the relative abundance of Gemmatimonadota, and increased abundance of Acidobacteria was observed in the BC treatment. In addition, the treatments of microplastics combined with biochar increased the relative abundance of Acidobacteria compared with the treatments adding microplastics alone. At the genus level, the top five genera were *Pedosphaeraceae*, *Dongia*, *TRA3-20*, *Sphingomonas*, and *Subgroup_10* among all treatments, and the average abundances were 2.1%, 2.4%, 3.0%, 2.8%, and 2.1%, respectively. There were significant differences in the relative abundances of seven of the top ten bacterial genera ($p < 0.05$) among the six treatments (Table S2). However, for the top 10 genera, there was no significant difference between low PE or high PE with CK, only the treatments of low PE + BC and high PE + BC significantly increased the relative abundance of *Subgroup_10* but significantly decreased abundance of *TRA3-20*, *SC-I-84*, and *Ellin6067* compared to CK. In addition, treatments of microplastics combined with biochar increased the relative abundance of *Subgroup_10*, *Bacillus*, and *Pseudomonas* but decreased the abundance of *TRA3-20*, *SC-I-84*, and *IMCC26256* compared with the treatments with microplastics alone.

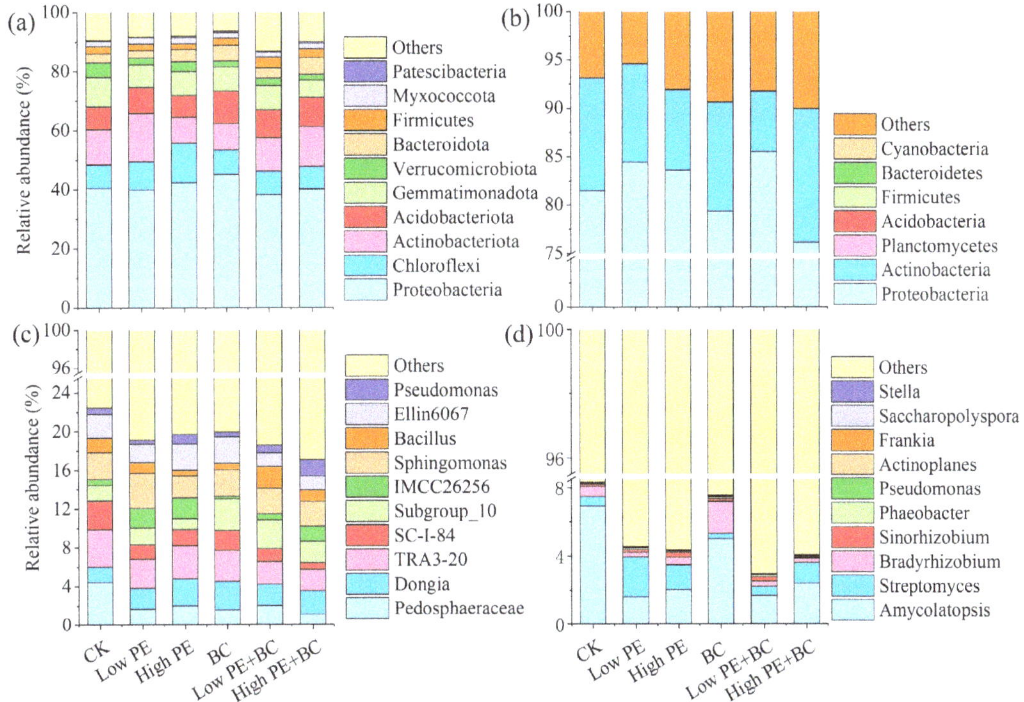

Figure 2. Community composition of total bacteria (top 10 phyla (**a**) and genera (**c**) in abundance) and *phoD-harboring* bacteria (phyla (**b**) and top 10 genera (**d**) in abundance) in soils under different treatments.

For *phoD-harboring* bacteria, the dominant phyla in all treatments were Proteobacteria and Actinobacteria at the phylum level, and the sum of their relative abundance was in the range of 89.9–94.6%. There was no significant difference in the relative abundance of Proteobacteria and Actinobacteria among other five treatments with CK (Table S1). At the genus level, *Amycolatopsis*, *Streptomyces*, and *Bradyrhizobium* were abundant in all treatments (at least one treatment > 1%). There was no significant difference in the relative abundances of *Amycolatopsis* or *Bradyrhizobium* among the other five treatments with CK, and only the low PE and high PE treatments significantly increased the relative abundance of *Streptomyces* compared with CK (Table S2). In addition, the treatments with microplastics combined with biochar increased the relative abundance of *Amycolatopsis* and *Bradyrhizobium* but decreased the abundance of *Streptomyces* compared with the treatments with microplastics alone.

3.3.3. Soil Microbial Function

FAPROTAX was used to predict the potential functions of total bacterial communities. In general, bacterial functions were mainly related to the cycling of C and N in the ecosystem, and there were obvious differences in the abundance of functional genes in different treatments (Figure 3). Compared to CK, the other five treatments increased the abundance of functional genes such as aerobic ammonia oxidation, denitrification, and iron respiration but decreased the aromatic compound degradation, methylotrophy, nitrate respiration, and nitrogen respiration. Moreover, high PE had a significantly higher abundance of functional genes of aerobic chemoheterotrophy, chemoheterotrophy, iron respiration, and photoheterotrophy but had a significantly lower abundance of functional genes of aromatic hydrocarbon degradation and chitinolysis than CK. In addition, compared to the treatments with microplastics alone, treatments with microplastics combined with biochar increased the abundance of aerobic ammonia oxidation, denitrification, nitrate respiration, and nitrogen respiration.

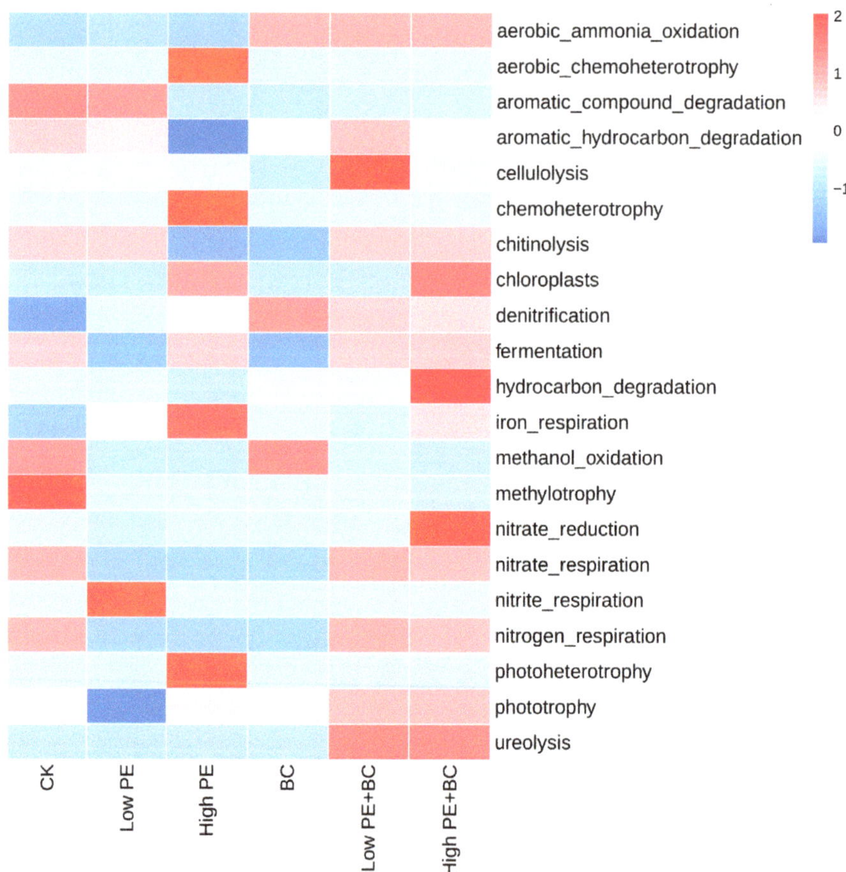

Figure 3. Heatmap of soil functional gene abundance in different treatments.

3.4. Correlations between Sugarcane Biomass, Soil Microbes, and Soil Physicochemical Properties

RDA was used to analyze the relationship between bacterial community structure and soil properties (Figure 4). For total bacteria (16S rRNA gene-based bacteria), pH, EC, SOC, TN, TP, TK, AN, AP, and AK together accounted for 53.1% of the bacterial community structure variation, with the first two axes accounting for 27.4% and 13.0% of the variation, respectively. The total bacterial (16S rRNA gene-based bacteria) community structure was mainly affected by pH, EC, and AK and was significantly negatively correlated with pH ($p < 0.05$). For *phoD-harboring* bacteria, the main physicochemical properties of the soil explained 40.3% of the bacterial community changes, and the first two axes explained 27.5% and 9.8% of the changes, respectively. The community structure of *phoD-harboring* bacteria was significantly negatively correlated with SOC, EC, and AK ($p < 0.05$).

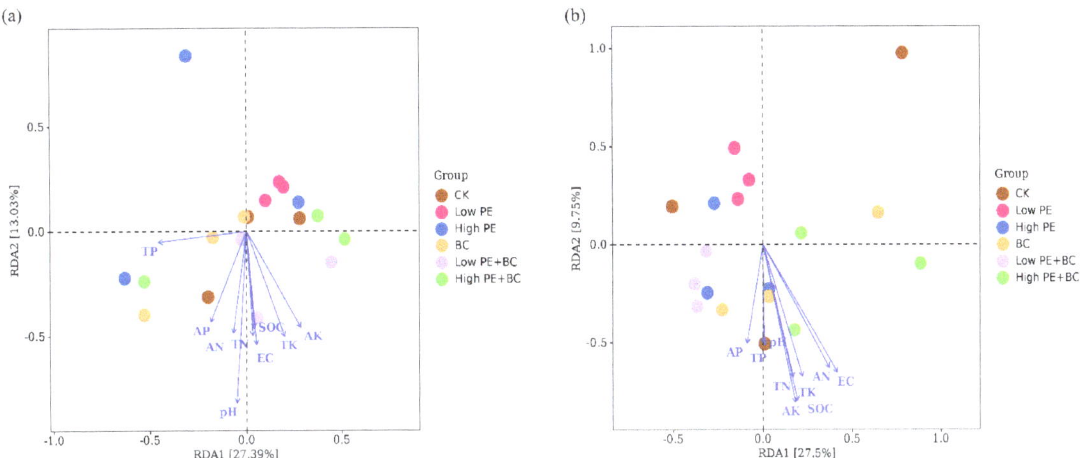

Figure 4. Redundancy analysis (RDA) of soil total bacteria (**a**), *phoD*-harboring bacteria (**b**), and soil physical and chemical properties.

The correlation between sugarcane biomass, soil bacterial community composition, functional gene abundance, and soil physical and chemical properties was further analyzed (Figure 5). For total bacteria (16S rRNA gene-based bacteria), the relative abundance of *Subgroup_10* was significantly and positively correlated with sugarcane AGB and BGB, soil pH, EC, SOC, and N, P, and K nutrient contents; the relative abundance of *IMCC26256* was significantly negatively correlated with the AGB and BGB of sugarcane and soil TP; and the relative abundance of *TRA3-20* was significantly negatively correlated with soil SOC, EC, TN, TK, AN, and AK ($p < 0.05$). For *phoD-harboring* bacteria, *Streptomyces* had a significant negative correlation with the AGB and BGB of sugarcane, soil pH, TN, AN, and AP, whereas *Bradyrhizobium* had a significant positive correlation with the AGB of sugarcane and soil AP. There was a significant positive correlation between *Saccharopolyspora* and soil TP ($p < 0.05$). For the abundance of functional genes, nitrate respiration and nitrogen respiration were positively correlated with soil pH and AP, whereas fermentation was negatively correlated with soil pH and AP. Denitrification and nitrite respiration were positively correlated with soil AN and EC, whereas methanol oxidation and methylotrophy were negatively correlated with AP. Aerobic ammonia oxidation was negatively correlated with AK and TK.

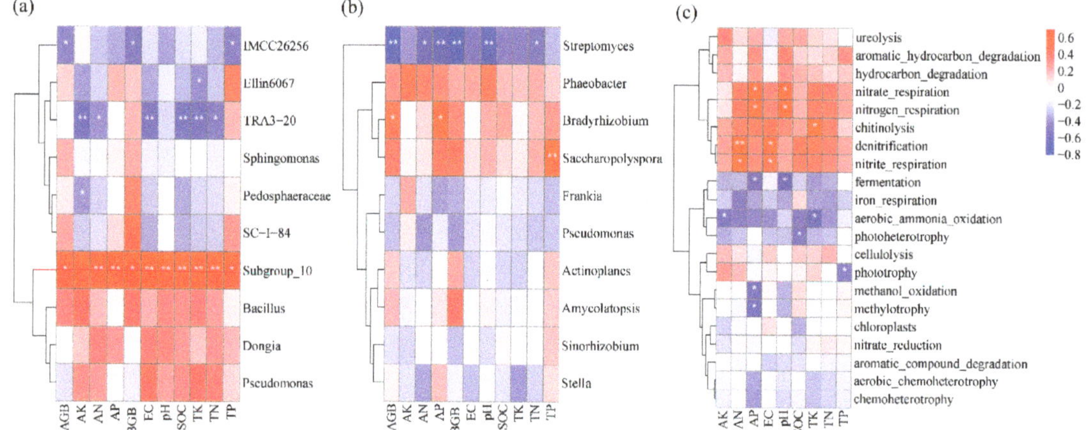

Figure 5. Correlation between soil total bacterial genera (**a**), *phoD-harboring* bacteria genera (**b**), ecosystem function (**c**), and soil physical and chemical properties. *, **, significant at $p \leq 0.05$, $p \leq 0.01$, respectively.

4. Discussion

4.1. Effects of Microplastics and Biochar on Sugarcane Biomass and Soil Physical and Chemical Properties

In this study, the addition of microplastics alone significantly decreased the AGB and BGB of sugarcane compared with the CK, and the decreasing amplitude in the high PE treatment was greater (Figure 1), indicating that the accumulation of microplastics would inhibit the biomass of sugarcane and that the degree of influence increased with the increasing concentration of microplastics. This is similar to the findings of many previous studies that the accumulation of microplastics can inhibit plant growth and development, thereby reducing plant [8] biomass [8,9]. Compared with the treatments with microplastics alone, the combination of microplastics with biochar significantly increased the total biomass of sugarcane, which confirmed that the addition of biochar could reduce the inhibition of microplastic accumulation on plant biomass [17,23]. Biochar addition has been found to be able to increase shoot dry matter production in different concentrations of PVC-contaminated soil [18]. At the same time, treatments with microplastics alone altered soil physicochemical properties, including the pH and total and available N and P contents (Table 1). Indeed, MPs like high-density PE have been proven to decrease soil pH through cation/proton exchange [24]. Microplastics have a high adhesion that can change plant root exudates and soil buffering capacity to co-mediate soil pH [9]. Soil microbial community composition is an important indicator of soil quality, and microbial activities also affect the recycling of N, P, and K nutrients [25]. Microplastic addition considerably changed microbial community and activities, which may be an important reason for the alteration in soil N and P content [26]. Many previous studies have reported the effects of PE microplastics on soil physical and chemical properties, but the results varied [19]. For example, the addition of PE microplastics reduced soil pH and organic matter content in red soil [27]. The discrepancy in the effects of PE microplastics on soil properties may be related to the concentration of microplastics, soil type, and climate [4]conditions [4,19]. Additionally, treatments with BC amendment had significantly higher soil pH, SOC, and nutrients than treatments with microplastics alone, which may have been due to the high pH, SOC, and nutrients contents of corn straw biochar itself [28]. In addition, the correlation analysis showed that the total biomass of sugarcane was significantly positively correlated with soil pH, SOC, and total and available N and P contents, indicating that the promotion of sugarcane biomass via biochar amendment may have been due to its improvement to soil quality [29].

4.2. Effects of Microplastics and Biochar on Bacterial Community Diversity and Composition

The diversity and composition of total bacteria (16S rRNA gene-based bacteria) were significantly affected by the addition of microplastics or biochar. Previous studies have also reported that microplastic accumulation or biochar amendment changed the diversity of bacteria [13,28]. The treatments with microplastics alone decreased the observed OTUs and the Chao1 and Shannon indices, whereas microplastics combined with biochar increased the three indices compared to CK (Table 2). From the perspective of the abundance and diversity of soil microbial communities, biochar addition can increase the abundance and diversity of soil total bacteria, thereby enhancing the stability and functional diversity of soil ecosystems [30,31]. For total bacteria, the other five treatments significantly decreased the relative abundance of Gemmatimonadota compared with CK (Figure 2 and Table S1), which may have been due to the inhibition of microplastics accumulation on the underground growth of sugarcane. Several bacteria in Gemmatimonadota form symbiotic relationships with plant roots [32], and the underground biomass of sugarcane treated with microplastics in this study was lower than that of CK, resulting in a decrease in the relative abundance of Gemmatimonadota. The BC treatment significantly increased the relative abundance of Acidobacteria compared with CK, which may have been related to the obvious increase in soil pH, SOC, and nutrients in BC [33]. At the genus level, compared with CK, only treatments with microplastics combined with biochar significantly increased the relative abundance of *Subgroup_10* and significantly decreased the abundance of *TRA3-20*. Correlation analysis showed that the relative abundance of *Subgroup_10* was significantly positively correlated with the AGB and BGB of sugarcane and the main soil physicochemical properties, whereas the relative abundance of *TRA3-20* was significantly negatively correlated with the majority of soil physicochemical properties. Additionally, compared with the treatment with microplastics alone, the treatment with microplastics combined with biochar increased the relative abundance of *Bacillus* and *Pseudomonas* but decreased the abundance of *IMCC26256*. *Bacillus* and *Pseudomonas* are important bacterial groups that can inhibit harmful bacteria and pathogens and promote plant growth. Therefore, the promoting effect of microplastics combined with biochar on sugarcane biomass may be related to changes in the total bacterial community composition in soil [34].

The diversity and composition of *phoD-harboring* bacteria in different treatments were also significantly different. The *phoD* gene of bacteria encodes alkaline phosphatase, which plays an important role in soil organic P decomposition [35]. Compared with CK, the observed OTUs and the Chao1 and Shannon indices of the other five treatments were significantly increased, and the index of the high PE treatment was the highest. In this study, the AP in the high PE treatment was the lowest, whereas the abundance and diversity of *phoD-harboring* bacteria were the highest, which may be explained by sugarcane requiring a certain amount of P provided by organic P decomposition for growth [36]. For *phoD-harboring* bacteria community composition, Proteobacteria and Actinobacteria were the two dominant phyla, both of which contained many currently known P solubilizing bacteria [37]. At the genus level, only the low PE and high PE treatments significantly increased the relative abundance of *Streptomyces* compared with CK (Figure 2 and Table S2). The higher relative abundance of *Streptomyces* at higher concentrations of microplastics (such as high PE) may have been due to its greater stress tolerance [38]. Moreover, compared with the treatment with microplastics alone, the treatment with microplastics combined with biochar increased the relative abundance of *Amycolatopsis* and *Bradyrhizobium*, both of which are plant growth-promoting bacteria [39]. Correlation analysis also confirmed that *Bradyrhizobium* was significantly positively correlated with sugarcane biomass and soil AP content, whereas *Streptomyces* was significantly negatively correlated with sugarcane biomass, soil pH, TN, AN, and AP ($p < 0.05$). These results confirm that biochar amendment could change the microbial composition and thus alleviate the inhibitory effect of microplastics on plant growth [40].

4.3. Effects of Microplastics and Biochar on Microbial Community Structure and Function

Changes in soil physical and chemical properties can cause changes in the soil microbial community structure and further change the function of soil microorganisms [41,42]. RDA showed that the community structure of total bacteria was mainly affected by pH, EC, TN, and AK, and the structure of the total bacterial community was significantly negatively correlated with pH ($p < 0.05$). Soil pH, EC, TN, and AK increased significantly with biochar addition, which in turn drove significant changes in the total bacterial community structure (Table 1 and Figure 4). Other studies also showed that the soil total bacterial community structure was affected by soil pH, SOC, AP, and other physical and chemical properties [37,43]. Soil pH and NO_3^- concentration played an important role in the formation of bacterial community structure in black soil [43], whereas the bacterial community structure on loess was significantly correlated with dissolved organic C, AP, and TP [37]. These results confirm that there were some differences in the factors affecting the structure of the total bacterial community in soil, which may have been due to differences in the soil types and fertilization management measures. The structure of *phoD-harboring* bacteria community was significantly negatively correlated with SOC, EC, and AK ($p < 0.05$). The results of different studies on the relationship between the structure of *phoD-harboring* bacteria community and soil physicochemical properties are quite [35] different [35,36]. Chen et al. [36] reported that soil pH and AP were related to changes in the community structure of *phoD-harboring* bacteria in soil, whereas Liu et al. [37] found that various physical and chemical properties had little effect on the structure of *phoD-harboring* bacteria community. These variable findings may imply that the structure of the *phoD-harboring* bacteria community is quite different under different soil types and climatic conditions [37].

Correlation analysis showed that AP and pH significantly affected soil microbial species and the abundance of functional genes (Figure 5). Soil pH was significantly correlated with *Subgroup_10* and *Streptomyces*, as well as nitrate respiration, nitrogen respiration, and fermentation. Similarly, found that pH was significantly correlated with nitrogen respiration and nitrate respiration [41]. Furthermore, there were significant differences in the abundance of functional genes among the different treatments. Compared with CK, the high PE treatments significantly changed the abundance of functional genes related to C cycling such as aerobic chemoheterotrophy, chemoheterotrophy, aromatic hydrocarbon degradation, and chitinolysis (Figure 3). The high functional genes related to C cycling in the high PE treatment may have been related to the changes in the content and form of C in the soil after adding a high concentration of microplastics, for microplastics are high-C polymers with a C content of more than 90% [44]. Moreover, compared with the treatment with microplastics alone, the treatment with microplastics combined with biochar increased the abundance of functional genes such as aerobic ammonia oxidation, denitrification, nitrate respiration, and nitrogen respiration. All of these functional genes participate in the soil N cycle and may affect the soil N content available to the plant [41]. The increase in soil AN content and the biomass of sugarcane in treatments with microplastics combined with biochar may also have been related to the alteration to functional genes after the biochar amendment [17].

5. Conclusions

The accumulation of polyethylene microplastics decreased both the aboveground biomass and underground biomass of sugarcane, soil pH, and nitrogen and phosphorus content, and the degree of influence increased with the increasing concentration of microplastics. In contrast, microplastics combined with biochar could alleviate the negative effects of microplastic accumulation on sugarcane biomass and soil quality. There were significant differences in the bacterial community alpha diversity indices and compositions among different treatments. Treatments with microplastics alone decreased the observed

OTUs and the Chao1 and Shannon indices of soil total bacteria (16s rRNA gene-based bacteria) while increasing the three indices in *phoD-harboring* bacteria compared to the control. Compared with microplastics alone, the treatments with microplastics combined with biochar increased the relative abundance of *Subgroup_10*, *Bacillus*, and *Pseudomonas* in soil total bacteria and *Amycolatopsis* and *Bradyrhizobium* in *phoD*-harboring bacteria, most of which can inhibit harmful bacteria and promote plant growth. In addition, compared with the treatments with microplastics alone, the treatments with biochar amendment increased the abundance of functional genes involved in the nitrogen cycle, such as aerobic ammonia oxidation, denitrification, nitrate respiration, and nitrogen respiration, which may have increased the soil nitrogen content available to the plant, thereby promoting the biomass of sugarcane. Overall, our results prove that biochar amendment may alleviate the negative effects of microplastic accumulation on sugarcane biomass by altering soil nutrients and microbial community structure and function.

Supplementary Materials: The following supporting information can be downloaded at: https://www.mdpi.com/article/10.3390/plants13010083/s1, Table S1: The relative abundances of the most abundant phyla in different treatments, Table S2: The relative abundances of the 10 most abundant genera in different treatments.

Author Contributions: Methodology, W.Z.; Data curation, W.Z. and D.C.; Writing—original draft, Q.W.; Writing—review & editing, J.T.; Supervision, J.A.; Funding acquisition, J.A. All authors have read and agreed to the published version of the manuscript.

Funding: This research was funded by [China Agricultural Research System of MOF and MARA] grant number [CARS-170203], [GuangDong Basic and Applied Basic Research Foundation] grant number [2022A1515110928] and [National Key Research and Development Program of China] grant number [2020YFD1000600].

Data Availability Statement: Data are contained within the article.

References

1. Kumar, A.; Mishra, S.; Pandey, R.; Yu, Z.G.; Kumar, M.; Khoo, K.S.; Thakur, T.K.; Show, P.L. Microplastics in terrestrial ecosystems: Unignorable impacts on soil characterises, nutrient storage and its cycling. *TrAC Trends Anal. Chem.* **2023**, *158*, 116869. [CrossRef]
2. Zhang, J.; Ren, S.; Xu, W.; Liang, C.; Li, J.; Zhang, H.; Li, Y.; Liu, X.; Jones, D.L.; Chadwick, D.R.; et al. Effects of plastic residues and microplastics on soil ecosystems: A global meta-analysis. *J. Hazard. Mater.* **2022**, *435*, 129065. [CrossRef] [PubMed]
3. Yang, L.; Zhang, Y.; Kang, S.; Wang, Z.; Wu, C. Microplastics in soil: A review on methods, occurrence, sources, and potential risk. *Sci. Total Environ.* **2021**, *780*, 146546. [CrossRef] [PubMed]
4. Chae, Y.; An, Y.-J. Current research trends on plastic pollution and ecological impacts on the soil ecosystem: A review. *Environ. Pollut.* **2018**, *240*, 387–395. [CrossRef] [PubMed]
5. Li, Y.-R.; Yang, L.-T. Sugarcane Agriculture and Sugar Industry in China. *Sugar Tech* **2015**, *17*, 1–8. [CrossRef]
6. Wu, Q.; Zhou, W.; Chen, D.; Cai, A.; Ao, J.; Huang, Z. Optimizing soil and fertilizer phosphorus management according to the yield response and phosphorus use efficiency of sugarcane in southern China. *J. Soil Sci. Plant Nutr.* **2020**, *20*, 1655–1664. [CrossRef]
7. Jacques, O.; Prosser, R.S. A probabilistic risk assessment of microplastics in soil ecosystems. *Sci. Total Environ.* **2021**, *757*, 143987. [CrossRef]
8. Yao, Y.; Lili, W.; Shufen, P.; Li, G.; Hongmei, L.; Weiming, X.; Lingxuan, G.; Jianning, Z.; Guilong, Z.; Dianlin, Y. Can microplastic mediate soil properties, plant growth and carbon/nitrogen turnover in the terrestrial ecosystem? *Ecosyst. Health Sustain.* **2022**, *8*, 2133638. [CrossRef]
9. De Souza Machado, A.A.; Lau, C.W.; Kloas, W.; Bergmann, J.; Bachelier, J.B.; Faltin, E.; Becker, R.; Görlich, A.S.; Rillig, M.C. Microplastics can change soil properties and affect plant performance. *Environ. Sci. Technol.* **2019**, *53*, 6044–6052. [CrossRef]
10. Gao, B.; Yao, H.; Li, Y.; Zhu, Y. Microplastic addition alters the microbial community structure and stimulates soil carbon dioxide emissions in vegetable-growing soil. *Environ. Toxicol. Chem.* **2021**, *40*, 352–365. [CrossRef]
11. Feng, X.; Wang, Q.; Sun, Y.; Zhang, S.; Wang, F. Microplastics change soil properties, heavy metal availability and bacterial community in a Pb-Zn-contaminated soil. *J. Hazard. Mater.* **2022**, *424*, 127364. [CrossRef] [PubMed]
12. Huang, Y.; Zhao, Y.; Wang, J.; Zhang, M.; Jia, W.; Qin, X. LDPE microplastic films alter microbial community composition and enzymatic activities in soil. *Environ. Pollut.* **2019**, *254*, 112983. [CrossRef] [PubMed]

13. Fei, Y.; Huang, S.; Zhang, H.; Tong, Y.; Wen, D.; Xia, X.; Wang, H.; Luo, Y.; Barceló, D. Response of soil enzyme activities and bacterial communities to the accumulation of microplastics in an acid cropped soil. *Sci. Total Environ.* **2020**, *707*, 135634. [CrossRef] [PubMed]

14. Zhou, J.; Gui, H.; Banfield, C.C.; Wen, Y.; Zang, H.; Dippold, M.A.; Charlton, A.; Jones, D.L. The microplastisphere: Biodegradable microplastics addition alters soil microbial community structure and function. *Soil Biol. Biochem.* **2021**, *156*, 108211. [CrossRef]

15. Chen, X.; Chen, X.; Zhao, Y.; Zhou, H.; Xiong, X.; Wu, C. Effects of microplastic biofilms on nutrient cycling in simulated freshwater systems. *Sci. Total Environ.* **2020**, *719*, 137276. [CrossRef] [PubMed]

16. Kuppusamy, S.; Thavamani, P.; Megharaj, M.; Venkateswarlu, K.; Naidu, R. Agronomic and remedial benefits and risks of applying biochar to soil: Current knowledge and future research directions. *Environ. Int.* **2016**, *87*, 1–12. [CrossRef] [PubMed]

17. Ran, T.; Li, J.; Liao, H.; Zhao, Y.; Yang, G.; Long, J. Effects of biochar amendment on bacterial communities and their function predictions in a microplastic-contaminated *Capsicum annuum* L. soil. *Environ. Technol. Innov.* **2023**, *31*, 103174. [CrossRef]

18. Khalid, A.R.; Shah, T.; Asad, M.; Ali, A.; Samee, E.; Adnan, F.; Bhatti, M.F.; Marhan, S.; Kammann, C.I.; Haider, G. Biochar alleviated the toxic effects of PVC microplastic in a soil-plant system by upregulating soil enzyme activities and microbial abundance. *Environ. Pollut.* **2023**, *332*, 121810. [CrossRef]

19. Wang, F.; Wang, Q.; Adams, C.A.; Sun, Y.; Zhang, S. Effects of microplastics on soil properties: Current knowledge and future perspectives. *J. Hazard. Mater.* **2022**, *424*, 127531. [CrossRef]

20. Wu, Q.; Zhou, W.; Lu, Y.; Li, S.; Shen, D.; Ling, Q.; Chen, D.; Ao, J. Combined Chemical Fertilizers with Molasses Increase Soil Stable Organic Phosphorus Mineralization in Sugarcane Seedling Stage. *Sugar Tech* **2022**, *25*, 552–561. [CrossRef]

21. Page, A.L. *Methods of Soil Analysis. Part 2. Chemical and Microbiological Properties*; American Society of Agronomy/Soil Science Society of America: Madison, WI, USA, 1982.

22. Sakurai, M.; Wasaki, J.; Tomizawa, Y.; Shinano, T.; Osaki, M. Analysis of bacterial communities on alkaline phosphatase genes in soil supplied with organic matter. *Soil Sci. Plant Nutr.* **2008**, *54*, 62–71. [CrossRef]

23. Hammer, E.C.; Forstreuter, M.; Rillig, M.C.; Kohler, J. Biochar increases arbuscular mycorrhizal plant growth enhancement and ameliorates salinity stress. *Appl. Soil Ecol.* **2015**, *96*, 114–121. [CrossRef]

24. Boots, B.; Russell, C.W.; Green, D.S. Effects of microplastics in soil ecosystems: Above and below ground. *Environ. Sci. Technol.* **2019**, *53*, 11496–11506. [CrossRef]

25. Das, P.P.; Singh, K.R.; Nagpure, G.; Mansoori, A.; Singh, R.P.; Ghazi, I.A.; Kumar, A.; Singh, J. Plant-soil-microbes: A tripartite interaction for nutrient acquisition and better plant growth for sustainable agricultural practices. *Environ. Res.* **2022**, *214*, 113821. [CrossRef] [PubMed]

26. Lian, J.; Liu, W.; Meng, L.; Wu, J.; Zeb, A.; Cheng, L.; Lian, Y.; Sun, H. Effects of microplastics derived from polymer-coated fertilizer on maize growth, rhizosphere, and soil properties. *J. Clean. Prod.* **2021**, *318*, 128571. [CrossRef]

27. Yan, Y.; Chen, Z.; Zhu, F.; Zhu, C.; Wang, C.; Gu, C. Effect of polyvinyl chloride microplastics on bacterial community and nutrient status in two agricultural soils. *Bull. Environ. Contam. Toxicol.* **2021**, *107*, 602–609. [CrossRef]

28. Kavitha, B.; Reddy, P.V.L.; Kim, B.; Lee, S.S.; Pandey, S.K.; Kim, K.-H. Benefits and limitations of biochar amendment in agricultural soils: A review. *J. Environ. Manag.* **2018**, *227*, 146–154. [CrossRef]

29. Song, D.; Xi, X.; Zheng, Q.; Liang, G.; Zhou, W.; Wang, X. Soil nutrient and microbial activity responses to two years after maize straw biochar application in a calcareous soil. *Ecotoxicol. Environ. Saf.* **2019**, *180*, 348–356. [CrossRef]

30. Bahram, M.; Hildebrand, F.; Forslund, S.K.; Anderson, J.L.; Soudzilovskaia, N.A.; Bodegom, P.M.; Bengtsson-Palme, J.; Anslan, S.; Coelho, L.P.; Harend, H.; et al. Structure and function of the global topsoil microbiome. *Nature* **2018**, *560*, 233–237. [CrossRef]

31. Hartmann, M.; Six, J. Soil structure and microbiome functions in agroecosystems. *Nat. Rev. Earth Environ.* **2023**, *4*, 4–18. [CrossRef]

32. Mujakić, I.; Piwosz, K.; Koblížek, M. Phylum Gemmatimonadota and its role in the Environment. *Microorganisms* **2022**, *10*, 151. [CrossRef] [PubMed]

33. Kalam, S.; Basu, A.; Ahmad, I.; Sayyed, R.Z.; El-Enshasy, H.A.; Dailin, D.J.; Suriani, N.L. Recent understanding of soil acidobacteria and their ecological significance: A critical review. *Front. Microbiol.* **2020**, *11*, 580024. [CrossRef] [PubMed]

34. Fan, K.; Delgado-Baquerizo, M.; Guo, X.; Wang, D.; Zhu, Y.-G.; Chu, H. Biodiversity of key-stone phylotypes determines crop production in a 4-decade fertilization experiment. *ISME J.* **2021**, *15*, 550–561. [CrossRef] [PubMed]

35. Wei, X.; Hu, Y.; Cai, G.; Yao, H.; Ye, J.; Sun, Q.; Veresoglou, S.D.; Li, Y.; Zhu, Z.; Guggenberger, G.; et al. Organic phosphorus availability shapes the diversity of *phoD*-harboring bacteria in agricultural soil. *Soil Biol. Biochem.* **2021**, *161*, 108364. [CrossRef]

36. Chen, X.; Jiang, N.; Condron, L.M.; Dunfield, K.E.; Chen, Z.; Wang, J.; Chen, L. Impact of long-term phosphorus fertilizer inputs on bacterial *phoD* gene community in a maize field, Northeast China. *Sci. Total Environ.* **2019**, *669*, 1011–1018. [CrossRef]

37. Liu, J.; Ma, Q.; Hui, X.; Ran, J.; Ma, Q.; Wang, X.; Wang, Z. Long-term high-P fertilizer input decreased the total bacterial diversity but not *phoD*-harboring bacteria in wheat rhizosphere soil with available-P deficiency. *Soil Biol. Biochem.* **2020**, *149*, 107918. [CrossRef]

38. Wang, J.; Li, Y.; Pinto-Tomás, A.A.; Cheng, K.; Huang, Y. Habitat Adaptation drives speciation of a *Streptomyces* species with distinct habitats and disparate geographic origins. *mBio* **2022**, *13*, e0278121. [CrossRef]

39. Luo, G.; Ling, N.; Nannipieri, P.; Chen, H.; Raza, W.; Wang, M.; Guo, S.; Shen, Q. Long-term fertilisation regimes affect the composition of the alkaline phosphomonoesterase encoding microbial community of a vertisol and its derivative soil fractions. *Biol. Fert. Soils* **2017**, *53*, 375–388. [CrossRef]

40. Palansooriya, K.N.; Sang, M.K.; Igalavithana, A.D.; Zhang, M.; Hou, D.; Oleszczuk, P.; Sung, J.; Ok, Y.S. Biochar alters chemical and microbial properties of microplastic-contaminated soil. *Environ. Res.* **2022**, *209*, 112807. [CrossRef]

41. Zhang, M.; Liang, G.; Ren, S.; Li, L.; Li, C.; Li, Y.; Yu, X.; Yin, Y.; Liu, T.; Liu, X. Responses of soil microbial community structure, potential ecological functions, and soil physicochemical properties to different cultivation patterns in cucumber. *Geoderma* **2023**, *429*, 116237. [CrossRef]

42. Ali, A.; Ghani, M.I.; Elrys, A.S.; Ding, H.; Iqbal, M.; Cheng, Z.; Cai, Z. Different cropping systems regulate the metabolic capabilities and potential ecological functions altered by soil microbiome structure in the plastic shed mono-cropped cucumber rhizosphere. *Agric. Ecosyst. Environ.* **2021**, *318*, 107486. [CrossRef]

43. Zhou, J.; Guan, D.; Zhou, B.; Zhao, B.; Ma, M.; Qin, J.; Jiang, X.; Chen, S.; Cao, F.; Shen, D.; et al. Influence of 34-years of fertilization on bacterial communities in an intensively cultivated black soil in northeast China. *Soil Biol. Biochem.* **2015**, *90*, 42–51. [CrossRef]

44. Rillig, M.C. Microplastic in terrestrial ecosystems and the soil? *Environ. Sci. Technol.* **2012**, *46*, 6453–6454. [CrossRef] [PubMed]

Article

Phylogeny and Genetic Divergence among Sorghum Mosaic Virus Isolates Infecting Sugarcane

Hui-Mei Xu [1,†], Er-Qi He [2,†], Zu-Li Yang [3], Zheng-Wang Bi [1], Wen-Qing Bao [1], Sheng-Ren Sun [4], Jia-Ju Lu [2,*] and San-Ji Gao [1,*]

1   National Engineering Research Center for Sugarcane, Fujian Agriculture and Forestry University, Fuzhou 350002, China; xhmxhm946946@163.com (H.-M.X.); sdwlbzw@163.com (Z.-W.B.); wenqingbao0706@163.com (W.-Q.B.)
2   Guizhou Institute of Subtropical Crops, Guizhou Academy of Agricultural Sciences, Xingyi 562400, China; heerqi003@163.com
3   Laibin Academy of Agricultural Sciences, Laibin 546100, China; yang_zuli@163.com
4   Institute of Nanfan & Seed Industry, Guangdong Academy of Sciences, Guangzhou 510316, China; ssr03@163.com
*   Correspondence: lujiaju82@163.com (J.-J.L.); gaosanji@fafu.edu.cn (S.-J.G.)
†   These authors contributed equally to this work.

Abstract: Sorghum mosaic virus (SrMV, the genus *Potyvirus* of the family *Potyviridae*) is a causal agent of common mosaic in sugarcane and poses a threat to the global sugar industry. In this study, a total of 901 sugarcane leaf samples with mosaic symptom were collected from eight provinces in China and were detected via RT-PCR using a primer pair specific to the SrMV coat protein (*CP*). These leaf samples included 839 samples from modern cultivars (*Saccharum* spp. hybrids) and 62 samples from chewing cane (*S. officinarum*). Among these, 632 out of 901 (70.1%) samples were tested positive for SrMV. The incidences of SrMV infection were 72.3% and 40.3% in modern cultivars and chewing cane, respectively. Phylogenetic analysis showed that all tested SrMV isolates were clustered into three clades consisting of six phylogenetic groups based on 306 *CP* sequences (this study = 265 and GenBank database = 41). A total of 10 SrMV isolates from South America (the United States and Argentina) along with 106 isolates from China were clustered in group D, while the remaining 190 SrMV isolates from Asia (China and Vietnam) were dispersed in five groups. The SrMV isolates in group F were limited to Yunnan province in China, and those in group A were spread over eight provinces. A significant genetic heterogeneity was elucidated in the nucleotide sequence identities of all SrMV CPs, ranging from 69.0% to 100%. A potential recombination event was postulated among SrMV isolates based on *CP* sequences. All tested SrMV *CP*s underwent dominant negative selection. Geographical isolation (South America vs. Asia) and host types (modern cultivars vs. chewing cane) are important factors promoting the genetic differentiation of SrMV populations. Overall, this study contributes to the global understanding of the genetic evolution of SrMV and provides a valuable resource for the epidemiology and management of the mosaic in sugarcane.

Keywords: genetic diversity; molecular evolution; mosaic disease; population structure; sorghum mosaic virus; sugarcane

Citation: Xu, H.-M.; He, E.-Q.; Yang, Z.-L.; Bi, Z.-W.; Bao, W.-Q.; Sun, S.-R.; Lu, J.-J.; Gao, S.-J. Phylogeny and Genetic Divergence among Sorghum Mosaic Virus Isolates Infecting Sugarcane. *Plants* **2023**, *12*, 3759. https://doi.org/10.3390/plants12213759

Academic Editor: Sergey Morozov

Received: 11 October 2023
Revised: 30 October 2023
Accepted: 31 October 2023
Published: 2 November 2023

1. Introduction

The *Potyviridae* is the largest family of known RNA viruses, having significant impact on agriculture and ecology [1–3]. There are currently 12 genera and 246 species in this virus family (https://ictv.global/taxonomy (accessed on 28 October 2023). The largest genus of plant viruses, the *Potyvirus*, belongs to the *Potyviridae* family and causes significant losses in a variety of crops around the world, such as maize, potatoes, sorghum, soybeans, sugarcane, and so on [4–7]. Mosaic is an important viral disease of sugarcane around the world, caused by a single or mixed infection of sugarcane mosaic virus (SCMV), sorghum mosaic virus (SrMV), sugarcane streak mosaic virus (SCSMV), and maize yellow mosaic virus (MaYMV) [8,9]. In addition to sugarcane, these viruses have a wide range of hosts such as maize, sorghum, and other grasses [8]. SCMV and SrMV (genus *Potyvirus*, family *Potyviridae*) are distributed globally, while SCSMV (genus *Poacevirus*, family *Potyviridae*) appears in Asia and Côte d'Ivoire and MaYMV (genus *Ampelovirus*, family *Closteroviridae*) occurs in Africa, Asia, as well as South America [9–12].

The potyviral genome encodes a long polyprotein that is processed by proteinases, giving rise to at least 10 mature proteins: P1 (protein 1 protease), HC-Pro (helper component protease), P3 (protein 3), PIPO (pretty interesting Potyviridae ORF), 6K1 (6 kDa peptide 1), CI (cylindrical inclusion), 6K2 (6 kDa peptide 2), NIa-Pro (nuclear inclusion a-protease), NIb (nuclear inclusion b, RNA-directed RNA polymerase), CP (capsid protein), as well as VPg (virus protein genome-linked) [5,13,14]. The potyviral CP participates in various biological functions such as coating and protection of the RNA genome, aphid transmission, as well as cell-to-cell and long-distance movement [4,15]. Meanwhile, this viral protein may also be involved in the regulation of CP stability and functional diversity during the viral life cycle through various post-translational modifications [4]. However, these biological functions in SrMV CP have not yet been identified. The CP-coding region of these potyvirus-encoded proteins is preferentially a targeted region used for viral genetic diversity and phylogeny analysis as well as disease diagnosis using molecular and serological approaches [5,8,16].

The genetic diversity of SrMV isolates has been explored using sequencing technology and virus taxonomy. In the 1990s, three strains of SrMV (H, I, and M) were identified, and these were distinguished from SCMV, johnsongrass mosaic virus (JGMV), and maize dwarf mosaic virus (MDMV) [17–19]. Perera et al. reported that the CP nucleotide identities ranged from 97.4% to 99.9% among SrMV isolates from Argentina [20]. In China, based on *CP* sequence analysis, Xu et al. (2008) revealed that obvious genetic diversity (76–100%) occurred among 18 SrMV isolates, while Luo et al. (2016) found that the nucleotide identities were 74.3–94.1% (nucleotide) and 84.7%–98.1% (amino acid) among 188 SrMV isolates worldwide [21]. Meanwhile, Zhou et al. (2014) demonstrated that SrMV-GX together with SrMV-XoS, SrMV-YH, and SrMV-H were grouped in the same evolutionary cluster based on the genomic sequence analysis, and they shared sequence identities of 80.9–95.4% and 90.4–98.4% at nucleotide and amino acid levels [22], respectively. Several phylogenetic groups of SrMV isolates were proposed based on viral CP and genome sequences. For example, two phylogenetic groups of SrMV were clustered based on host origins, i.e., modern cultivars (*Saccharum* hybrids spp.) vs. noble cane (*S. officinarum*) [21,23]. Three and six phylogenetic groups of SrMV were proposed by Zhang et al. (2015) [24] and Luo et al. (2016) [25], respectively, in China.

Evolutionary driving forces, population structure, and differentiation among SrMV isolates have been investigated. For instance, insertion/deletion mutations, negative selection, and frequent gene flow were proposed to contribute to the genetic divergence and population structure of SrMV isolates [25]. No obvious recombination event was found in the *CP* gene region of all tested SrMV isolates [24,25]. High rates of mutation and recombination between potyvirus strains result in the creation of new viral strains. These novel isolates show a high degree of pathogenicity in a variety of host species and cultivars, which is posing a challenge to global crop production [4,26,27]. Importantly, it is critical to distinguish between SrMV strains when breeding resistant sugarcane genotypes. However, these research aspects of SrMV remain unclear. In China, SrMV is one of main viruses infecting sugarcane, particularly modern commercial cultivars, followed by SCSMV [8,25,28]. Therefore, in this study we extensively analyze the occurrence, distribution, genetic variation, and population differentiation of SrMV infecting sugarcane based on viral *CP* fragments. These findings offer insights into the virus's prevalence in China's sugarcane-growing regions and crucial recommendations for the management and prevention of mosaic disease.

2. Results

2.1. Detection of SrMV Using RT-PCR

The SrMV was detected using RT-PCR in 632 of 901 (70.1%) leaf samples. The incidences of SrMV-positive were 72.3% and 40.3% in modern cultivars and chewing cane, respectively. Subsequently, 265 representative *CP* fragments (approximately 850 bp) were selected for a further sequence analysis.

2.2. Phylogenetic Relationship among SrMV Isolates

A phylogenetic analysis showed that all the 306 SrMV isolates (this study = 265 and GeneBank library = 41) were clustered into three clades (I, II, and III), including six different groups (A–F) with 4–120 isolates. Clades I and II consisted of two (A and B) and three (C–E) groups, respectively. Clades III included a unique group F. Moreover, 39.2% and 37.9% of SrMV isolates were assigned to groups A and D, respectively. Apart from 106 SrMV isolates from China, 10 isolates from the United States and Argentina were clustered into group D. The remaining SrMV isolates from Asia (China and Vietnam) were clustered in six groups (Figure 1). Notably, the 18 SrMV isolates from chewing cane were distributed in four groups (SrMV-A, -D, -E, and -F). The frequency of SrMV phylogroups over eight Chinese sugarcane-planting provinces is shown in Figure 2. The SrMV isolates from groups A and D were observed in eight provinces, while the SrMV isolates from group B were found in seven provinces except Guangdong (GD). Additionally, the SrMV isolates from group E were found in six provinces except Sichuan (SC) and Yunnan (YN) provinces. Group C was present in four provinces including Fujian (FJ), Guangxi (GX), Hainan (HN), and Sichuan, but group F only occurred in Yunnan province.

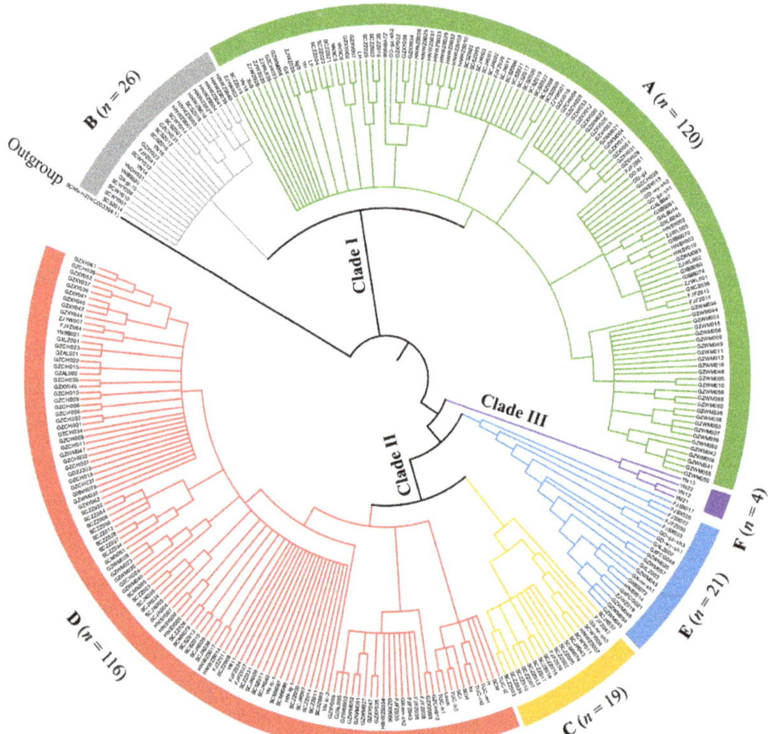

Figure 1. Phylogenetic tree based on nucleotide sequences of coat protein (*CP*) from 306 sorghum mosaic virus (SrMV) isolates. All tested SrMV CP sequences included 265 isolates from this study

plus 41 isolates from the GenBank database. A sequence of sugarcane mosaic virus (SCMV) isolate SCMV-HZ (GenBank accession no. NC_003398) was used as outgroup. The number (*n*) of isolates in each phylogroup is in parentheses.

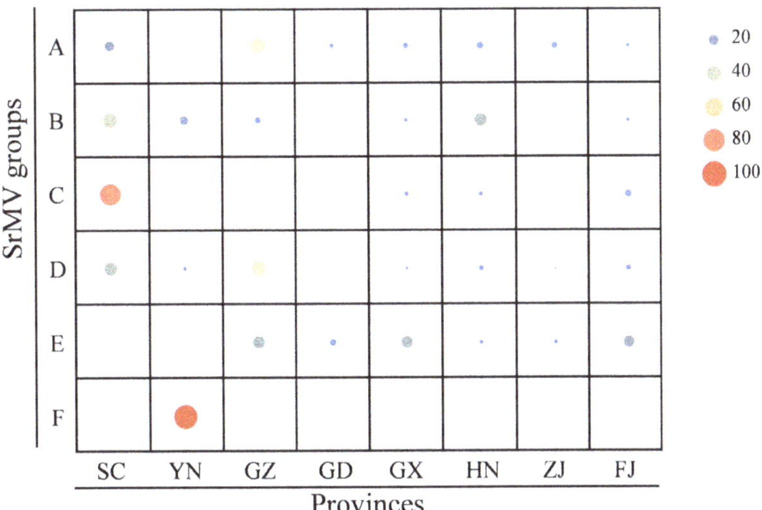

Figure 2. Distribution and frequency (%) of the SrMV phylogroups (A–F) over eight sugarcane planting provinces in China. FJ = Fujian (*n* = 99), GD = Guangdong (*n* = 45), GX = Guangxi (*n* = 173), GZ = Guizhou (*n* = 170), HN = Hainan (*n* = 51), SC = Sichuan (*n* = 181), YN = Yunnan (*n* = 81), and ZJ = Zhejiang (*n* = 101). The size of the circles represents the SrMV phylogroup frequency (%), wherein larger circles correspond to a higher frequency.

2.3. Sequence Identities between SrMV Populations

The sequence identities of 265 SrMV isolates obtained in this study ranged from 70.3 to 100% (nucleotide) and from 73.8 to 100% (amino acid). In each phylogenetic group, the minimum sequence identities of 73.2% (nucleotide) and 80.5% (amino acid) were observed between SCJK003 (MZ419743) and other isolates in group D (Table 1). Among six phylogenetic groups, nucleotide sequence identities ranged between 69.0% (between groups D and E) and 97.5% (between groups C and D), while amino acid sequence identities were 72.8% (between groups A and D) and 100% (between groups C and D). Notably, obvious divergence was exhibited between clade I and the other two clades (II and III), as evidenced by lower nucleotide sequence identities (<85%) among SrMV isolates, except those between the SCJK003 isolate (group D) and 111 SrMV isolates in group A. In addition, the nucleotide and amino acid sequence identities between geographical groups were between 71.3–100% and 74.9–100%, respectively (Table S1). Meanwhile, nucleotide and amino acid identities between host origin groups were shared by 71.3–100% and 74.9–100%, respectively (Table S2).

To further investigate the variation among SrMV *CP* sequences, 12 representative CP amino acid sequences (two sequences in each phylogroup) were aligned. At least four Insertion/deletion (InDel) at the N-terminal and 26 mutation sites were exhibited in SrMV CP sequences (Figure S1). It is noteworthy that no deletion, but a unique site mutation, was present among these CP amino acid sequences in the SrMV-F group as compared to other groups.

Table 1. Percentage identities (%) of nucleotide (low-left) and amino acid (up-right) sequences of SrMV coat protein within and between phylogenetic groups [a].

Group	A ($n = 120$)	B ($n = 26$)	C ($n = 19$)	D ($n = 116$)	E ($n = 21$)	F ($n = 4$)
A	85.3–100 (81.1–100)	77.8–97.7	73.2–89.2	72.8–90.0	73.8–91.8	77.5–90.3
B	79.3–94.0	89.2–100 (88.5–99.6)	76.0–89.2	76.4–90.0	77.1–92.2	81.4–91.0
C	70.7–80.7	69.1–81.3	87.8–100 (86.1–99.5)	79.2–100	81.4–97.5	88.5–95.7
D	70.1–90.7	69.3–84.6	71.7–97.5	80.5–100 (73.2–100)	75.0–98.5	82.5–96.4
E	72.5–81.8	69.4–82.5	78.9–92.4	69.0–92.6	86.7–100 (82.4–100)	84.6–96.0
F	71.6–80.3	72.2–81.2	80.3–87.3	75.2–87.4	78.0–87.3	97.1–99.6 (95.6–99.5)

[a] Nucleotide sequence identities (%) of SrMV CP within phylogenetic groups are shown in parentheses.

2.4. Genetic Recombination Events among SrMV Isolates

Genetic recombination events were identified using RDP4 based on 306 *CP* sequences of SrMV. A significant recombination event was found as supported by four algorithms ($p < 0.05$). The potential recombinant isolate was FJSX017 (MZ419585) from Fujian province, China. The recombinant was derived from the major parent SCH (U07219) from the United States and a minor unknown parent (Table 2).

Table 2. Recombination signals detection among 306 SrMV isolates based on the *CP* genes.

Recombinant	Potential Parents		Detection Method [a]						
	Main Parent	**Minor Parent**	**R**	**G**	**B**	**M**	**C**	**S**	**T**
FJSX017 (MZ419585)	SCH (U07219)	Unknown	-	-	-	+	+	+	+

[a] Seven algorithms include RDP (R), GENECONV (G), Booscan (B), Maximum Chisquare (M), Chimaera (C), Sister Scan (S), and 3Seq (T); +, significant ($10^{-6} < p \le 0.05$); -, non-significant ($p > 0.05$).

2.5. Neutrality Test and Selection Pressure on SrMV Populations

Nucleotide diversity (π) showed that SrMV CP sequences in the Asian population had a higher genetic variation ($\pi = 0.12160$), while the sequences in the American population had a lower genetic variation ($\pi = 0.02200$). However, the π values of the SrMV CP sequences in modern cultivars and chewing cane were 0.12062 and 0.10265, respectively, indicating a higher genetic variation of SrMV CP in both host origins. A neutrality test showed that Tajima's D values for four SrMV populations were all negative, suggesting that the SrMV population exhibited a trend of expansion. Conversely, Tajima's D values for neutrality tests were not statistically significant ($p > 0.10$) in all cases. Meanwhile, the ratios of dN/dS ranged from 0.070 to 0.078 (less than 1) among four populations, suggesting that the SrMV CP gene was under a negative selection (Table 3).

Table 3. Genetic variation and population genetic parameters between different populations based on SrMV *CP* sequences.

Population	π	Tajima's D [a]	dN/dS [b]
Total ($n = 306$)	0.12186	−0.28177 (ns)	0.077
Asia ($n = 296$)	0.12160	−0.25049 (ns)	0.078
South America ($n = 10$)	0.02200	−1.25048 (ns)	0.077
Modern cultivar ($n = 288$)	0.12062	−0.28584 (ns)	0.078
Chewing cane ($n = 18$)	0.10265	−0.61461 (ns)	0.070

[a] ns, non-significant. [b] dN/dS, the ratio of nonsynonymous (dN) and synonymous (dS) substitution.

2.6. Genetic Differentiation and Gene Flow between SrMV Populations

Geographic (Asia vs. America) and host (modern cultivars vs. chewing cane) origins showed considerable genetic differentiation as detected by three permutation-based statistical tests (Ks*, Z*, and Snn) that reached significant levels ($p < 0.05$). The Fst values were >0.33, and the Nm values were <1.0 between geographical groups (Asia vs. America), suggesting that the gene flow between two populations was not frequent. Conversely, the Fst values were <0.33, and the Nm values were >1.0 between host origins (modern cultivars vs. chewing cane), indicating that the gene flow between two populations was frequent (Table 4).

Table 4. Tests of genetic differentiation and gene flow among SrMV groups based on geographical origins and host types [a].

Comparison	Ks* (*p*-Value)	Z* (*p*-Value)	Snn (*p*-Value)	Fst	Nm
Asia (*n* = 296) vs. South America (*n* = 10)	4.14107 (0.0000 ***)	9.68253 (0.0000 ***)	0.99183 (0.0000 ***)	0.44293	0.63
Modern cultivar (*n* = 288) vs. chewing cane (*n* = 18)	4.15947 (0.0000 ***)	9.71312 (0.0000 ***)	0.92157 (0.0300 *)	0.15628	2.70

[a] Asterisks: *, $0.01 < p < 0.05$; ***, $p < 0.001$.

3. Discussion

The crop productivity is affected by a wide range of adverse environmental factors, including biotic and abiotic stress [29]. Mosaic can cause losses ranging from 17% to 50% in susceptible varieties [8]. A survey of the occurrence and distribution of causal agents is an important step for prevention and control for this disease. However, SrMV is often mixed with other viruses causing mosaic diseases, and, therefore, distinguishing the species or strain of viruses causing mosaic disease is nearly impossible through a visual observation [8,25,30,31]. In this study, the RT-PCR technology was employed to accurately identify SrMV. A higher SrMV detection rate was found in modern cultivars than chewing cane. A lower SrMV detection rate existed in chewing cane because of the lower vulnerability of the host to SrMV pathogenesis [21,32]. In addition to cultivar resistance, vector populations and their vagility being subjected to ecosystem simplification also affects virus infection rates [33]. However, this difference in interaction between the virus and sugarcane host need to be further explored. In addition to SrMV, SCSMV is another main causal agent of mosaic in sugarcane modern cultivars in China [8,21,22].

According to our findings, the SrMV isolates from China and South America were grouped together in the SrMV-D group, while the SrMV isolates from Asia were distributed throughout the six phylogroups. Compared to a previous study by Luo et al. (2016), a large number of sugarcane samples was used in this study, but no new phylogroup was proposed [25]. Nonetheless, more phylogroups were discovered in some specific provincial regions in China. For example, Luo et al. (2016) proposed that only one phylogroup (SrMV-G1) occurred in Guizhou province, while the results of the current study indicate that four phylogroups (SrMV-A, -B, -D, and -E) were in this region [25]. The possible reason is that more leaf samples with mosaic were analyzed in this study. The low sequence identities among the SrMV isolates were indicative of high genetic divergence. The viral species demarcation in *Potyviridae* family is typically based on the sequence identity of the CP-coding region with a threshold of <76% (nucleotide) and <82% (amid acid) [34]. Therefore, even if these isolates are in line with the threshold of viral species in this family, more research is required to determine whether SrMV isolates from Yunnan Province clustered in phylogroup F belong to a unique quasispecies or species. New viral species will be considered following investigations based on full genome sequences (genomic feature and phylogeny), together with additional data about biological characteristics such as host range and vector [5].

A high rate of viral genome mutation aids in the creation of novel strains, including resistance-breaking isolates [4]. Our data showed that there are at least four InDels in the N-terminal of SrMV CPs and numerous site mutations across the viral CP sequence. Notably, an obvious feature of CP amino acid sequences in the SrMV-F group is no deletion, but a unique mutation site is present compared to other groups. It is unclear whether these different SrMV isolates in different phylogroups are associated with the variation of viral pathogenicity. Additionally, recombination is a major driving force in the evolution of potyviruses [35,36], but this evolutionary force seems to be an uncommon mechanism of speciation [35]. Our data showed that there was a potential recombination event in all tested SrMV *CP* sequences. However, no recombination was found in previous studies by Zhang et al. (2015) [24] and Luo et al. (2016) [25]. Natural selection is another important evolutionary mechanism and driving force for viral population variation, and purification selection accelerates the elimination of harmful mutations in genes as well as the formation of a stable population genetic structure [37]. Notably, all potyvirus genomes undergo a negative selection, with certain genes such as *HC-Pro*, *CP*, *Nia*, and *NIb* being more strongly selected than others [35]. In this study, the tested SrMV *CP* was subjected to negative selection.

The genetic makeup of viral populations is significantly influenced by geographic isolation [37]. However, modern travel and trade have grown to be significant factors in the transmission of viruses and the swapping of their hosts [35]. Sugarcane is a vegetative propagated crop and frequent exchange of germplasm resources or plant settings between Asian countries, which likely resulted in the absence of obvious population divergence of SrMV within the Asian population. Similarly, Wang et al. (2017) also demonstrated that there was no obvious geographic difference among SrMV isolates [38]. But, to some extent, SrMV populations in China were linked to their geographical origins [24]. Here, our data showed that geographic isolation plays a significant role in the divergence of SrMV isolates between Asia and South America. The host type is another crucial factor leading to the genetic differentiation of plant viruses [33,37]. Based on the phylogenetic analysis of 18 Chinese SrMV isolates, they were divided into two virus populations associated with host types (moder cultivar and chewing cane) [21]. Our data revealed that the phylogenetic grouping of SrMV isolates was not related to two host sources. On the other hand, these SrMV isolates were strongly differentiating the populations of chewing cane and modern cultivars, according to genetic differentiation analysis. A large-scale study of SrMV samples, host sources, and sugarcane-planting regions should be carried out to further explore SrMV population differentiation. Overall, various driving forces contribute to form different SrMV populations or quasispecies.

4. Materials and Methods

4.1. Collection and Distribution of Leaf Samples

A total of 901 leaf samples with mosaic were collected from eight sugarcane-planting provinces from 2017 to 2020, including 839 samples from modern cultivars (*Saccharum* spp. hybrids, Sh) and 62 samples from chewing cane (*S. officinarum*, So). Distribution of leaf samples in different provinces: Guangdong (Sh = 45), Guangxi (Sh = 173), Guizhou (Sh = 165 and So = 5), Fujian (Sh = 90 and So = 9), Hainan (Sh = 51), Sichuan (Sh = 181), Yunnan (Sh = 81), and Zhejiang (Sh = 53 and So = 48). All leaf samples were scrubbed and disinfected with 75% alcohol and stored at $-80\ ^\circ$C for further molecular detection.

4.2. RT-PCR Detection

Total RNA was extracted from leaf samples using the TRIzol® Reagent (Invitrogen, Carlsbad, CA, USA). After the quality and quantity of total RNA were checked, these RNA samples were used for molecular detection using RT-PCR [39]. The HiScript II 1st Strand cDNA Synthesis Kit (Novozan, Nanjing, China) was used to synthesize cDNA from each RNA sample (1.0 μg) with the reverse transcription primer Oligo (dT) $_{23}$VN. The set of SrMV-specific primers SrMV-F (5′-ACAGCAGAWGCAACRGCACAAGC-3′) and SrMV-R

(5′-CTCWCCGACATTCCCATCCAAGCC-3′) was used for PCR amplification [39]. The PCR amplification in a 25 μL volume included 1 μL cDNA, 12.5 μL Premix Taq (Ex Taq Version 2.0 plus dye) (TaKaRa, Dalian, China), and 1 μL of each primer (10 μmol/L). The PCR was performed in the following conditions: 94 °C for 5 min; followed by 35 cycles at 94 °C for 30 s, 52 °C for 30 s, and 72 °C for 1 min; a final extension at 72 °C for 10 min. The PCR products were analyzed via a gel electrophoresis on 1.0% agarose gels.

4.3. Cloning and Sequencing of RT-PCR Fragments

The target fragments from partial SrMV-positive PCR products were eluted using a Gel Extraction Kit (OMEGA Bio-Tek, Norcross, GA, USA). The purified PCR fragments were ligated into the pMD19-T vector (TaKaRa) and then transformed into *Escherichia coli* DH5α competent cells. Three positive colonies from each leaf sample were sent to Sangon Biotech Co., Ltd. (Shanghai, China) for sequencing. The inserted fragments were sequenced bidirectionally using the M13 universal primers.

4.4. Sequence Alignment and Phylogenetic Analysis

A total of 306 CP sequences (this study = 265 and GenBank database = 41) trimming the primer pair sequences were used for sequence alignment and phylogenetic analysis, including 288 sequences from modern cultivars and 28 sequences from chewing cane (Table S3). Sequence alignment was carried out using the ClustalW algorithm implemented in MEGA 10.1.8 software [40] The Neighbor-joining (NJ) method was used to construct the phylogenetic tree, and the robustness of the nodes of the phylogenetic tree was assessed from 1000 bootstrap replicates. A sequence of SCMV isolate SCMV-HZ (NC_003398) was used as outgroup. Sequence identity analysis was conducted using BioEdit 7.1.9 software [41].

4.5. Genetic Recombination Analysis

Sequence recombination analysis was performed using seven different recombination algorithms (RDP, GENECONV, Chimaera, MaxChi, Bootscane, SISCAN, and 3Seq) implemented in RDP4 (Recombination Detection Program version 4) software [42]. Only recombination events that were detected by more than four algorithms ($p < 0.05$) were considered significant.

4.6. Evaluation of Population Genetic Parameters

All population genetic parameters of SrMV *CP* sequences based on different geographical origins (South America vs. Asia) and sugarcane hosts (modern cultivar vs. chewing cane) were calculated using DnaSP version 5.10.01 software [43]. Genetic parameters included nucleotide diversity (π) [44] and Tajima's D [45]. Three statistical test values (Ks*, Z*, and Snn) were used to evaluate the genetic differentiation between SrMV populations. |Fst| > 0.33 or Nm < 1 indicates that gene flow between populations is not frequent, while |Fst| < 0.33 or Nm > 1 suggests frequent gene flow. The selection pressure on SrMV CP in each population was evaluated by calculating the ratio of nonsynonymous (dN) and synonymous (dS) substitutions in nucleotide sequences. Positive, neutral, and negative selections were indicated by dN/dS ratios >1, =1, and <1.

5. Conclusions

In this study, the molecular divergence and population structure of SrMV isolates infecting sugarcane (modern cultivars and ancient chewing cane) were investigated based on the *CP* fragment sequences. A high incidence (70.1%) of the samples was tested positive for SrMV through an RT-PCR assay. Based on 306 SrMV *CP* sequences, three clades including six phylogenetic groups were proposed. High genetic diversity was present among all tested SrMV isolates based on *CP* sequence identities ranging from 69.0% to 100%. The SrMV-A and -D groups dispersed in eight sugarcane planting regions/provinces, but SrMV-F only occurred in Yunnan province, China. Our data suggested that numerous evolutionary driving forces such as nucleotide mutant, gene recombination, and purifying selection as well as geographical and host isolation contributed to form different SrMV populations around the world. These findings enrich the information of the genetic diversity of this virus. However, the molecular divergence and genetic population of SrMV at the complete genome level remain unclear. In addition, the molecular mechanism of the interaction between this virus and host sugarcane is largely unknown. Therefore, these research aspects need to be further explored.

Supplementary Materials: The following supporting information can be downloaded at: https://www.mdpi.com/article/10.3390/plants12213759/s1. Table S1 Percent identities (%) of nucleotide (low-left) and amino acid (up-right) sequences of SrMV *CP* within and between different geographical origins. Table S2 Percent identities (%) of nucleotide (low-left) and amino acid (up-right) sequences of SrMV *CP* within and between host types. Table S3 Information of sequences from sorghum mosaic virus (SrMV) isolates worldwide. Figure S1. Insertion/deletion (InDel) and site mutation among amino acid sequences of SrMV CP. Twelve representative CP sequences (two sequences in each phylogroup) were selected for the alignment by DNAMAN version 6 software. Insertion/deletion (InDel) is showed in red boxes. A unique site mutation of SrMV CP between SrMV-F and other groups is marked with a solid triangle.

Author Contributions: Conceptualization: S.-J.G.; experimentation: H.-M.X., E.-Q.H., Z.-L.Y., Z.-W.B., W.-Q.B. and S.-R.S.; writing—review and editing: H.-M.X., E.-Q.H. and J.-J.L.; methodology: S.-R.S. and S.-J.G.; project administration: S.-J.G.; funding acquisition: S.-J.G.; supervision: J.-J.L. and S.-J.G. All authors have read and agreed to the published version of the manuscript.

Funding: This research was supported by the earmarked fund for China Agriculture Research System (grant no. CARS-170302).

Data Availability Statement: All data supporting the findings of this study are available within the paper and its Supplementary File.

Acknowledgments: We thank Mei-Ting Huang and Hua-Ying Fu for the help of leaf sample collection.

Conflicts of Interest: The authors declare no conflict of interest.

References

1. Pasin, F.; Daròs, J.A.; Tzanetakis, I.E. Proteome expansion in the *Potyviridae* evolutionary radiation. *FEMS Microbiol. Rev.* **2022**, *46*, fuac011. [CrossRef]
2. Adams, M.J.; Lefkowitz, E.J.; King, A.M.; Carstens, E.B. Ratification vote on taxonomic proposals to the international committee on taxonomy of viruses (2014). *Arch. Virol.* **2014**, *159*, 2831–2841. [CrossRef] [PubMed]
3. Sanjuan, R.; Agudelo-Romero, P.; Elena, S.F. Upper-limit mutation rate estimation for a plant RNA virus. *Biol. Lett.* **2009**, *5*, 394–396. [CrossRef] [PubMed]
4. Yang, X.; Li, Y.; Wang, A. Research advances in *Potyviruses*: From the laboratory bench to the field. *Annu. Rev. Phytopathol.* **2021**, *59*, 1–29. [CrossRef] [PubMed]
5. Inoue-Nagata, A.K.; Jordan, R.; Kreuze, J.; Li, F.; Lopez-Moya, J.J.; Makinen, K.; Ohshima, K.; Wylie, S.J.; ICTV Report Consortium. ICTV virus taxonomy profile: *Potyviridae* 2022. *J. Gen. Virol.* **2022**, *103*, 001738. [CrossRef] [PubMed]
6. Chen, J.; Chen, J.; Adams, M.J. Characterisation of potyviruses from sugarcane and maize in China. *Arch. Virol.* **2002**, *147*, 1237–1246. [CrossRef] [PubMed]
7. Sztuba-Solińska, J.; Urbanowicz, A.; Figlerowicz, M.; Bujarski, J.J. RNA-RNA recombination in plant virus replication and evolution. *Annu. Rev. Phytopathol.* **2011**, *49*, 415. [CrossRef]
8. He, E.; Bao, W.; Sun, S.; Hu, C.; Chen, J.; Bi, Z.; Xie, Y.; Lu, J.; Gao, S. Incidence and distribution of four viruses causing diverse mosaic diseases of sugarcane in China. *Agronomy* **2022**, *12*, 302. [CrossRef]
9. Desalegn, B.; Kebede, E.; Legesse, H.; Fite, T. Sugarcane productivity and sugar yield improvement: Selecting variety, nitrogen fertilizer rate, and bioregulator as a first-line treatment. *Heliyon* **2023**, *9*, e15520. [CrossRef]
10. Domingo, E.; Garcia-Crespo, C.; Lobo-Vega, R.; Perales, C. Mutation rates, mutation frequencies, and proofreading-repair activities in RNA virus genetics. *Viruses* **2021**, *13*, 1882. [CrossRef]
11. Elena, S.F.; Fraile, A.; Garcia-Arenal, F. Evolution and emergence of plant viruses. *Adv. Virus Res.* **2014**, *88*, 161–191.
12. Forgia, M.; Navarro, B.; Daghino, S.; Cervera, A.; Gisel, A.; Perotto, S.; Aghayeva, D.N.; Akinyuwa, M.F.; Gobbi, E.; Zheludev, I.N. Hybrids of RNA viruses and viroid-like elements replicate in fungi. *Nat. Commun.* **2023**, *14*, 2591. [CrossRef] [PubMed]
13. Ha, C.; Revill, P.; Harding, R.M.; Vu, M.; Dale, J.L. Identification and sequence analysis of potyviruses infecting crops in Vietnam. *Arch. Virol.* **2008**, *153*, 45–60. [CrossRef]

14. Hema, M.; Joseph, J.; Gopinath, K.; Sreenivasulu, P.; Savithri, H.S. Molecular characterization and interviral relationships of a flexuous filamentous virus causing mosaic disease of sugarcane (*Saccharum officinarum* L.) in India. *Arch. Virol.* **1999**, *144*, 479–490. [CrossRef] [PubMed]

15. Valli, A.; García, J.A.; López-Moya, J.J. Potyviruses (*Potyviridae*). In *Encyclopedia of Virology*, 4th ed.; Bamford, D.H., Zuckerman, M., Eds.; Academic Press: Oxford, UK, 2021; pp. 631–641.

16. Han, Z.; Yang, C.; Xiao, D.; Lin, Y.; Wen, R.; Chen, B.; He, X. A rapid, fluorescence switch-on biosensor for early diagnosis of sorghum mosaic virus. *Biosensors* **2022**, *12*, 1034. [CrossRef] [PubMed]

17. Shukla, D.D.; Frenkel, M.J.; Mckern, N.M.; Ward, C.W.; Jilka, J.; Tosic, M.; Ford, R.E. Present status of the sugarcane mosaic subgroup of potyviruses. *Arch. Virol. Suppl.* **1992**, *5*, 363–373.

18. Yang, Z.N.; Mirkov, T.E. Sequence and relationships of sugarcane mosaic and sorghum mosaic virus strains and development of RT-PCR-based RFLPs for strain discrimination. *Phytopathology* **1997**, *87*, 932–939. [CrossRef]

19. Grisham, M.P.; Pan, Y.B. A genetic shift in the virus strains that cause mosaic in Louisiana sugarcane. *Plant Dis.* **2007**, *91*, 453–458. [CrossRef] [PubMed]

20. Perera, M.F.; Filippone, M.P.; Ramallo, C.J.; Cuenya, M.I.; Garcia, M.L.; Ploper, L.D.; Castagnaro, A.P. Genetic diversity among viruses associated with sugarcane mosaic disease in Tucuman, Argentina. *Phytopathology* **2009**, *99*, 38–49. [CrossRef]

21. Xu, D.L.; Zhou, G.H.; Shen, W.K.; Deng, H.H. Genetic diversity of sorghum mosaic virus infecting sugarcane. *Acta Agron. Sin.* **2008**, *34*, 1916–1920. [CrossRef]

22. Zhou, L.; Xu, Z.; Tang, Y.; Lv, B.; Li, J.; Si, H.; Wen, R.; Chen, B. Complete genome sequence analysis of sorghum mosaic virus from sugarcane in Guangxi. *Genom. Appl. Biol.* **2014**, *33*, 953–960.

23. Hudson, R.R.; Boos, D.D.; Kaplan, N.L. A statistical test for detecting geographic subdivision. *Mol. Biol. Evol.* **1992**, *9*, 138–151.

24. Zhang, Y.; Huang, Q.; Yin, G.; Jia, R.; Lee, S.; Xiong, G.; Yu, N.; Pennerman, K.K.; Liu, Z.; Zhang, S. Genetic diversity of viruses associated with sugarcane mosaic disease of sugarcane inter-specific hybrids in China. *Eur. J. Plant Pathol.* **2015**, *143*, 351–361. [CrossRef]

25. Luo, Q.; Ahmad, K.; Fu, H.Y.; Wang, J.; Gao, S.J. Genetic diversity and population structure of sorghum mosaic virus infecting *Saccharum* spp. hybrids. *Ann. Appl. Biol.* **2016**, *169*, 398–407. [CrossRef]

26. Lefeuvre, P.; Martin, D.P.; Elena, S.F.; Shepherd, D.N.; Roumagnac, P.; Varsani, A. Evolution and ecology of plant viruses. *Nat. Rev. Microbiol.* **2019**, *17*, 632–644. [CrossRef]

27. Liang, S.S.; Luo, Q.; Chen, R.K.; Gao, S.J. Advances in researches on molecular biology of viruses causing sugarcane mosaic. *J. Plant Prot.* **2017**, *44*, 363–370.

28. Lu, G.; Wang, Z.; Xu, F.; Pan, Y.B.; Grisham, M.P.; Xu, L. Sugarcane mosaic disease: Characteristics, identification and control. *Microorganisms* **2021**, *9*, 1984. [CrossRef]

29. Javed, T.; Gao, S.J. WRKY transcription factors in plant defense. *Trends Genet.* **2023**, *39*, 787–801. [CrossRef]

30. Muhammad, K.; Herath, V.; Ahmed, K.; Tahir, M.; Verchot, J. Genetic diversity and molecular evolution of sugarcane mosaic virus, comparing whole genome and coat protein sequence phylogenies. *Arch. Virol.* **2022**, *167*, 2239–2247. [CrossRef]

31. Opedal, O.H.; Armbruster, W.S.; Hansen, T.F.; Holstad, A.; Pelabon, C.; Andersson, S.; Campbell, D.R.; Caruso, C.M.; Delph, L.F.; Eckert, C.G.; et al. Evolvability and trait function predict phenotypic divergence of plant populations. *Proc. Natl. Acad. Sci. USA* **2023**, *120*, e2091739176. [CrossRef]

32. Wang, K.L.; Deng, Q.Q.; Dou, Z.M.; Wang, M.Q.; Chen, J.W.; Shen, W.K. Molecular detection of viral diseases in chewing cane (*Saccharum officinarum*) from southern China. *Int. J. Agric. Biol.* **2018**, *20*, 655–660. [CrossRef]

33. Fraile, A.; Garcia-Arenal, F. Environment and evolution modulate plant virus pathogenesis. *Curr. Opin. Virol.* **2016**, *17*, 50–56. [CrossRef] [PubMed]

34. Adams, M.J.; Antoniw, J.F.; Fauquet, C.M. Molecular criteria for genus and species discrimination within the family *Potyviridae*. *Arch. Virol.* **2015**, *150*, 459–479. [CrossRef] [PubMed]

35. Gibbs, A.J.; Hajizadeh, M.; Ohshima, K.; Jones, R. The potyviruses: An evolutionary synthesis is emerging. *Viruses* **2020**, *12*, 132. [CrossRef]

36. He, Z.; Dong, Z.; Gan, H. Genetic changes and host adaptability in sugarcane mosaic virus based on complete genome sequences. *Mol. Phylogenet. Evol.* **2020**, *149*, 106848. [CrossRef] [PubMed]

37. Sun, S.; Chen, J.; He, E.; Huang, M.; Fu, H.; Lu, J.; Gao, S. Genetic variability and molecular evolution of maize yellow mosaic virus populations from different geographic origins. *Plant Dis.* **2021**, *105*, 896–903. [CrossRef] [PubMed]

38. Wang, X.-Y.; Li, W.-F.; Huang, Y.-K.; Zhang, R.-Y.; Shan, H.-L.; Yin, J.; Luo, Z.-M. Molecular detection and phylogenetic analysis of viruses causing mosaic symptoms in new sugarcane varieties in China. *Eur. J. Plant Pathol.* **2017**, *148*, 931–940. [CrossRef]

39. Xie, Y.; Wang, M.; Xu, D.; Li, R.; Zhou, G. Simultaneous detection and identification of four sugarcane viruses by one-step RT-PCR. *J. Virol. Methods* **2009**, *162*, 64–68. [CrossRef] [PubMed]

40. Stecher, G.; Tamura, K.; Kumar, S. Molecular evolutionary genetics analysis (MEGA) for macOS. *Mol. Biol. Evol.* **2020**, *37*, 1237–1239. [CrossRef] [PubMed]

41. Hall, T.A. Bioedit: A user-friendly biological sequence alignment editor and analysis program for windows 95/98/nt. *Nucleic Acids Symp. Ser.* **1999**, *41*, 95–98.

42. Martin, D.P.; Murrell, B.; Khoosal, A.; Muhire, B. Detecting and analyzing genetic recombination using RDP4. *Methods Mol. Biol.* **2017**, *1525*, 433–460. [PubMed]

43. Librado, P.; Rozas, J. DnaSP v5: A software for comprehensive analysis of DNA polymorphism data. *Bioinformatics* **2009**, *25*, 1451–1452. [CrossRef] [PubMed]
44. Nei, M.; Tajima, F. DNA polymorphism detectable by restriction endonucleases. *Genetics* **1981**, *97*, 145–163. [CrossRef] [PubMed]
45. Tajima, F. The effect of change in population size on DNA polymorphism. *Genetics* **1989**, *123*, 597–601. [CrossRef] [PubMed]

Article

Transcriptomic Analysis Reveals Candidate Genes in Response to Sorghum Mosaic Virus and Salicylic Acid in Sugarcane

Genhua Zhou [1,2,†], Rubab Shabbir [1,2,†], Zihao Sun [1,2,†], Yating Chang [1,2], Xinli Liu [1,2] and Pinghua Chen [1,2,*]

1   Key Laboratory of Sugarcane Biology and Genetic Breeding, Ministry of Agriculture, National Engineering Research Center for Sugarcane, Fujian Agriculture and Forestry University, Fuzhou 350002, China; mszhougenhua@163.com (G.Z.); rubabshabbir28@gmail.com (R.S.); jeremy_szh@126.com (Z.S.); cyating0302@163.com (Y.C.); 18350065738@163.com (X.L.)

2   Key Laboratory of Ministry of Education for Genetics, Breeding and Multiple Utilization of Crops, College of Agriculture, Fujian Agriculture and Forestry University, Fuzhou 350002, China

*   Correspondence: 000q010027@fafu.edu.cn

†   These authors contributed equally to this work.

Abstract: Sorghum mosaic virus (SrMV) is one of the most prevalent viruses deteriorating sugarcane production. Salicylic acid (SA) plays an essential role in the defense mechanism of plants and its exogenous application has been observed to induce the resistance against biotic and abiotic stressors. In this study, we set out to investigate the mechanism by which sorghum mosaic virus (SrMV) infected sugarcane responds to SA treatment in two sugarcane cultivars, i.e., ROC22 and Xuezhe. Notably, significantly low viral populations were observed at different time points (except for 28 d in ROC22) in response to post-SA application in both cultivars as compared to control based on qPCR data. Furthermore, the lowest number of population size in Xuezhe (20 copies/μL) and ROC22 (95 copies/μL) was observed in response to 1 mM exogenous SA application. A total of 2999 DEGs were identified, of which 731 and 2268 DEGs were up- and down-regulated, respectively. Moreover, a total of 806 DEGs were annotated to GO enrichment categories: 348 biological processes, 280 molecular functions, and 178 cellular components. GO functional categorization revealed that DEGs were mainly enriched in metabolic processes, extracellular regions, and glucosyltransferase activity, while KEGG annotation revealed that DEGs were mainly concentrated in phenylpropanoid biosynthesis and plant-pathogen interaction suggesting the involvement of these pathways in SA-induced disease resistance of sugarcane in response to SrMV infection. The RNA-seq dataset and qRT-PCR assay showed that the transcript levels of *PR1a*, *PR1b*, *PR1c*, *NPR1a*, *NPR1b*, *PAL*, *ICS*, and *ABA* were significantly up-regulated in response to SA treatment under SrMV infection, indicating their positive involvement in stress endorsement. Overall, this research characterized sugarcane transcriptome during SrMV infection and shed light on further interaction of plant-pathogen under exogenous application of SA treatment.

Keywords: sugarcane; sorghum mosaic virus; salicylic acid; transcriptomics

Citation: Zhou, G.; Shabbir, R.; Sun, Z.; Chang, Y.; Liu, X.; Chen, P. Transcriptomic Analysis Reveals Candidate Genes in Response to Sorghum Mosaic Virus and Salicylic Acid in Sugarcane. *Plants* **2024**, *13*, 234. https://doi.org/10.3390/plants13020234

Academic Editor: Won Kyong Cho

Received: 17 October 2023
Revised: 23 December 2023
Accepted: 24 December 2023
Published: 14 January 2024

1. Introduction

Environmental stressors brought on by extreme climate change are becoming more prevalent worldwide and have also affected the geographic distribution and incidence of plant pathogens and related diseases, which by the end of this century could push about one-third of the world's food crop productivity outside of the safe climatic range [1]. Various biotic stresses are major factors that can impede the whole growth and development period of sugarcane, resulting in 10–15% yield losses worldwide [1]. Plants recognize the first layer of conserved pathogen-associated molecular patterns (PAMPs) through the pattern-recognition receptors (PRRS) to activate pattern-triggered immunity (PTI) in response to various stresses [2,3]. In order to activate effector-triggered immunity (ETI), plants selectively detect effectors using polymorphic nucleotide-binding domain and leucine-rich repeat (NB-LRR) proteins expressed by the majority of resistance genes [2–4].

Numerous transcriptomic approaches have been undertaken in sugarcane infected with various biotic and abiotic stresses. For pathogen infection, several studies have been undertaken on sugarcane infected by sugarcane yellow leaf virus (SCYLV) [5], sugarcane streak mosaic virus (SCSMV) [6], *Sporisorium scitamineum* [7], *Acidovorax avenae* subsp. *avenae* (Aaa) [8] and *Xanthomonas albilineans* (*Xa*) [9]. Comparative transcriptome analysis of resistant and susceptible sugarcane cultivars (LCP85-384 and ROC20) evidenced the upregulated expression of numerous *WRKYs* genes especially for *WRKY33* alleles against *Xa* infection [9]. The transcriptional regulatory co-expression network revealed 16 and 25 hub genes in *Saccharum spontaneum* (SES208) that were enriched for possible involvement in MAPK signaling, plant-pathogen interactions, and hormonal signaling against resistance to leaf scald [10].

Several studies have revealed that the exogenous application of SA regulates ROS homeostasis, GR (glutathione reductase), and APX (ascorbic acid-peroxidase) activity and ultimately increases the resistance against biotic and abiotic stressors [11]. The SA application also induces the PR-proteins against *V. dahlia* in *S. melongena* [12] which in turn regulate resistance against *P. palmivora* by increasing the activity of PAL (phenylalanine ammonialyase), POD (peroxidase), and CAT (catalase) in rubber trees [13]. Another research revealed that SA application could increase the APX and POD activity to enhance resistance against *tomato yellow leaf curl virus* [14]. An increasing number of studies revealed that SA pathway PR1 proteins play a pivotal role in the response of plants to pathogen infection. For instance, ectopic expression of *Mangifera indica MiPR1A* in transgenic *Arabidopsis* attributed to enhanced disease resistance to *Colletotrichum gloeosporioides* pathogen [15]. Overexpression and virus-induced gene silencing of *Triticum aestivum TaPR1-7* and *Solanum lycopersicum SlPR1* exhibited decreased pathogenesis caused by *Puccinia striiformis* f. sp. *tritici* and *Ralstonia solanacearum*, respectively [16,17]. The SA has the advantages of high efficiency, non-toxicity, broad-spectrum, environmental safety, simple use, and low cost, and has become a research hotspot in the field of resistance induction to plant diseases [18,19].

Single and/or mixed viral disease pathogens such as sugarcane mosaic virus (SCMV), sorghum mosaic virus (SrMV), sugarcane streak mosaic virus (SCSMV), and maize yellow mosaicvirus (MaYMV) causing significant loss to the global sugar industry. These viruses have a wide range of hosts. i.e., maize, sorghum, and other grasses in addition to sugarcane. SrMV (genus: Potyvirus, family: Potyviridae) is a causal agent of common mosaic in sugarcane. Additionally, SrMV symptoms include chlorotic streaks on leaves, stunted growth, necrosis, and death of the plant [20].

However, only gene profiles of sugarcane responses to a single stressor alone have been considered. To our knowledge, no report relates to an integrated transcriptome analysis of SrMV infection and SA in sugarcane. Therefore, we investigated the molecular responses in sugarcane cultivar Xuezhe under SrMV infection and SA application based on the RNA-seq platform by Illumina NGS technology. Genes and pathways involved in response to the coupling effect of SrMV infection and SA application as part of a defense strategy against this pathogen were identified. These potential genes serve as helpful genetic resources for sugarcane breeding as well as provide insights into the signaling cascades activated by SrMV infection under SA enrichment.

2. Results

2.1. SA Application Inhibited SrMV from Infecting Sugarcane Plants

Based on qPCR data, significantly low viral populations were observed at different time points (except for 28 d in ROC22) in response to post-SA application in both cultivars as compared to control (no SA application). The lowest number of population size in Xuezhe (20 copies/μL) and ROC22 (95 copies/μL) was observed in response to 1 mM exogenous SA application as compared to other SA treatments and controls. In contrast to the aforementioned results, the highest number of population size was identified in

ROC22 (1209 copies/μL) in response to a higher dose of SA (4 mM SA) at 28 d post-SrMV infection. No variance in viral population size was observed at 14 d post-SrMV infection in response to different SA treatments in ROC22. Overall, the results indicated the positive involvement of exogenous SA application in decreased population size of SrMV in both sugarcane cultivars (Figure 1).

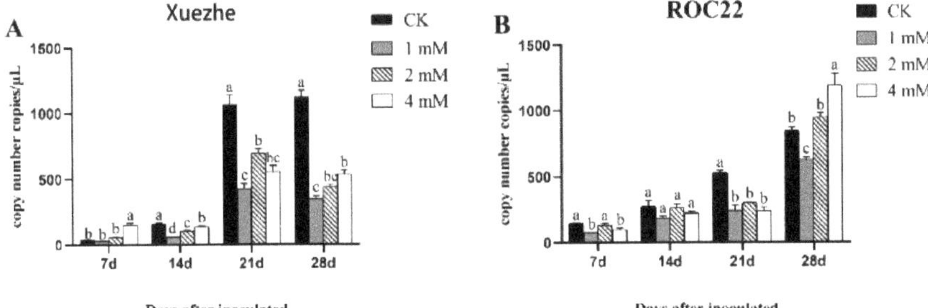

Figure 1. Virus populations in two sugarcane cultivars post SA application determined by qPCR assay: (**A**,**B**) represent the viral copy numbers of Xuezhe and ROC22, respectively. SA solution with concentrations of 0 mM (CK), 1 mM, 2 mM, and 4 mM was prepared and applied to sugarcane seedlings following SrMV inoculation. The letters on the bar graph indicate significant differences among different treatment means ($p < 0.05$), the same letters indicate no significant differences, and different letters indicate significant differences. Values are means ± standard errors.

2.2. Transcriptome Sequencing and Assembly

To investigate the genes and pathways involved in sugarcane response to SrMV infection and SA treatment, a total of 12 cDNA libraries were screened (Table 1). The number of raw reads and clean data after quality control ranged from 40,056,842 to 47,349,582 and 37,546,314 to 45,324,754 for the Xuezhe cultivar, respectively. The GC content, Q20, and Q30 of 12 libraries was more than 56%, 97%, and 93% respectively. Moreover, the mapping rate ranged between 78.52% and 89.37% (Table 1).

Table 1. Summary of transcriptome sequencing and assembly of sugarcane in response to SrMV infection and SA application.

Sample	Raw Reads	Clean Reads	Q20 (%)	Q30 (%)	GC (%)	Mapping Rate (%)
CK-1	44,914,726	43,314,264	97.5	93.34	51.76	85.01
CK-2	40,068,588	37,546,314	97.5	93.28	49.83	87.59
CK-3	41,815,628	40,116,578	97.36	93.24	56.06	86.90
CK-4	46,929,116	45,112,170	97.48	93.22	50.38	81.90
CK-5	43,052,022	41,398,032	97.6	93.42	48.49	87.37
CK-6	44,478,236	42,657,658	97.47	93.26	51	85.09
SA-1	41,941,044	40,182,148	97.43	93.31	53.92	81.29
SA-2	44,570,072	42,837,078	97.36	93.14	54.66	83.18
SA-3	45,719,490	44,294,824	97.39	93.19	53.73	78.52
SA-4	44,841,752	43,183,362	97.28	93.01	55.88	89.37
SA-5	40,056,842	38,165,778	97.38	93.25	55.25	84.26
SA-6	47,349,582	45,324,754	97.41	93.34	56.06	87.88

Note: Q20%, and Q30%: respectively refer to the percentage of bases with sequencing quality above 99% and 99.9% of the total bases; GC%: is the percentage of the sum of G and C bases corresponding to the quality control data to the total bases. SA solution with a concentration of 1 mM was prepared and applied to sugarcane seedlings following SrMV inoculation with 6 biological replicates (SA-1 to SA-6).

2.3. Identification, Functional Annotation of DEGs, Gene Ontology (GO), and Kyoto Encyclopedia of Genes and Genomes (KEGG) Analysis

A total of 2999 DEGs were identified in treatment comparison with control of which 731 DEGs were upregulated while 2268 DEGs were downregulated. Go analysis revealed that the top 10 upregulated and downregulated transcripts were assigned to one of the three main GO categories: cellular component (CC), biological process (BP), and molecular function (MF). A total of 806 DEGs were annotated, including 348 BP, 280 MF, and 178 CC. The DEGs were further distributed to 30 main GO groups. Notably, 204 DEGs (51 for each ontology) were attributed to the cellular glucan metabolic process, glucan metabolic process, cellular polysaccharide metabolic process, and polysaccharide metabolic process of the BP category. Similarly, 39 and 36 DEGs were attributed to the extracellular region and cell periphery of the CC category. Additionally, glucosyltransferase activity, oxidoreductase, antioxidant activity, and peroxidase activity had 52, 43, 43, and 42 DEGs, respectively, in the MF category (Figure 2).

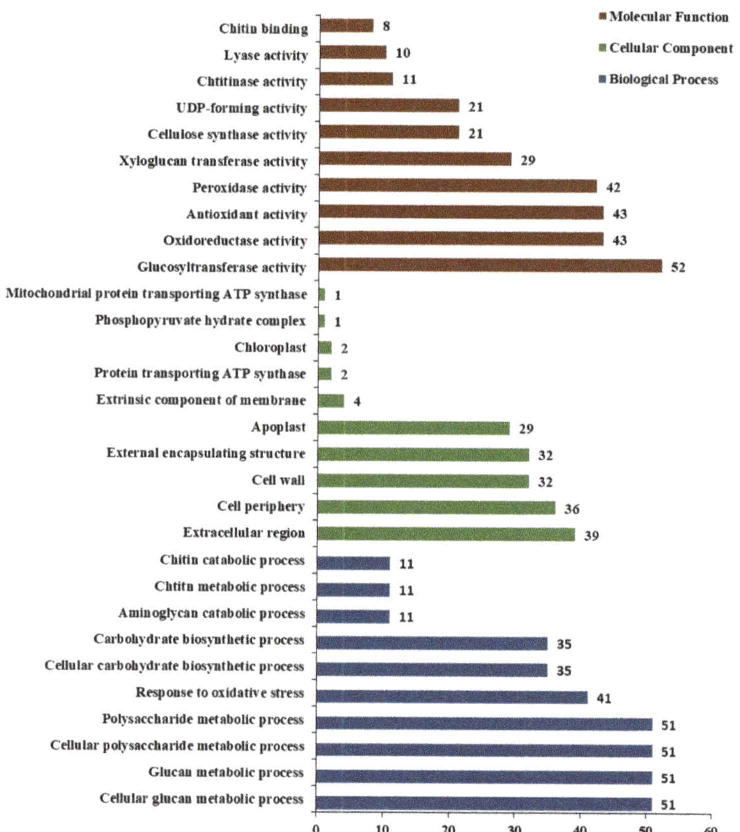

Figure 2. *Differentially expressed genes GO enrichment analysis.* Note: The abscissa is the GO term, and the ordinate is the significance level of GO term enrichment, which is represented by −log10(padj), and different colors represent different functional categories.

To further study the biological pathways triggered by SrMV infection and SA application, DEGs were annotated by KEGG enrichment analysis. Overall, the most significant 20 KEGG pathways were selected (Figure 3). The results showed that DEGs are mainly concentrated in phenylpropanoid biosynthesis, plant-pathogen interaction, benzoxazinoid biosynthesis, fatty acid elongation, thiamine metabolism, alpha-linolenic acid metabolism, and zeatin biosynthesis. The enrichment of DEGs was most significant in phenylpropanoid biosynthesis and plant-pathogen interaction, suggesting that that pathway may be involved in SA-induced disease resistance of sugarcane (Figure 3).

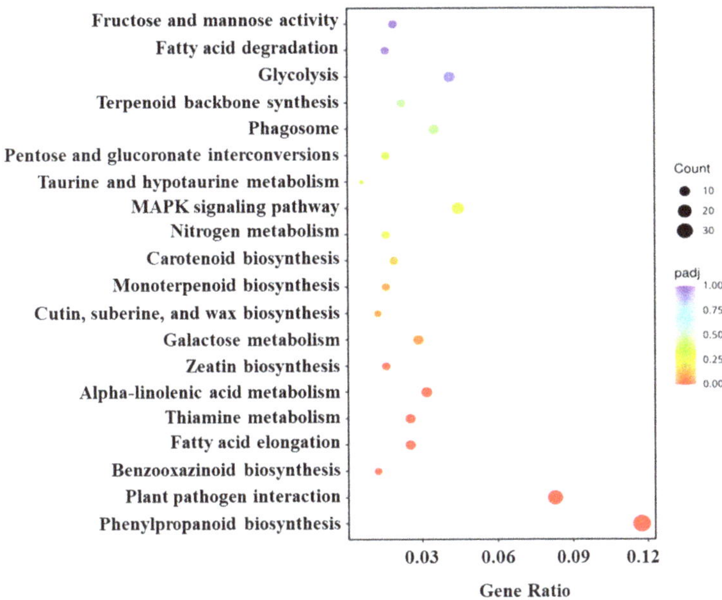

Figure 3. KEGG pathway classification and functional enrichment of differentially expressed genes in sugarcane response to SrMV infection and SA treatment.

2.4. Transcript Profiling of Candidate Genes

Based on transcript expression of genes in RNA-seq data, 8 DEGs were selected for validation by qRT-PCR assay (Figures 4 and 5). The characterization of genes responding to SrMV infection and SA was determined by qRT-PCR assay. The transcript levels of 3 *PR1s* (*PR1a*, *PR1b*, and *PR1c*), 2 *NPR1s* (*NPR1a* and *NPR1b*), *PAL*, *ICS*, and *ABA* were significantly up-regulated after the SA treatment in response to SrMV.

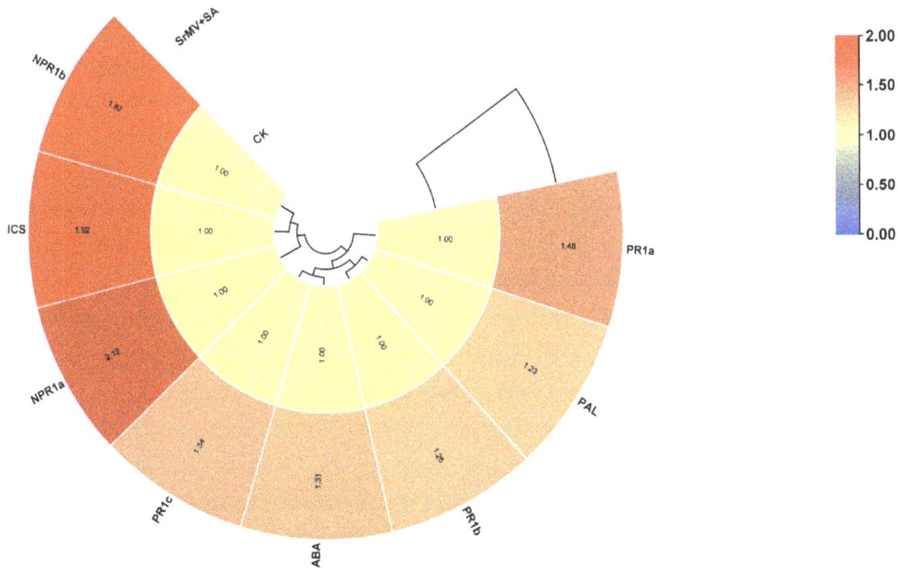

Figure 4. Transcript expression of selected candidate genes from RNA-seq dataset.

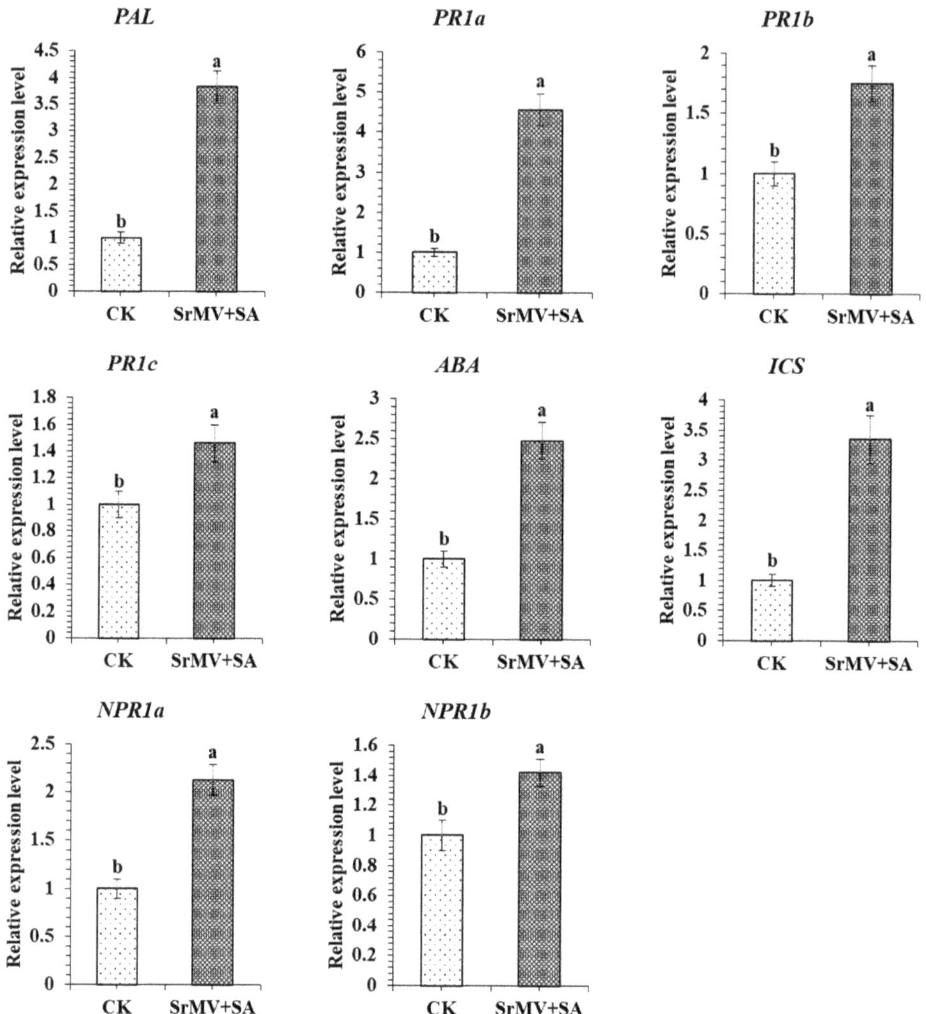

Figure 5. Transcript profiling of candidate genes by qRT-PCR assay. Values for relative expression levels are means ± standard errors. Different lowercase letters above the bars are significantly different at a 5% probability level.

3. Discussion

Climate change is predicted to have significant effects on sugarcane production around the globe, particularly in developing countries because of relatively low adaptive capacity, high vulnerability to natural hazards, limited gene pool, and poor forecasting systems as well as preventive/mitigating measures. Sugarcane production may have been negatively affected and will continue to be considerably affected by increases in the frequency and intensity of extreme environmental conditions such as biotic and abiotic stresses [1–9]. Over the last decade, the advent of next-generation sequencing (NGS) such as transcriptomics, metabolomics, genomics, and proteomics have emerged as promising tools for exploring genetic and molecular basis to identify networks of regulatory candidate genes and biosynthetic cascades under the ongoing climate change scenario in sugarcane [6–9]. The combination of sugarcane omics technologies through identification and modifications in transcripts, metabolites, DNA, and proteins enables a deeper understanding of the genetic mechanisms underlying the complex architecture of many phenotypic/genotypic traits of sugarcane relevance [7–9]. This study also employed a transcriptomic approach to identify key and candidate genes responding to SrMV infection coupled with SA application.

3.1. SA Pathway and Regulatory Genes

SA plays an important role in all aspects of plant immunity and can effectively promote plant growth and development [2]. The effects of SA on plant growth and development are dependent on the concentration, plant growth conditions, and different developmental stages. Usually, high SA concentration negatively regulates plant growth while the application of optimal SA concentration shows beneficial effects. Exogenous application of SA could significantly promote plant growth and increase plant dry yield under both salt stress and no-stress conditions, and the effect was more obvious under salt stress [21]. Exogenous application of SA could significantly reduce the infection effect of the virus and effectively increase plant height and leaf number [22]. Similar findings were reported by Zagier et al. [23] to control fig mosaic virus by exogenous application of SA. Based on qPCR results, low viral populations were observed at different time points (except for 28 d in ROC22) in response to post-SA application in both cultivars as compared to the control in this study. Similarly, Suharti et al. [24] found that after SA induction, *Xanthomonas oryzae* infection was inhibited and plant biochemical resistance increased, especially tannins and phenols. However, salicylic acid did not affect the growth of rice plants which might be caused by species differences.

The SA signaling pathway is involved in dramatic transcriptional reprogramming against environmental stimuli in plants via several important genes such as *NPR1/NPR3* and disease process-related proteins PR1/5, etc. SA signals play a key role in triggering defense responses to stress and activating plant systemic acquired resistance (SAR). In SA signal transduction, up-regulated expression of SAR-related genes leads to increased plant disease resistance. Zhang et al. [25] used RNA-seq to study the differential expression of 32 SA signal-related genes, including *NPR1*, *TRXs*, *NPR3*, *NIMINs*, *WRKYs*, *TGAs*, *SABP2* and *MESs* genes. NPR1 is a major regulator in the SA signaling pathway, controlling multiple immune responses including SAR. In NPR1 mutants, SA-mediated PR gene expression and pathogen resistance were eliminated. In the inactivated state, *NPR1* exists in the cytoplasm as a disulfide-bonded oligomer. After induction by exogenous SA, the cytoplasmic TRX catalyzes the REDOX of *NPR1* from an oligomeric form to a monomer form. The monomer form of *NPR1* can then enter the nucleus and regulate downstream transcription factors, such as TGA and WRKY. Plant WRKY TFs are involved in regulating disease-resistance responses at multiple levels, such as directly or indirectly regulating downstream target genes, activating, or inhibiting the expression of other TFs, and regulating WRKY family members and self-genes using multiple feedforward and feedback methods. WRKYs have been shown to play a major positive or negative regulatory role in SA-dependent defense responses in plants, with more than half of *Arabidopsis* WRKY genes induced or inhibited when treated with SA [26]. For example, transcriptional regulator *AtWRKY70* signals through the SA-mediated pathway and thereby regulates a subset of genes common to senescence and plant defense [27]. SA-induced *AtWRKY18* positively regulates the expression of the PR gene and resistance to *Pseudomonas syringae* as well as *AtWRKY18* enhances the developmental regulatory defense response dependent on *NPR1* [28]. A recent study by Chu et al. [29] reported the involvement of *ScPR1* as a positive regulator of defense response against *Aaa* infection in sugarcane. Previously, it has been found that *NPR3* binds to *NPR1* and TGAs to regulate the defense response of developing flowers to pathogens [30]. Notably, SA-mediated genes (*ScNPR3*, *ScTGA4*, *ScPR1*, and *ScPR5*), participated in the response of sugarcane to infection by *Xa* strains, but sugarcane responses to two strains differing in pathogenicity through differential modulation of SA and ROS [31].

Transcription factors are ubiquitous regulatory proteins in plants, which play a regulatory role in plant growth and development, biological and abiotic stress response, hormone response, and metabolism [32]. In this study, by further analysis of DEGs, it was found that WRKY and GST transcription factor families were highly expressed after SA treatment. WRKY transcription factor is one of the ten major transcription factor families in plants. TFs are activated by different pathways of signal transduction and can directly or indirectly combine with *cis*-acting elements to modulate the transcription efficiency of target genes, which play key regulators in crop genetic improvement. Interestingly, WRKY TFs are widely involved in plant disease resistance response, plant growth, development and physiological processes, and regulatory networks affecting plant secondary metabolic pathways [33,34]. When plants suffer from stress, the antioxidant activity in the plant is enhanced, among which GST is one of the main antioxidant enzymes, which can degrade harmful substances produced in plants and reduce cell damage. In this study, the large expression of the GST transcription factor can promote the production of this enzyme and thus enhance the plant's defense against disease. In conclusion, the large expression of WRKY and GST transcription factors may be the main response mode of SrMV resistance induced by SA in sugarcane in this experiment, which can be further studied in the future. In this study, it was also found that the ABA pathway genes were also significantly upregulated after the treatment. ABA has been reported to play a role in abiotic and biological stress through bidirectional JA and SA crosstalk [35].

3.2. Metabolic Adjustments

Phenylalanine ammonia-lyase converts phenylalanine into trans-cinnamic acid to synthesize SA in a distinct enzymatic pathway. The SA signaling pathway can be triggered both by PTI and ETI. Moreover, plants have developed a multitude of multiple genes regulated defense responses such as PAMP by modulating different metabolic pathways i.e., galactose-, phenylpropanoid biosynthesis-, and amino acids-metabolisms [33]. These responses also include activation of pathogenesis-related proteins (PR), production of defense markers (reactive oxygen species), and signaling dependent on the phytohormones [30,31]. Under *Aaa* infection, increased expression of several genes involved in the phenylpropanoid biosynthesis pathway would be a defense strategy employed by sugarcane against pathogen stress [8]. In this study, we found that phenylpropanoid biosynthesis pathway genes were significantly enriched under SrMV infection coupled with SA as compared to control, suggesting that SA signaling cascade may also exist in sugarcane to positively regulate SrMV pathogenesis.

4. Conclusions and Prospects

This study provides the first report on a transcriptome dataset of 12 cDNA libraries in response to SrMV infection and SA enrichment in sugarcane. Most of the DEGs were annotated for cellular glucan metabolic process, glucan metabolic process, cellular polysaccharide metabolic process, and polysaccharide metabolic process. KEGG pathway analysis revealed that phenylpropanoid biosynthesis and plant-pathogen interaction pathways contributed to sugarcane response to SrMV infection coupled with SA supply. In this study, 8 up-regulated genes were selected for qRT-RCR validation: 3 *PR1*, 2 *NPR1*, 1 *PAL*, 1 *ICS*, and 1 *ABA*. The results showed that the expression profiles of several genes were consistent with the expression trend of the transcriptome. These results provide data support for further understanding of the SA regulatory network in sugarcane and the molecular mechanism of SA regulating defense and stress response at the transcriptional level and lay a foundation for revealing the systematic gene functions and protein interactions in the SA signaling pathway in sugarcane under SrMV pathogenesis. Additionally, this study also lays the foundation for the identification of new gene resources in response to SrMV infection coupled with SA application.

5. Materials and Methods

5.1. Plant Materials and Virus Detection

The experimental materials were selected from the SrMV-sensitive sugarcane variety Xuezhe and the SrMV-resistant sugarcane variety ROC22, both of which were provided by the National Sugarcane Germplasm Resource Nursery in Kunming, Yunnan Province, China. Sugarcane stems were cut into single-budded setts followed by imbibition in water for 24 h at room temperature and incubation at 32 °C till seedlings emerged and grew up to a height of 3–5 cm. The seedlings were transplanted to nutrient soil in an incubator with the following specifications: 28 °C temperature, 16 h of light/8 h of darkness, and 6000 Lx of light intensity [18]. SA solution with concentrations of 0 mM (CK), 1 mM, 2 mM, and 4 mM was prepared and applied to sugarcane seedlings soon after SrMV inoculation. The samples of Xuezhe and ROC22 seedlings inoculated with SrMV were taken at 7, 14, 21, and 28 days after inoculation. The SrMV content in sugarcane leaves at different periods was determined using the detection method established by the standard curve (attachment) constructed in the earlier stage of the laboratory (Figure S1).

5.2. RNA Extraction and cDNA Library Construction

The healthy seedlings of sugarcane cultivar Xuezhe at the 3–5 leaf stage were selected for infection by SrMV. The leaf cut method was employed to infect the leaves with SrMV for 28 days, and each treatment had 6 biological repeats. SA solution with a concentration of 1 mM was prepared and applied to sugarcane seedlings following SrMV inoculation. The plant leaf samples were used for transcriptome analysis. RNA samples were sent to Novogene Biological Company, Beijing, China, for transcriptome sequencing analysis. According to the NEBNext® Ultra ™ RNA Library Prep Kit for Illumina® (San Diego, CA, USA) manual of the kit, cDNA libraries were constructed. Finally, the sequencing was performed through the Illumina HiSeq platform.

5.3. Quality Control and Alignment of Sequencing Data

Trimmatic (0.39) [36] was used to remove a small number of reads with sequencing connectors, sequencing errors, and low-quality sequences from the original data. Q20, Q30, and GC contents were calculated from the clean data. Genome (https://download.cncb.ac.cn/gwh/Plants/Saccharum_officinarum_LApurple_GWHBEII00000000/GWHBEII00000000.genome.fasta.gz (accessed on 12 March 2022)) and annotation (https://download.cncb.ac.cn/gwh/Plants/Saccharum_officinarum_LApurple_GWHBEII00000000/GWHBEII00000000.gff.gz (accessed on 12 March 2022)) files LA purple from the genome website were downloaded and by using HISAT2 (2.0.5) [37] an index of the reference genome to obtain location information was calculated.

5.4. Screening and Functional Enrichment Analysis of Differentially Expressed Genes

StringTie (1.3.3b) [38] was used to predict new genes, and the number of fragments per thousand base length from a gene in each million reads of each gene was calculated according to the gene length. DESeq2 (1.20.0) [39] was used for differential expression analysis based on the negative binomial distribution model. The fragments per kilobase of transcript per million fragments mapped (FPKM) value for each transcript was calculated and then transformed to log2 (Fold Change) values for the generation of a heatmap with TBtools v0.6655. DEGs were screened with a p-value ≤ 0.05. ClusterProfiler (3.8.1) [40] software was used to conduct GO (Gene Ontology) and KEGG (Kyoto Encyclopedia of Genes and Genomes) analysis of DEGs. The criterion for significantly different expression was established at $|\log_2(\text{fold change})| > 1.0$ and p-value < 0.005. Following transcriptome assembly, BLASTx with an E-value cutoff of 10^{-5} was used to assign functional categories in the GO and KEGG databases. Wallenius non-central hyper-geometric distribution from the GOseq R package (v.4.2.3) was used to analyze the GO enrichment of differential expression genes (DEGs). GO terms with a p-value of 0.005 or lower showed substantial enrichment. KOBAS 2.0 was used for the KEGG enrichment analysis of DEGs with a p-value < 0.05.

5.5. Quantitative Real-Time PCR Analysis

The RNA from leaf samples was extracted by using a megazol reagent (Invitrogen, Waltham, MA, USA) according to the manufacturer's instructions. The extracted sample RNA was reversely transcribed into cDNA for qRT-PCR verification. A total of 8 differential genes were selected for verification by qRT-PCR. The qRT-PCR was carried out using ChamQ Universal SYBR qPCR master mix (Vazyme, Nanjing, China) on a QuantStudio® Real-Time PCR system (Applied Biosystems, Waltham, MA, USA). Additionally, IDT online software (https://sg.idtdna.com/PrimerQuest/Home/Index (accessed on 12 March 2022)) was used to design gene-specific primers for qRT-PCR (Table S1). The reaction mixture comprised 10 µL of 2× SYBR PremixEx TaqTM (Takara, San Jose, CA, USA), 2 µL of each forward and reverse primer, 1 µL of cDNA, and 6.6 µL of ddH$_2$O to make a final volume of 20 µL. For each sample, three biological and three technical replicates were used. Glyceraldehyde 3-phosphate dehydrogenase (GAPDH) was used as a reference gene and $2^{-\Delta\Delta Ct}$ method was used to determine the gene expression.

5.6. Statistical Analysis

The means of different treatments were compared using the least significance difference (LSD) test at a 5% probability level ($p \leq 0.05$) with a statistical software package Statistix 8.1. This means having different letters above the bars are significantly different at a 5% probability level.

Supplementary Materials: The following supporting information can be downloaded at https://www.mdpi.com/article/10.3390/plants13020234/s1.

Author Contributions: Conceptualization, P.C.; methodology, P.C.; software, Z.S.; validation, P.C. and R.S.; formal analysis, G.Z. and Z.S.; investigation, G.Z.; resources, P.C.; data curation, P.C.; writing—original draft preparation, Z.S. and R.S.; writing—review and editing, G.Z., Y.C., X.L. and R.S.; visualization, R.S.; supervision, P.C.; project administration, P.C.; funding acquisition, P.C. All authors have read and agreed to the published version of the manuscript.

Funding: This research was funded by the National Natural Science Foundation, China grant number 31070330 and The APC was funded by KH180187A.

Data Availability Statement: Data are contained within the article and Supplementary Materials.

Conflicts of Interest: The authors declare no conflict of interest.

References

1. Kummu, M.; Heino, M.; Taka, M.; Varis, O.; Viviroli, D. Climate change risks pushing one-third of global food production outside the safe climatic space. *One Earth* **2021**, *4*, 720–729. [CrossRef] [PubMed]
2. Javed, T.; Gao, S. WRKY transcription factors in plant defense. *Trends Genet.* **2023**, *39*, 787–801. [CrossRef] [PubMed]
3. Yuan, M.; Ngou, B.P.M.; Ding, P.; Xin, X.F. PTI-ETI crosstalk: An integrative view of plant immunity. *Curr. Opin. Plant Biol.* **2021**, *62*, 102030. [CrossRef] [PubMed]
4. Javed, T.; Shabbir, R.; Ali, A.; Afzal, I.; Zaheer, U.; Gao, S.J. Transcription factors in plant stress responses: Challenges and potential for sugarcane improvement. *Plants* **2020**, *9*, 491. [CrossRef] [PubMed]
5. Shabbir, R.; Zhaoli, L.; Yueyu, X.; Zihao, S.; Pinghua, C. Transcriptome Analysis of Sugarcane Response to Sugarcane Yellow Leaf Virus Infection Transmitted by the Vector Melanaphis sacchari. *Front. Plant Sci.* **2022**, *13*, 921674. [CrossRef]
6. Dong, M.; Cheng, G.; Peng, L.; Xu, Q.; Yang, Y.; Xu, J. Transcriptome analysis of sugarcane response to the infection by Sugarcane steak mosaic virus (SCSMV). *Trop. Plant Biol.* **2017**, *10*, 45–55. [CrossRef]
7. McNeil, M.D.; Bhuiyan, S.A.; Berkman, P.J.; Croft, B.J.; Aitken, K.S. Analysis of the resistance mechanisms in sugarcane during *Sporisorium scitamineum* infection using RNA-seq and microscopy. *PLoS ONE* **2018**, *13*, e0197840. [CrossRef]
8. Chu, N.; Zhou, J.-R.; Fu, H.-Y.; Huang, M.-T.; Zhang, H.-L.; Gao, S.-J. Global Gene Responses of Resistant and Susceptible Sugarcane Cultivars to Acidovorax avenae subsp. avenae Identified Using Comparative Transcriptome Analysis. *Microorganisms* **2020**, *8*, 10. [CrossRef]
9. Ntambo, M.S.; Meng, J.-Y.; Rott, P.C.; Henry, R.J.; Zhang, H.-L.; Gao, S.-J. Comparative Transcriptome Profiling of Resistant and Susceptible Sugarcane Cultivars in Response to Infection by Xanthomonas albilineans. *Int. J. Mol. Sci.* **2019**, *20*, 6138. [CrossRef]
10. Ma, Y.; Yu, H.; Lu, Y.; Gao, S.; Fatima, M.; Ming, R.; Yue, J. Transcriptome analysis of sugarcane reveals rapid defense response of SES208 to Xanthomonas albilineans in early infection. *BMC Plant Biol.* **2023**, *23*, 52. [CrossRef]

11. Wang, L.; Li, S. Salicylic acid-induced heat or cold tolerance in relation to Ca^{2+} homeostasis and antioxidant systems in young grape plants. *Plant Sci.* **2006**, *170*, 685–694. [CrossRef]

12. Mahesh, H.M.; Murali, M.; Pal, M.A.C.; Melvin, P.; Sharada, M.S. Salicylic acid seed priming instigates defense mechanism by inducing PR-Proteins in *Solanum melongena* L. upon infection with *Verticillium dahliae* Kleb. *Plant Physiol. Biochem.* **2017**, *117*, 12–23. [CrossRef] [PubMed]

13. Deenamo, N.; Kuyyogsuy, A.; Khompatara, K.; Chanwun, T.; Ekchaweng, K.; Churngchow, N. Salicylic acid induces resistance in rubber tree against *Phytophthora palmivora*. *Int. J. Mol. Sci.* **2018**, *19*, 1883. [CrossRef] [PubMed]

14. Li, T.; Huang, Y.; Xu, Z.S.; Wang, F.; Xiong, A.S. Salicylic acid-induced differential resistance to the Tomato yellow leaf curl virus among resistant and susceptible tomato cultivars. *BMC Plant Biol.* **2019**, *19*, 173. [CrossRef] [PubMed]

15. Liu, Y.; Luo, C.; Liang, R.; Lan, M.; Yu, H.; Guo, Y.; Chen, S.; Lu, T.; Mo, X.; He, X. Genome-wide identification of the mango CONSTANS (CO) family and functional analysis of two MiCOL9 genes in transgenic Arabidopsis. *Front. Plant Sci.* **2022**, *13*, 1028987. [CrossRef] [PubMed]

16. Liu, R.; Lu, J.; Xing, J.; Xue, L.; Wu, Y.; Zhang, L. Characterization and functional analyses of wheat TaPR1 genes in response to stripe rust fungal infection. *Sci. Rep.* **2023**, *13*, 3362. [CrossRef]

17. Chen, N.; Shao, Q.; Xiong, Z. Isolation and characterization of a pathogenesis-related protein 1 (SlPR1) gene with induced expression in tomato (*Solanum lycopersicum*) during *Ralstonia solanacearum* infection. *Gene* **2023**, *855*, 147105. [CrossRef]

18. Yuan, W.; Jiang, T.; Du, K.; Chen, H.; Cao, Y.; Xie, J.; Li, M.; Carr, J.P.; Wu, B.; Fan, Z.; et al. Maize phenylalanine ammonia-lyases contribute to resistance to *Sugarcane mosaic virus* infection, most likely through positive regulation of salicylic acid accumulation. *Mol. Plant Pathol.* **2019**, *20*, 1365–1378. [CrossRef]

19. Ling, H.; Huang, N.; Wu, Q.; Su, Y.; Peng, Q.; Ahmed, W.; Gao, S.; Su, W.; Que, Y.; Xu, L. Transcriptional insights into the sugarcane-sorghum mosaic virus interaction. *Trop. Plant Boil.* **2018**, *11*, 163–176. [CrossRef]

20. Xu, H.M.; He, E.Q.; Yang, Z.L.; Bi, Z.W.; Bao, W.Q.; Sun, S.R.; Lu, J.J.; Gao, S.J. Phylogeny and genetic divergence among sorghum mosaic virus isolates infecting sugarcane. *Plants* **2023**, *12*, 3759. [CrossRef]

21. Jini, D.; Joseph, B. Physiological mechanism of salicylic acid for alleviation of salt stress in rice. *Rice Sci.* **2017**, *24*, 97–108. [CrossRef]

22. Gunes, A.; Inal, A.; Alpaslan, M.; Eraslan, F.; Bagci, E.G.; Cicek, N. Salicylic acid induced changes on some physiological parameters symptomatic for oxidative stress and mineral nutrition in maize (*Zea mays* L.) grown under salinity. *J. Plant Physiol* **2007**, *164*, 728–736. [CrossRef] [PubMed]

23. Zagier, S.; Fadhal, F.A.; Alisawi, O. Control Fig Mosaic Virus By plant extracts with salicylic acid. *IOP Conf. Ser. Earth Environ. Sci.* **2021**, *790*, 012058. [CrossRef]

24. Suharti, W.S.; Leana, N.W.A. Induce resistance of rice plants against bacterial leaf blight by using salicylic acid application. *IOP Conf. Ser. Earth Environ. Sci.* **2021**, *746*, 012003.

25. Zhang, X.; Dong, J.; Liu, H.; Wang, J.; Qi, Y.; Liang, Z. Transcriptome sequencing in response to salicylic acid in Salvia miltiorrhiza. *PLoS ONE* **2016**, *11*, e0147849. [CrossRef]

26. Canet, J.V.; Dobón, A.; Roig, A.; Tornero, P. Structure-function analysis of *npr1* alleles in Arabidopsis reveals a role for its paralogs in the perception of salicylic acid. *Plant Cell Environ.* **2010**, *33*, 1911–1922. [CrossRef]

27. Ülker, B.; Shahid, M.; Somssich, I.E. The WRKY70 transcription factor of Arabidopsis influences both the plant senescence and defense signaling pathways. *Planta* **2007**, *226*, 125–137. [CrossRef]

28. Dong, J.; Chen, C.; Chen, Z. Expression profiles of the Arabidopsis WRKY gene superfamily during plant defense response. *Plant Mol. Boil.* **2003**, *51*, 21–37. [CrossRef]

29. Chu, N.; Zhou, J.R.; Rott, P.C.; Li, J.; Fu, H.Y.; Huang, M.T.; Zhang, H.L.; Gao, S.J. ScPR1 plays a positive role in the regulation of resistance to diverse stresses in sugarcane (*Saccharum* spp.) and *Arabidopsis thaliana*. *Ind. Crops. Prod.* **2022**, *180*, 114736. [CrossRef]

30. Shi, Z.; Maximova, S.; Liu, Y.; Verica, J.; Guiltinan, M.J. The salicylic acid receptor NPR3 is a negative regulator of the transcriptional defense response during early flower development in Arabidopsis. *Mol. Plant* **2013**, *6*, 802–816. [CrossRef]

31. Zhao, J.Y.; Chen, J.; Shi, Y.; Fu, H.Y.; Huang, M.T.; Rott, P.C.; Gao, S.J. Sugarcane responses to two strains of *Xanthomonas albilineans* differing in pathogenicity through a differential modulation of salicylic acid and reactive oxygen species. *Front. Plant Sci.* **2022**, *13*, 1087525. [CrossRef] [PubMed]

32. Chen, C.; Chen, Z. Potentiation of developmentally regulated plant defense response by AtWRKY18, a pathogen-induced Arabidopsis transcription factor. *Plant Physiol.* **2002**, *129*, 706–716. [CrossRef] [PubMed]

33. Wang, H.; Cheng, X.; Yin, D.; Chen, D.; Luo, C.; Liu, H.; Huang, C. Advances in the Research on Plant WRKY Transcription Factors Responsive to External Stresses. *Curr. Issues Mol. Biol.* **2023**, *45*, 2861–2880. [CrossRef] [PubMed]

34. Phukan, U.J.; Jeena, G.S.; Shukla, R.K. WRKY transcription factors: Molecular regulation and stress responses in plants. *Front. Plant Sci.* **2016**, *7*, 760. [CrossRef]

35. Yang, L.; Miao, L.; Gong, Q.; Guo, J. Advances in studies on transcription factors in regulation of secondary metabolites in Chinese medicinal plants. *Plant Cell Tissue Organ Cult. (PCTOC)* **2022**, *151*, 1–9. [CrossRef]

36. Bolger, A.M.; Lohse, M.; Usadel, B. Trimmomatic: A flexible trimmer for Illumina sequence data. *Bioinformatics* **2014**, *30*, 2114–2120. [CrossRef]

37. Kim, D.; Paggi, J.M.; Park, C.; Bennett, C.; Salzberg, S.L. Graph-based genome alignment and genotyping with HISAT2 and HISAT-genotype. *Nat. Biotech.* **2019**, *37*, 907–915. [CrossRef]

38. Pertea, M.; Pertea, G.M.; Antonescu, C.M.; Chang, T.C.; Mendell, J.T.; Salzberg, S.L. StringTie enables improved reconstruction of a transcriptome from RNA-seq reads. *Nat. Biotech.* **2015**, *33*, 290–295. [CrossRef]
39. Love, M.I.; Huber, W.; Anders, S. Moderated estimation of fold change and dispersion for RNA-seq data with DESeq2. *Genome Boil.* **2014**, *15*, 550. [CrossRef]
40. Wu, T.; Hu, E.; Xu, S.; Chen, M.; Guo, P.; Dai, Z.; Feng, T.; Zhou, L.; Tang, W.; Zhan, L.I.; et al. ClusterProfiler 4.0: A universal enrichment tool for interpreting omics data. *Innovation* **2021**, *2*, 100141. [CrossRef]

Article

Relationship between Sugarcane *eIF4E* Gene and Resistance against *Sugarcane Streak Mosaic Virus*

Hongli Shan, Du Chen, Rongyue Zhang, Xiaoyan Wang, Jie Li, Changmi Wang, Yinhu Li and Yingkun Huang *

Sugarcane Research Institute, Yunnan Academy of Agricultural Science, Yunnan Key Laboratory of Sugarcane Genetic Improvement, Kaiyuan 661699, China; shhldlw@163.com (H.S.); lbxnyzhfwzx@163.com (D.C.); rongyuezhang@hotmail.com (R.Z.); xiaoyanwang402@sina.com (X.W.); lijie0988@163.com (J.L.); wcmlucky@163.com (C.W.); liyinhu93@163.com (Y.L.)
* Correspondence: huangyk64@163.com; Tel.: +86-873-7227017; Fax: +86-873-7222489

Abstract: Sugarcane mosaic disease, mainly caused by *Sugarcane streak mosaic virus* (*SCSMV*), has serious adverse effects on the yield and quality of sugarcane. Eukaryotic translation initiation factor 4E (*eIF4E*) is a natural resistance gene in plants. The *eIF4E*-mediated natural recessive resistance results from non-synonymous mutations of the eIF4E protein. In this study, two sugarcane varieties, CP94-1100 and ROC22, were selected for analysis of their differences in resistance to *SCSMV*. Four-base missense mutations in the ORF region of *eIF4E* resulted in different conserved domains. Therefore, the differences in resistance to *SCSMV* are due to the inherent differences in *eIF4E* of the sugarcane varieties. The coding regions of *eIF4E* included 28 SNP loci and no InDel loci, which were affected by negative selection and were relatively conserved. A total of 11 haploids encoded 11 protein sequences. Prediction of the protein spatial structure revealed three non-synonymous mutation sites for amino acids located in the cap pocket of eIF4E; one of these sites existed only in a resistant material (Yuetang 55), whereas the other site existed only in a susceptible material (ROC22), suggesting that these two sites might be related to the resistance to *SCSMV*. The results provide a strong basis for further analysis of the functional role of *eIF4E* in regulating mosaic resistance in sugarcane.

Keywords: eukaryotic translation initiation factor 4E (eIF4E); *sugarcane streak mosaic virus* (*SCSMV*); resistance difference; amino acid variation; non-synonymous mutation

Citation: Shan, H.; Chen, D.; Zhang, R.; Wang, X.; Li, J.; Wang, C.; Li, Y.; Huang, Y. Relationship between Sugarcane *eIF4E* Gene and Resistance against *Sugarcane Streak Mosaic Virus*. *Plants* **2023**, *12*, 2805. https://doi.org/10.3390/plants12152805

Academic Editor: Sergey Morozov

Received: 1 June 2023
Revised: 15 July 2023
Accepted: 25 July 2023
Published: 28 July 2023

1. Introduction

Sugarcane (*Saccharum officinarum* L.) is an important sugar and energy crop in China and the world and is the raw material for 78% of world sugar production. In sugarcane production, sugarcane mosaic disease (SMD) is one of the most common and serious viral diseases, named sugarcane cancer, which seriously affects the yield and quality of sugarcane. *Sugarcane streak mosaic virus* (*SCSMV*), which has been identified as a major virus causing SMD in recent years, belongs to the genus *Poacevirus* in the *Potyviridae* family [1–4]. The genome of *SCSMV* is composed of a sense single-stranded RNA, about 9.7–10.0 kb in size. Eleven functional proteins are produced from the N-terminal to C-terminal by hydrolase cutting, including Protein 1 (P1), Helper component-proteinase (HC-Pro), Protein 3 (P3), P3N-PIPO (P3 N-terminal fused with Pretty Interesting *Potyviridae* ORF), the first protein of 6 kD (6K1), Cylindrical inclusion, the second protein of 6 kD (6K2), Viral protein genome-linked (VPg), the protease of Nuclear Inclusion protein a (NIa-Pro), Nuclear Inclusion protein b (NIb), and Coat protein (CP). CP is highly conserved, and its structure affects the ability for virus assembly and aphid transmission [5,6]. VPg protein performs a variety of functions in the process of virus infection [7].

The process of viral translation from mRNA to protein in eukaryotes involves more than 12 eukaryotic translation initiation factors (eIFs). Among these, the eukaryotic translation initiation factor, 4E (eIF4E), is one of the most important and it plays an important role in viral translation and the synthesis of proteins [8]. One of the key links in the establishment of systemic infection in hosts by viruses in the *Potyviridae* family is using eIF4E of the host to complete the translation. Thus, eIF4E is an indispensable host factor for potato virus Y genome translation. In most single-stranded RNA viruses, the VPg is encoded by the virus-like cap structure which interacts with the host factor, eIF4E, to initiate the translation of the viral genome [9,10]. When eIF4E mutates, its interaction with the virus VPg is eliminated or weakened, and the infectivity of the virus is lost [11,12]. Lin et al. [13] found four predominant mutations in *eIF4E1* of tobacco-resistant sources with PVY resistance. The mutations included single base insertion, substitution, and partial or complete deletion of gene fragments. Plant disease resistance induced by *eIF4E* mutants has been observed in a variety of plants. The difference between *eIF4E* in susceptible plants and *eIF4E* in resistant plants was only the difference of a single or a few amino acids in their protein sequence, but it has led to plants resistant to multiple viruses or multiple strains of a virus.

Breeding or utilizing resistant varieties is the most economical and effective means of controlling sugarcane mosaic disease. Using plant-derived genes to conduct plant antiviral research enhances biosafety [14]. Previous studies have shown that *eIF4E*-mediated natural recessive resistance is derived from the non-synonymous mutation of the eIF4E protein. Therefore, in cultivating resistant varieties, it is necessary to explore the relationship between the mutation site of *eIF4E* and disease resistance in the host. However, in sugarcane, different varieties show different levels of resistance to sugarcane mosaic disease in the field. There is no report of resistance caused by the sugarcane resistance gene, or pathogenicity change caused by the natural variation of the virus. The polymorphism of the *eIF4E* coding region in sugarcane is unknown. There are two sugarcane varieties with significant differences in resistance against sugarcane mosaic disease in the field. They are used as samples in this study. Sequence analysis was performed for both the pathogenic virus and translation initiation factor eIF4E of the host. The aim was to determine whether the difference in resistance was caused by the differences in *eIF4E* sequences between the sugarcane varieties. The non-synonymous mutation sites of the eIF4E amino acid in susceptible and resistant sugarcane varieties were explored by cloning the coding sequences of *eIF4E* in 11 different resistant materials and analyzing the polymorphisms in *eIF4E* in resistant and susceptible sugarcane varieties. The results provide guidance for the molecular breeding of sugarcane resistance to mosaic disease with *eIF4E* as the target.

2. Results

2.1. Differences in Resistance of Sugarcane Varieties against Sugarcane Mosaic Disease

Many years of field investigation of naturally occurring sugarcane diseases have shown that four resistant varieties, including CP94-1100, and seven susceptible varieties, including ROC22, have significant and stable resistance against sugarcane mosaic disease (Table 1). Of these, the incidence of ROC22 mosaic disease in different sugarcane-growing areas in Yunnan (China) reached 100%, and the whole plant had serious mosaic infection with a coverage area of 100%. The CP94-1100 sugarcane grew normally, the whole plant was relatively strong, and the leaves were clean, having no mosaic symptoms (Figure 1). CP94-1100 and ROC22 were selected as representatives of different resistant varieties, for the study of the source of resistance.

Table 1. Samples and phenotype of resistance for *sugarcane streak mosaic virus* in sugarcane.

Number	Sample	Phenotype	Number	Sample	Phenotype
1	CP 94-1100	R	7	Guitang 08-1180	S
2	Funong 30	R	8	ROC 10	S
3	Yuetang 55	R	9	Q 124	S
4	CP 89-2377	R	10	Funong 02-6427	S
5	Funong 09-2201	S	11	ROC 22	S
6	Yunrui 10-701	S			

Figure 1. Symptoms of sugarcane mosaic disease in different varieties of sugarcane in Yunnan Province, China. From left to right, varieties are CP94-1100 (**a**), and ROC22 (**b**).

Highly resistant to sugarcane mosaic disease (R) and highly susceptible to sugarcane mosaic disease (S).

2.2. Detection and Sequence Analysis of SCSMV-CP and SCSMV-VPg in Different Resistant Sugarcane Varieties

SCSMV was found in different resistant sugarcane varieties (CP94-1100 and ROC22) through the detection of the *SCSMV-CP* and *SCSMV-VPg*, indicating that both varieties were infected by *SCSMV*. Two *SCSMV-CP* sequences obtained using RT-PCR were submitted to the GenBank database with the accession numbers ON505212 (CP94-1100) and ON505213 (ROC22). The sequence length was 938 bp, and the homologies with the published *SCSMV-CP* sequence (accession number JF488065) were 99.9% and 99.6%, respectively. The nucleotide sequence and coding proteins were basically the same. The *SCSMV-VPg* sequences were submitted to the GenBank database with the accession numbers MZ605863 and MZ605884 for CP94-1100 and ROC22, respectively. The sequence length of *SCSMV-VPg* was 594 bp, and the homologies were 99.7% and 99.0% with the published gene sequence of *SCSMV-VPg* (accession number JF488065). The nucleotide sequence and coding protein were basically the same (Table 2), indicating that the *SCSMV* isolates of the two varieties were the same. The differences in the symptoms in sugarcane were attributed to the differences in sugarcane varieties.

Table 2. Sequence analysis of SCSMV and eIF4E for sugarcane CP94-1100 and ROC22.

Varieties	CP		VPg		eIF4E	
	Nucleotide	Amino Acid	Nucleotide	Amino Acid	Nucleotide	Amino Acid
ROC22	938	240	594	198	663	220
CP94-1100	938	240	594	198	663	220
Sequence identities/%	99.5	99.2	98.2	99.0	99.1	98.2

Note: *Sequence identities*—The homologies of the varieties ROC22 and CP94-1100 on nucleotide sequence and amino acid sequence, respectively.

The standard curve for detecting SCSMV virus accumulation was Y = 2.3421X + 23.155, $R^2 \geq 0.998$, E $\geq$ 98.8%. It met the requirements of an ideal standard curve ($R^2 \geq 0.98$, 85% $\leq$ E $\leq$ 110) [15]. RT-qPCR showed the accumulation of SCSMV in susceptible varieties was greater than that in resistant varieties. The accumulation of SCSMV was lowest in CP94-1100 and highest in ROC22 (Figure 2).

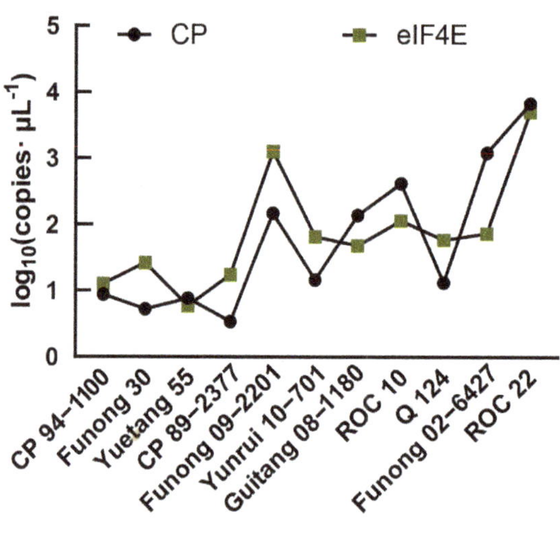

Figure 2. Quantification of accumulation levels for SCSMV and eIF4E.

2.3. Cloning and Sequence Analysis of eIF4E from Different Varieties

2.3.1. Detection of eIF4E from Different Varieties

The *eIF4E* sequences of two sugarcane varieties (CP94-1100 and ROC22) were cloned, sequenced, and blasted. The homology with the coding region sequence of the published *eIF4E* (accession number KX757017) was 99.4% and 99.2%. The obtained sequence was confirmed as the coding region sequence of sugarcane *eIF4E*, named *SceIF4E-R* and *SceIF4E-S* and submitted to the GenBank database, with the accession numbers MT680884 (CP94-1100) and MT680894 (ROC22). The nucleotide and amino acid sequences of the *eIF4E* in CP94-1100 and ROC22 were significantly different, and the nucleotide sequence in the coding region was 663 bp, with a consistency of 99.1%. The amino acid sequence was 220 aa, and the consistency was 98.2% (Table 2). This result indicated that the amino acid properties of eIF4E varied greatly in the two sugarcane varieties with different levels of resistance to *SCSMV*. These results show that the difference in resistance was derived from the variation in *eIF4E*.

The standard curve for detecting eIF4E accumulation was Y = 2.2802X + 24.615, $R^2 \geq 0.994$, E $\geq$ 1.03%. It met the requirements of an ideal standard curve ($R^2 \geq 0.98$, 85% $\leq$ E $\leq$ 110) [15]. RT-qPCR showed the accumulation of eIF4E in susceptible varieties was greater than that in resistant varieties. The accumulation of eIF4E was the least in CP94-1100 and the most in ROC22 (Figure 2).

2.3.2. Bioinformatic Analysis of Coding Region Sequence of eIF4E in Different Varieties

Bioinformatic analysis of the *eIF4E* sequences of two sugarcane varieties with different levels of resistance (CP94-1100 and ROC22) showed that there were base differences between the two varieties at 60 bp, 180 bp, 328 bp, 383 bp, 496 bp, and 661 bp in the ORF region. Of these, four sites, 60 bp, 180 bp, 328 bp, and 496 bp, led to the missense mutation of amino acids (Figure 3). These results indicate that the protein character differed between the varieties. The physicochemical properties of the proteins encoded by *SceIF4E-R* and

SceIF4E-S were predicted online. The results show that the molecular formulas of SceIF4E-R and SceIF4E-S were $C_{1093}H_{1652}N_{306}O_{328}S_7$ and $C_{1092}H_{1649}N_{307}O_{330}S_7$, respectively. The relative molecular weights were 24.55 kDa and 24.58 kDa, respectively. The theoretical isoelectric points were 5.70 for both, and the instability coefficients were 29.95 and 32.29, respectively. The total average hydrophilicity values were -0.617 and -0.662, respectively, indicating that both proteins were stable and hydrophilic. The SceIF4E-R and SceIF4E-S proteins were both present in the cytoplasm, which had no transmembrane structure and signal peptide. These results indicate that SceIF4E is neither a membrane protein nor a secretory protein (Figure 4). The results of protein secondary and tertiary structure prediction show that SueIF4E-R was composed of a random coil (53.18%), alpha-helix (28.18%), extension strand (14.55%), and bate turn (4.09%), while SceIF4E-S was composed of a random coil (50.45%), alpha-helix (28.64%), extension strand (16.36%), and bate turn (4.55%) (Figure 5). Conservative domain analysis showed that both SceIF4E-R and SceIF4E-S had an IF4E conserved domain, belonging to the IF4E family of proteins; however, SceIF4E-R also contained a CDC 33 or PTZ00040 domain (Figure 6). These domains suggested that SceIF4E-R and SceIF4E-S proteins had the same function as the IF4E family of proteins, but the functions of SceIF4E-R and SceIF4E-S proteins differed.

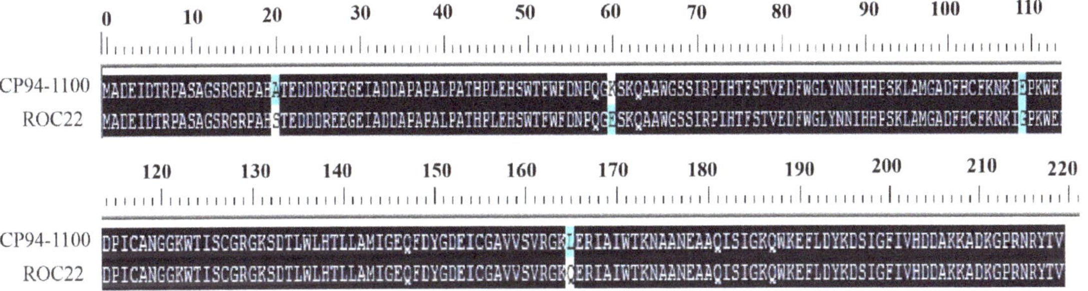

Figure 3. Alignment analysis of amino acid sequences of eIF4E in sugarcane.

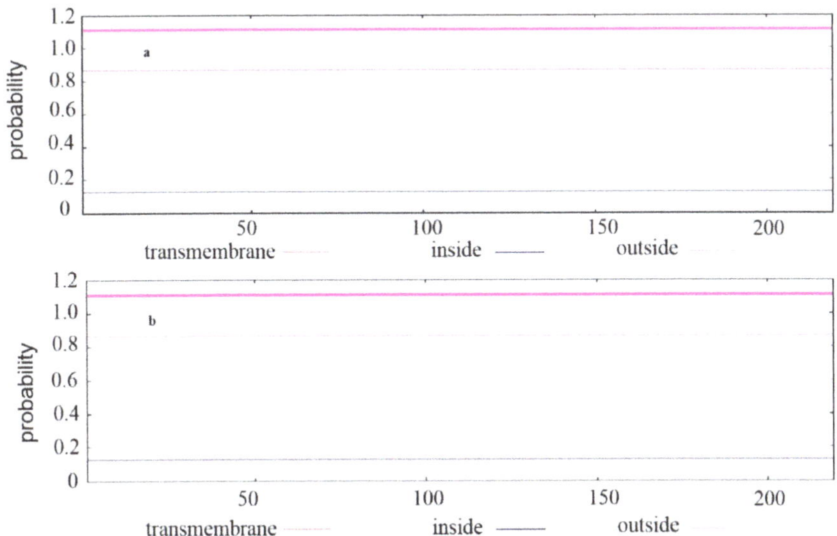

Figure 4. Prediction and analysis of transmembrane domains of SceIF4E. SceIF4E-R (**a**) and SceIF4E-S (**b**).

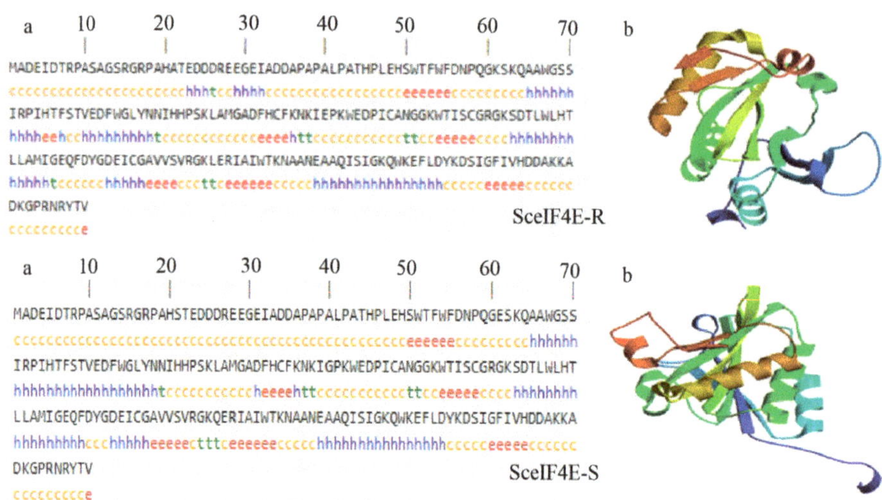

Figure 5. Protein secondary and tertiary structures of SceIF4E. Secondary structure (**a**): orange c represents random coil, red e represents extended strand, blue h represents alpha helix, and green t represents beta turn; and tertiary structure (**b**). Blue h: alpha helix; red e: extended strand; orange c: random coil; green t: bate turn.

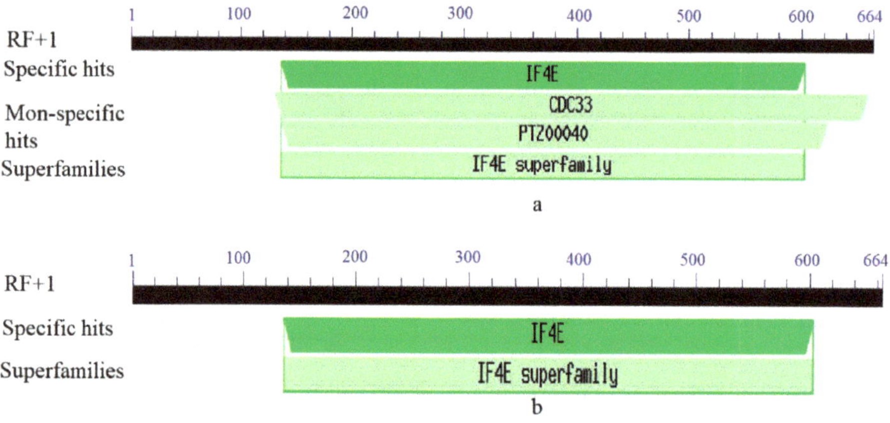

Figure 6. Conserved domain of SceIF4E. SceIF4E-R (**a**) and SceIF4E-S (**b**).

To further clarify the functions of *SceIF4E-R* and *SceIF4E-S* and understand their phylogenetic relationships with eIF4E genes in other plants, the amino acid sequences of *eIF4E* in 10 plants were downloaded from the NCBI database (https://www.ncbi.nlm.nih.gov/, accessed on 11 June 2022) for analysis and construction of the phylogenetic tree. The 10 plants were sugarcane (*Saccharum hybrid* cultivars), sorghum (*Sorghum bicolor*), maize (*Zea mays*), Chenopodium album (*Panicum miliaceum*), millet (*Setaria italica*), *Saccharum spontaneum*, rice (*Oryza sativa*), barley (*Hordeum vulgare*), pineapple (*Ananas comosus*), and sweet potato (*Ipomoea triloba*). The phylogenetic tree was divided into three groups. The first group contained only pineapple *eIF4E*. The second group included *SceIF4E-R*, *SceIF4E-S*, and the *eIF4E* of eight *Gramineae* plants. The third group contained sweet potato *eIF4E* (Figure 7). The phylogenetic tree showed that all *SceIF4E* clustered in the same branch, including the *eIF4E* of sugarcane (*S. hybrid* cultivar, ASU92319) and *S. spontaneum* (AWA44706), and *SceIF4E-R* had the closest genetic relationship with *S. spontaneum* (AWA44706).

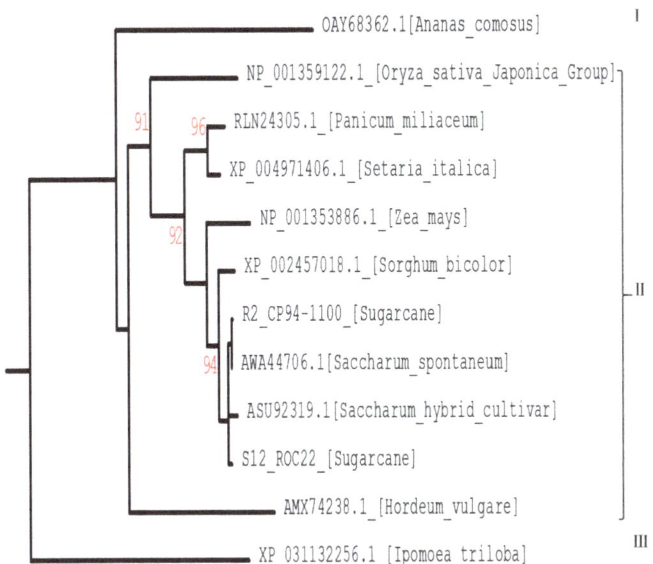

Figure 7. Phylogenetic tree of eIF4E protein in sugarcane and ten plants. Red numbers on branches indicate bootstrap values based on 1000 replicates (values < 90 are not shown).

2.4. Polymorphism Analysis of the Coding Region Sequence of eIF4E

2.4.1. Cloning of the Coding Region Sequence of eIF4E in Different Resistant Varieties

The coding region of *eIF4E* from eleven sugarcane materials comprising four resistant varieties, including CP94-1100, and seven susceptible varieties, including ROC22, was amplified using RT-PCR. Electrophoresis showed that the main bands of the PCR amplification products of all samples were clear, with no obvious heterozygous band. The size of the bands was about 660 bp, which was consistent with the expected size (Figure 8). BLAST results showed that the highest homology with sugarcane *eIF4E1* (KX757017) was 99.4%. The sequence similarity with sorghum *eIF4E1* (XM_002456973) was 97.6% and that with maize *eIF4E1* (NM_001366957) was 94.9%, indicating that the coding region sequence of sugarcane *eIF4E* was successfully obtained. It was submitted to GenBank with the accession number MT680884-MT680894.

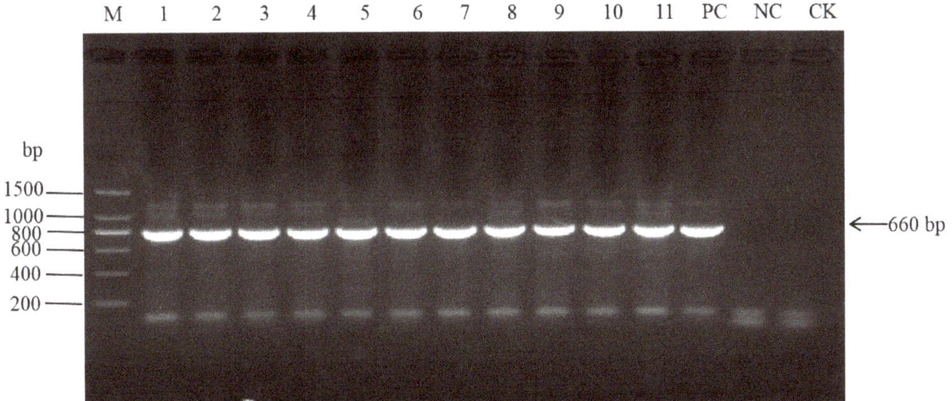

Figure 8. Detection of *eIF4E1* by PCR in different sugarcane varieties. From left to right, marker (M), samples from Nos. 1–11 (Lanes 1–11), positive control (PC), negative control (NC), blank control (CK).

2.4.2. Nucleotide Sequence Polymorphism of the Coding Region of eIF4E

The *eIF4E* coding region sequences of 11 sugarcane materials were analyzed with respect to the open reading frame (ORF) length, insertion–deletion (InDel), single nucleotide polymorphism (SNP), nucleotide diversity index (Pi), haplotype number, and diversity (Table 3). The ORF length of 11 materials was 663 bp, forming 11 haplotypes. Each haplotype contained one material, and the haplotype diversity was 1.000. There were 28 polymorphic loci, which were SNP loci with simplified information, and no InDel locus was found. The frequency of SNP loci was 1SNP/23.7 bp. Among the twenty-eight polymorphic variation sites, eight sites were the same in the *SCSMV* materials with high resistance and high susceptibility, six variation sites only existed in high resistance to the *SCSMV* materials, and fourteen variation sites only existed in high susceptibility to *SCSMV* materials. The Pi value of all tested materials was 0.00932. The Pi value of highly resistant material (0.00930) was less than that of highly susceptible material (0.00970) (Table 3), indicating that the genetic variation of highly resistant material was less than that of highly susceptible material.

Table 3. Analysis of polymorphism of *eIF4E* cDNA in sugarcane.

	Length	No. of Haplotypes	Haplotype Diversity (Hd ± SD)	Nucleotide Diversity (Pi ± SD)	No. of Polymorphic Sites (S)	Specific Polymorphic Sites
All samples	663	11	1.000 ± 0.039	0.00932 ± 0.0011	28	SNP3; SNP101; SNP172; SNP242; SNP383; SNP416; SNP419; SNP496
Highly resistant to SCSMV	663	4	1.000 ± 0.177	0.00930 ± 0.0020	14	SNP181; SNP193; SNP261; SNP406; SNP451; SNP614
Highly susceptible to SCSMV	663	7	1.000 ± 0.076	0.00970 ± 0.00153	22	SNP46; SNP60; SNP180; SNP304; SNP328; SNP415; SNP433; SNP437; SNP448; SNP493; SNP501; SNP554; SNP644; SNP662

The distribution of polymorphic sites in the *eIF4E* nucleotide sequence of the tested materials is shown in Figure 9. There were multiple peaks in the previous sequence, indicating that there were many regions where the sequence changed and there may be multiple hot mutation sites. The peak appears in the region of about 400 bp, indicating that the region around 400 bp was the enrichment region of nucleotide change.

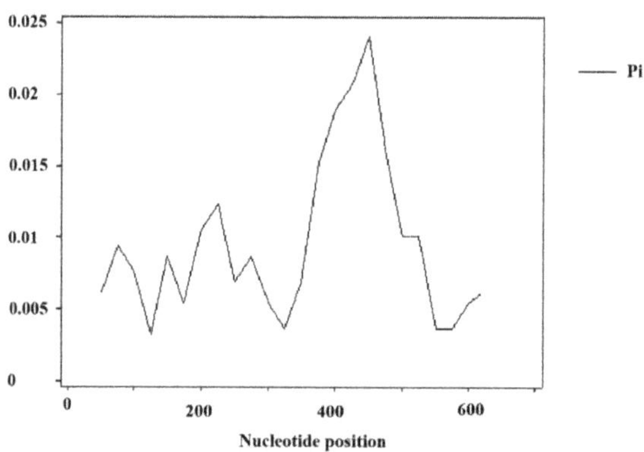

Figure 9. Distribution of the polymorphic sites of *eIF4E* in sugarcane.

2.4.3. Amino Acid Sequence Analysis of eIF4E Coding Region

The amino acid sequences corresponding to the coding region of *eIF4E* were predicted using the ORF finder (http://www.nibi.nlm.nih.gov/, accessed on 12 July 2022) and the interspecies variation of amino acid sequences of 11 sugarcane materials was analyzed using the DNAMAN software. The results show that all the samples encoded 220 amino acids. The interspecies variation analysis of amino acids showed that 203 amino acids were synonymous mutations, and 17 amino acids were non-synonymous mutations. The base mutations were all substitutions, and there was no insertion or deletion. A total of 11 proteins were encoded. Eighteen base mutations in 28 SNPs resulted in 17 amino acid non-synonymous mutations, which were SNP46, SNP60, SNP71, SNP172, SNP180, SNP193, SNP261, SNP304, SNP328, SNP406, SNP415, SNP416, SNP433, SNP448, SNP451, SNP493, SNP496, and SNP501. SNP415 and SNP416 resulted in amino acid non-synonymous mutations at the same site (Table 4). Of all the amino acid non-synonymous mutation sites, only four (SNP193, SNP261, SNP406, and SNP451) occurred in highly resistant *SCSMV* materials and only ten (SNP46, SNP60, SNP180, SNP304, SNP328, SNP415, SNP433, SNP448, SNP493, and SNP501) occurred in highly susceptible *SCSMV* materials. To evaluate the natural selection pressure of the eIF4E gene, the ratio of synonymous mutation (Ks) to non-synonymous mutation (Ka) (Ka/Ks) of *eIF4E* sequences obtained from all materials was analyzed, yielding a value of 0.3727 (less than 1), indicating that the eIF4E gene was subjected to negative selection pressure as a whole.

Table 4. The types of SNP mutation of *eIF4E* gene in sugarcane.

Number	Mutation	Polymorphic Type	Amino Acid Variations	Mutation Type	Material
1	SNP46	G/A	GLY15Asp	Non-synonymous mutation	Hap7
2	SNP60	G/T	Ala20Ser	Non-synonymous mutation	Hap7; Hap10; Hap11
3	SNP71	G/C	Glu23Asp	Non-synonymous mutation	Hap1; Hap2; Hap3; Hap4; Hap5; Hap6; Hap7; Hap8; Hap9; Hap10; Hap11
4	SNP101	T/C	Asp33Asp	Synonymous mutation	Hap3; Hap8
5	SNP172	A/C	Gln58Pro	Non-synonymous mutation	Hap1; Hap2; Hap3; Hap4; Hap5; Hap6; Hap7; Hap8; Hap9; Hap10; Hap11
6	SNP180	A/G	Lys60Glu	Non-synonymous mutation	Hap11
7	SNP181	A/G	Ser61Ser	Synonymous mutation	Hap2
8	SNP193	C/G	Ala64Val	Non-synonymous mutation	Hap3
9	SNP242	G/A	Glu80Glu	Synonymous mutation	Hap2; Hap4; Hap7; Hap9
10	SNP261	A/G	Asn87Asp	Non-synonymous mutation	Hap2
11	SNP304	T/C	Phe101Ser	Non-synonymous mutation	Hap9
12	SNP328	A/G	Glu109Gly	Non-synonymous mutation	Hap11
13	SNP383	T/C	Cys127Cys	Synonymous mutation	Hap1; Hap7
14	SNP406	T/C	Leu135Pro	Non-synonymous mutation	Hap2
15	SNP415	A/G	His138Arg	Non-synonymous mutation	Hap5
16	SNP416	C/T	His138Arg	Non-synonymous mutation	Hap4; Hap9
17	SNP419	T/C	Th139Thr	Synonymous mutation	Hap3; Hap5
18	SNP433	T/C	Ile144Thr	Non-synonymous mutation	Hap5
19	SNP437	C/T	Gly145Gly	Synonymous mutation	Hap5
20	SNP448	A/G	Asp149Gly	Non-synonymous mutation	Hap8
21	SNP451	A/G	Tye150Cys	Non-synonymous mutation	Hap2
22	SNP493	A/G	Lys164Arg	Non-synonymous mutation	Hap7
23	SNP496	A/T A/G	Gln165Leu Gln165Arg	Non-synonymous mutation	Hap1; Hap7; Hap8
24	SNP501	A/G	Arg167Gly	Non-synonymous mutation	Hap6
25	SNP554	T/C	Ile184Ile	Synonymous mutation	Hap8
26	SNP614	C/T	Asp204Asp	Synonymous mutation	Hap3
27	SNP644	G/A	Arg214Arg	Synonymous mutation	Hap9
28	SNP662	G/A	Val220Val	Synonymous mutation	Hap11

2.5. Spatial Structure of eIF4E Protein

Wheat eIF4E protein (2idr.1.A) was selected as the model protein using SWISS-MODEL software to predict the spatial structures of 11 proteins encoded by *eIF4E*. The spatial structures of 11 proteins showed high similarity. Previous studies showed that the cap structure of eIF4E played an indispensable role in protein function [16,17]. The key regions of the cap structure are shown in the A and B regions of Figure 9. Region A is located in the 57–73 amino acid site range of eIF4E, and region B is located in the 107–118 amino acid site range. In this study, the amino acid variation sites, 58, 60, and 64 sites, were all located in region A. Of these, the 58 amino acid variation site was the variation site of all materials relative to the control sequence, KX757017. The 64 amino acid variation site was only found in Protein 3 encoded by the eIF4E gene of Yuetang 55, an extremely resistant material, and the amino acid variation was Ala64Val. The 60 amino acid variation site was only found in Protein 11 encoded by the eIF4E gene of ROC22, an extremely susceptible material, and the amino acid variation was Lys60Glu (Tables 4 and 5, Figure 10).

Table 5. Amino acid variation of eIF4E protein sequences between KX757017 and sugarcane used in this study.

Varieties	Haplotype	Protein	Amino Acid Variation Sites																
			15	20	23	58	60	64	87	101	109	135	138	144	149	150	164	165	167
KX757017			G	A	E	Q	K	A	N	F	E	L	H	L	D	Y	K	Q	R
CP 94-1100	Hap1	Protein-1	—	—	D	P	—	—	—	—	—	—	—	—	—	—	—	L	—
Funong 30	Hap2	Protein-2	—	—	D	P	—	—	D	—	—	P	—	—	—	C	—	—	—
Yuetang 55	Hap3	Protein-3	—	—	D	P	—	V	—	—	—	—	—	—	—	—	—	—	—
CP 89-2377	Hap4	Protein-4	—	—	D	P	—	—	—	—	—	—	—	—	—	—	—	—	—
Funong 09-2201	Hap5	Protein-5	—	—	D	P	—	—	—	—	—	—	R	T	—	—	—	—	—
Yunrui 10-701	Hap6	Protein-6	—	S	D	P	—	—	—	—	—	—	—	—	—	—	—	—	G
Guitang 08-1180	Hap7	Protein-7	D	—	D	P	—	—	—	—	—	—	—	—	—	—	R	L	—
ROC 10	Hap8	Protein-8	—	—	D	P	—	—	—	—	—	—	—	—	G	—	—	R	—
Q 124	Hap9	Protein-9	—	—	D	P	—	—	—	S	—	—	—	—	—	—	—	—	—
Funong 02-6427	Hap10	Protein-10	—	S	D	P	—	—	—	—	—	—	—	—	—	—	—	—	—
ROC 22	Hap11	Protein-11	—	S	D	P	E	—	—	—	G	—	—	—	—	—	—	—	—

The same amino acids with KX75701 (—).

Figure 10. Protein 3D structure and mutation sites of *eIF4E*-coding proteins.

3. Discussion

SCSMV is one of the main viral pathogens of sugarcane mosaic disease. Since the first detection of SCSMV in Yunnan in 2011 [18], it has spread rapidly, with strong pathogenicity, and has become widespread in all sugarcane areas in China, causing serious harm. A detection rate of 100% has revealed it as the main pathogen of sugarcane mosaic disease in China [19,20]. SCSMV is a member of the potato virus Y family, and the nucleotide sequence variation of its cp gene is highly limited [21]. The VPg protein encoded by the SCSMV genome is a key protein in plants infected by the virus [22]. Therefore, the CP and VPg proteins of the virus are essential for successfully infecting the host. In this study, the cp and vpg sequences of SCSMV in different varieties infected by SCSMV were analyzed to determine differences in the pathogenicity of SCSMV among different varieties. The results show that the nucleotide and amino acid sequences of the cp and vpg genes of SCSMV belonging to the same isolate were highly consistent (more than 99.0%) in two sugarcane varieties with different resistances. These results indicate that the differences in resistance between the varieties were not caused by the differences in SCSMV. The results of this study are consistent with those reported by Wang et al. [20] and Zhang et al. [23]. The Chinese isolates of SCSMV had obvious geographical variations in characteristics, but there was no obvious genetic differentiation within the population. Quantification of SCSMV accumulation levels in different varieties show that the accumulation of SCSMV in susceptible varieties was greater than that in resistant varieties. Furthermore, the eIF4E accumulation levels in different varieties were quantified. It was also found that the accumulation level of eIF4E in susceptible varieties was also greater than that in resistant varieties. These results indicated that eIF4E may contribute to the accumulation of SCSMV in sugarcane.

The virus has a simple structure, and lacks a cap structure that is required to start translation. After the virus enters a plant, the VPg encoded by the virus interacts with the eIF4E of the plant to complete virus replication and infection [24]. Therefore, *eIF4E* is considered to be a recessive resistance gene against viruses in many plants. The genome translation efficiency of viruses in the *Potyviridae* family (including potato virus Y and barley yellow mosaic virus) depends on the interaction level of VPg and eIF4E [25,26]. Studies have shown that the introduction of *eIF4E* can enable the replication of wheat mosaic virus (WYMV) in barley protoplasts [27]. Therefore, the resistance of different sugarcane varieties is considered to be caused by differences in the translation efficiency of the *eIF4E* sequences of the varieties. The results of this study show that the amino acid properties of *eIF4E* were quite different between the two sugarcane varieties with different levels of resistance to *SCSMV*, suggesting that the ability of eIF4E to recruit subsequent translation initiation factors and ribosome subunits differed, leading to the difference in resistance derived from the variation of *eIF4E*. Further bioinformatics analysis of the *eIF4E*-ORF of two sugarcane varieties with different levels of resistance (CP94-1100 and ROC22) showed that the physicochemical properties of the two proteins were basically the same, but there were four-base missense mutations in the ORF region, resulting in different conservative domains. Although SceIF4E-R and SceIF4E-S belonged to the IF4E family of proteins and had the same function as the IF4E family of proteins, SceIF4E-R contained CDC 33 or PTZ00040 domains, indicating that SceIF4E-R may have other functions. In phylogenetic analysis, SceIF4E-R and SceIF4E-S clustered with eight *Gramineae* crops, in different groups from pineapple and sweet potato, indicating that the eIF4E had obvious species differences and high homology within species. The results of this study were consistent with the results of Geng et al. [28] on *eIF4E* sequence analysis of wheat varieties. SceIF4E-R is closely related to *S. spontaneum*, indicating that they have similar functions. Huang et al. [29] analyzed the phenotypic genetic diversity of the agronomic traits of *S. spontaneum* resources at home and abroad. The results show that *S. spontaneum* is highly resistant to mosaic disease. Li et al. [30] pointed out that *S. spontaneum* was a promising resistant germplasm for breeding sugarcane varieties which were resistant to mosaic disease. In this study, SceIF4E-R was cloned from the resistant germplasm, CP94-1100, and was closely related to *S. spontaneum*, indicating that SceIF4E-R may have the function of resistance to mosaic disease.

Polymorphism analysis of 11 sugarcane varieties with extremely high resistance and high susceptibility to *SCSMV* showed that the coding region of *eIF4E* was 663 bp in length, with 28 SNPs and no InDel locus. The SNP frequency was 1 SNP/23.7 bp, slightly lower than that of the sugarcane tillering key gene, *HTD2* [31], and higher than that of the tropical sucrose synthase gene *SuSy* (1SNP/108 bp) [32]. The SNP frequency of 1 SNP/23.7 bp was much higher than the polymorphism of cabbage (1 SNP/150 bp) [33] and soybean (1 SNP/272 bp) [34]. SNP is a gene sequence polymorphism at the genome level caused by a single nucleotide variation in a closely related population or conspecific individual. It is caused by single base conversion, transversion, and single base insertion, and deletion. SNP located in the non-coding region does not affect the amino acid sequence of its protein, and SNP located in the gene coding region will change the amino acid sequence of its protein, thus changing the function of the protein [35]. To sum up, SNP frequency is related to plant species, gene types, and functions.

Synonymous and non-synonymous mutations in the coding region of genes reflect variation in genes, and the ratio of their mutation rates (the Ka/Ks value) reflect the effect and direction of gene selection. A Ka/Ks value of more than 1 indicates that the gene is affected by positive selection and belongs to the rapid evolution gene; Ka/Ks = 1 indicates that synonymous mutation and non-synonymous mutation frequency are the same and the population is not affected by selection pressure; a Ka/Ks value less than 1 indicates that the gene is affected by negative selection and is a relatively conservative gene [36]. In this study, the Ka/Ks value of the coding region of *eIF4E* was less than 1, indicating that the eIF4E gene was affected by negative selection and was a relatively conservative gene, consistent with the results of Ruffel et al. [37] on the conservation of the eIF4E amino acid sequence in higher plants.

In plants, cloning and utilizing natural resistance genes have important applications in the cultivation of disease-resistant varieties. The absence or mutation of *eIF4E* in host plants can lead to recessive resistance and hinder infection by viruses in the family *Potyviridae*. Studies have shown that the natural recessive resistance mediated by *eIF4E* is derived from the non-synonymous mutation of the eIF4E protein [38,39]. Stein et al. [40] conducted a sequence analysis on *eIF4E* in *Barley Yellow Mosaic Virus*-resistant barley and found that the mutated amino acid residues, namely, Ser57, Lysl18, Thrl20, Asnl60, Glnl61, Ser205, Asp206, and Gly208, were located on both sides of the cap-binding pocket of the elF4E. The analysis of a three-dimensional structure model of the Arabidopsis eIF (iso) 4E protein revealed that mutant amino acid residues (G1y43, Lysl02, Thrl04, Lys149, Glnl50, Ser194, Asp195, and Thr197) corresponding to barley were found near the cap-binding pocket. A related analysis showed that the tryptophan at positions 46 and 92 in Arabidopsis eIF (iso) 4E was located in the hat recognition region. When these two tryptophans were mutated to leucine, eIF (iso) 4E lost its ability to bind to virus VPg and methylguanosine [41]. Related studies showed that the combination of VPg and eIF4E occurred near the cap pocket [16,17]. These indicate the mutation sites of *eIF4E* were primarily concentrated near the cap pocket. In this study, 11 tested materials were divided into 11 haplotypes, encoding 11 proteins. A total of 28 SNPs were detected in the coding region, of which 18 SNP loci could cause 17 amino acid sites with non-synonymous mutations. The three-dimensional structure analysis of the protein showed that the two amino acid non-synonymous mutation sites were located near the cap pocket. The nucleotides at the 193 bp site of Yuetang 55 were mutated from C to G, and the encoded 64-position alanine was mutated to valine. The nucleotide at the 180 bp site of ROC22 was mutated from A to G, and the encoded 60-lysine was mutated to glutamic acid. It is these two sites that may be the resistance sites of the eIF4E gene to *SCSMV* or *SCSMV* sensitivity in sugarcane. The next step is to study whether the protein formed by the mutation in these two sites can bind to VPg and develop the marker to marker-assisted selection for breeding *SCSMV*-resistant cultivars.

4. Materials and Methods

4.1. Tested Material

Eleven varieties with stable and universal resistance against sugarcane mosaic disease in the field [42] served as samples. Susceptible varieties were denoted S, while resistant samples were denoted R (Table 1). Healthy leaves of the resistant varieties (without mosaic symptoms) were collected, and the mosaic leaves of susceptible varieties (with typical mosaic symptoms) were stored at $-80\ ^\circ$C.

4.2. Design and Synthesis of Primers

SCSMV-CP-F/R primers and SCSMV-qCP-F/R were designed to amplify the *cp* sequence of *SCSMV* according to the whole genome sequence of SCSMV-JP2 (accession number JF488065) reported in the GenBank database. SCSMV-VPg-4F/4R primers were designed to amplify the *vpg* sequence and partial Nia-pro nucleotide sequence of SCSMV-GN12 (accession number KT257273) reported in the GenBank database. eIF4E-5F/5R primers and eIF4E-qF/qR were designed according to the reported sequence of the sugarcane translation initiation factor, *eIF4E* (accession number KX757017), in the GenBank to amplify the *eIF4E* sequence. The primer sequences (Table 6) were synthesized by Shanghai Sangon Biotechnology Co., Ltd. (Shanghai, China).

Table 6. Primers used in this study.

Primers	Sequence (5′ to 3′)	Product Size	Target Gene
SCSMV-CP-F	ACAAGGAACGCAGCCACCT	938 bp	*CP*
SCSMV-CP-R	ACTAAGCGGTCAGGCAAC		
SCSMV-VPg-4F	GGGAAGAAGCGTCGAACTCA	712 bp	*vpg* + Partial Nia-Pro
SCSMV-VPg-4R	CAACACACCAACTCTGCGTG		
eIF4E-5F	ATGGCCGACGAGATCGACAC	660 bp	*eIF4E*
eIF4E-5R	GCAACTCCTCGGCAAATACAG		
SCSMV-qCP-F	AACAACAACGAGTCAAGCTG	143 bp	SCSMV
SCSMV-qCP-R	AGAGATGAGAGCTTGTGGTG		
eIF4E-qF	GGCGAACAATTCGACTATGG	93 bp	*eIF4E*
eIF4E-qR	TCTGAGCAGCTTCATTAGCA		

4.3. Extraction of Total RNA from Plants

Total RNA was extracted from 0.2 g of tested material using a TransZol Plant kit (TransGen Biotech, Co., Ltd., Beijing, China). The specific method was implemented according to the kit instructions, and the RNA precipitation was dissolved in sterile water treated with 30 μL of DEPC. The total RNA was submitted as a template to synthesize cDNA using TransScript One-Step gDNA Removal and cDNA Synthesis SuperMix kit (TransGen Biotech Co., Ltd.) according to the instructions.

4.4. Gene Cloning

cDNA was used as the template, and PCR amplification was carried out using primers SCSMV-CP-F/R, SCSMV-VPg-4F/4R, and eIF4E-5F/5R. PCR amplification of *SCSMV-CP* was performed in a 25 μL reaction mixture, including 9.5 μL of ddH$_2$O, 12.5 μL of 2 × Easy Taq PCR SuperMix (TransGen Biotech Co., Ltd., Beijing, China), 2.0 μL of cDNA template, and 0.5 μL of upstream and downstream primers each (20 μg/μL). In the PCR amplification of *SCSMV-VPg* and *eIF4E*, the cDNA template and upstream and downstream primers were 1.0 μL. The thermal cycling conditions of *SCSMV-CP* were as follows: 5 min at 94 $^\circ$C followed by 35 cycles for 30 s at 94 $^\circ$C, 30 s at 50 $^\circ$C, and 1 min at 72 $^\circ$C, with a final extension for 10 min at 72 $^\circ$C. The thermal cycling conditions of *SCSMV-VPg* and *eIF4E* were as follows: 5 min at 94 $^\circ$C followed by 35 cycles for 30 s at 94 $^\circ$C, 30 s at 60 $^\circ$C, and 1 min at 72 $^\circ$C, with a final extension for 10 min at 72 $^\circ$C. The amplified product (10 μL) was analyzed through electrophoresis on 1% agarose gel stained with ethidium bromide. The PCR products were purified according to the manufacturer's instructions on the DNA agarose gel purification kit (Tiangen Biotech Co., Ltd., Beijing, China). The purified PCR products were cloned with pEASY-T5 Zero Cloning Kit (TransGen Biotech Co., Ltd., Beijing, China) and transformed into *Escherichia coli* cells Trans1-T1 (TransGen Biotech Co., Ltd., Beijing, China). Six positive clones per sample were sequenced.

4.5. SCSMV Accumulation Detection

The positive CP gene plasmid with correct sequencing was extracted. The plasmid DNA concentration was determined by nucleic acid protein analyzer. The copies of plasmid concentration were calculated according to the follow formula [43]. Then, the plasmid was used as a standard sample. The ChamQ Universal SYBR qPCR Master Mix kit (Vazyme, Nanjing Biological Company) was used to perform qPCR with primers SCSMV-qCP-F/R or eIF4E-qF/qR. The standard sample was successively diluted into 5 plasmid samples with 10-times gradient, and then to make the standard curve by qPCR. The DNA concentration of all samples was uniformly quantified to 500 ng/µL as template DNA for qPCR. The obtained Ct values were used to calculate the copies of SCSMV according to regression equation of the standard curve.

$$\text{Copies (copies·µL}^{-1}) = [6.02 \times 10^{23} \text{ (copies·mol}^{-1}) \times \text{DNA concentration (ng·µL}^{-1}) \times 10^{-9}]/[\text{base number (bp)} \times 660 \text{ (ng·mol}^{-1})]$$

4.6. Sequence Analysis

To identify homology, the sequences were compared with the published sequences in GenBank using the BLAST function on the NCBI website. DNAMAN software was used for sequence multiple alignments. The nucleotide sequence and amino acid sequence differences between the samples and the control KX757017 were analyzed. Single nucleotide polymorphism (SNP) and insertion–deletion (Indel) sites were analyzed. The open reading frame (ORF) of *eIF4E* was searched using the online tool, ORF Finder, and the amino acid sequence was deduced. The conserved domain of the eIF4E protein was analyzed using the online tool, Conserved Domain Search. ProtScale (https://web.expasy.org/protscale/, accessed on 25 May 2022) and ProtParam (https://web.expasy.org/, accessed on 25 May 2022) in ExPASy server ProtParam/) were used to analyze the amino acid residues, theoretical molecular weight, isoelectric point, instability index, hydrophilicity, and hydrophobicity of the encoded protein. The signal peptide and transmembrane region of amino acid sequence were analyzed using SignalP 5.0 Server (https://services.healthtech.dtu.dk/service.php/SignalP-5.0, accessed on 26 May 2022) and TMHMM Server V2.0 (https://services.healthtech.dtu.dk/service.php/TMHMM-2.0, accessed on 27 May 2022). Protein secondary structure was predicted using SOPMA (https://npsa-prabi.ibcp.fr/cgi-bin/npsa_automat.pl/page=npsa_sopma.html, accessed on 1 July 2022), and the tertiary structure was predicted using SWISS-MODEL (https://swissmodel.expasy.org/interactive, accessed on 16 July 2022). Subcellular localization was predicted using Plant-mPLoc (http://www.csbio.sjtu.edu.cn/cgi-bin/PlantmPLoc.cgi, accessed on 10 June 2022). The eIF4E amino acid sequences of other species were downloaded from the NCBI database, and the sequence analysis and multiple sequence alignment were performed using DNAMAN and ClustalW software. The phylogenetic tree was constructed using the neighbor-joining method and Kimura's two-parameter model as implemented in the MEGA version 6.0 [44]. The bootstrap value was 1000 replicates. Haplotype diversity (Hd), nucleotide diversity (Pi), standard deviation (SD), number of polymorphic loci (S), and Ka/Ks were calculated using the Dnasp v5.1 software.

Author Contributions: H.S. completed the writing, review and editing of the first draft of the paper. D.C., R.Z. and X.W. conducted the study. J.L. and C.W. verified the results. Y.H. carried out the data analysis, literature references searching, and the writing of some parts of the paper. Y.L. was the initiator and leader of the project and manages paper modification. All authors have read and agreed to the published version of the manuscript.

Funding: This work was funded by the Yunnan Province Technology Innovation Talent Training Object Project (Grant No. 202005AD160012); The National Science Foundation of China (Grant No. 31701490); China Agriculture Research System of MOF and MARA (Grant No. CARS-170303); Yunling Industry and Technology Leading Talent Training Program (Grant No. 2018LJRC56) and the Yunnan Province Agriculture Research System (Grant No. YNGZTX-4-92).

Data Availability Statement: All authors agree with MDPI Research Data Policies.

Acknowledgments: We are grateful to the Yunnan Province Technology Innovation Talent Training Object Project (202005AD160012); The National Science Foundation of China (31701490); China Agriculture Research System of MOF and MARA (CARS-170303); Yunling Industry and Technology Leading Talent Training Program (2018LJRC56); and the Yunnan Province Agriculture Research System (YNGZTX-4-92).

Conflicts of Interest: The authors declare no conflict of interest.

References

1. Hema, M.; Sreenivasulu, P.; Savithri, H.S. Taxonomic position of *Sugarcane streak mosaic virus* in the family *Potyviridae*. *Arch. Virol.* **2002**, *147*, 1997–2007. [CrossRef]
2. King, A.M.Q.; Adams, M.J.; Carstens, E.B.; Lefkowitz, E. Family Potyviridae. In *Virus Taxonomy: Classification and Nomenclature of Viruses*; Ninth Report of the International Committee on Taxonomy of Viruses; Elsevier Academic Press: San Diego, CA, USA, 2012; pp. 1069–1090.
3. Rabenstein, F.; Seifers, D.L.; Schubert, J.; French, R.C.; Stenger, D.C. Phylogenetic relationships, strain diversity and biogeography of tritimoviruses. *J. Gen. Virol.* **2002**, *83*, 895–906. [CrossRef] [PubMed]
4. Tatineni, S.; Ziem, A.D.; Wegulo, S.N.; French, R. Triticum mosaic virus: A distinct member of the family Potyviridae with an unusually long leader sequence. *Phytopathology* **2009**, *99*, 943–950. [CrossRef]
5. Hong, Y.; Levay, K.; Murphy, J.F.; Klein, P.G.; Shaw, J.G.; Hunt, A.G. A potyvirus polymerase interacts with the viral coat protein and VPg in the yeast cells. *Virology* **1995**, *214*, 159–166. [CrossRef] [PubMed]
6. Puustinen, P.; Makinen, K. Uridylylation of the potyvirus VPg by viral replicase NIb correlates with the nucleotide binding capacity of VPg. *J. Biol. Chem.* **2004**, *279*, 38103–38110. [CrossRef] [PubMed]
7. Lopez-Moya, J.J.; Wang, R.Y.; Pirone, T.P. Context of the coat protein DAG motif affects Potyvirus transmissibility by aphids. *J. Gen. Virol.* **1999**, *80*, 3281–3288. [CrossRef]
8. Gazo, B.M.; Murphy, P.; Gatchel, J.R.; Browning, K.S. A novel interaction of cap-binding protein complexes eukaryotic initiation factor (eIF) 4F and eIF(iso)4F with a region in the 3′-untranslated region of satellite tobacco necrosis virus. *J. Korean Soc. Appl. Biol.* **2004**, *279*, 13584–13592. [CrossRef]
9. Léonard, S.; Plante, D.; Wittmann, S.; Daigneault, N.; Fortin, M.G.; Laliberte, J.F. Complex formation between potyvirus VPg and translation eukaryotic initiation factor 4E correlates with virus infectivity. *J. Virol.* **2000**, *74*, 7730–7737. [CrossRef]
10. Wittmann, S.; Chatel, H.; Fortin, M.G.; Laliberté, J.F. Interaction of the viral protein genome linked of turnip mosaic potyvirus with the translational eukaryotic initiation factor (iso)4E of *Arabidopsis thaliana* using the yeast two-hybrid system. *Virology* **1997**, *234*, 84–92. [CrossRef]
11. Robaglia, C.; Caranta, C. Translation initiation factors: A weak link in plant RNA virus infection. *Trends Plant Sci.* **2006**, *11*, 40–45. [CrossRef]
12. Wang, A.M.; Krishnaswamy, S. Eukaryotic translation initiation factor 4e-mediated recessive resistance to plant viruses and its utility in crop improvement. *Mol. Plant Pathol.* **2012**, *13*, 795–803. [CrossRef]
13. Lin, S.F.; Wang, R.G.; Ren, X.L.; Li, Z.Q.; Yuan, Y.; Long, M.J.; Zhang, J.S.; Wang, Z.L. Genotypic identification of Potato virus Y resistance in tobacco germplasm resources. *Acta Tabacaria Sin.* **2021**, *27*, 37–44. (In Chinese) [CrossRef]
14. Yakupjan, H.; Asigul, I.; Wang, Y.J.; Liu, Y.L. Advances in genetic engineering of plant virus resistance. *Chin. J. Biotechnol.* **2015**, *31*, 976–994. [CrossRef]
15. Green, M.R.; Sambrook, J. *Molecular Cloning: A Laboratory Manual*, 4th ed.; He, F.C., Ed.; Translator; Science Press: Beijing, China, 2017; pp. 509–545.
16. Monzingo, A.F.; Dhaliwal, S.; Dutt-Chaudhuri, A.; Lyon, A.; Sadow, J.H.; Hoffman, D.W.; Robertus, J.D.; Browning, K.S. The structure of eukaryotic translation initiation factor-4E from wheat reveals a novel disulfide bond. *Plant Physiol.* **2007**, *14*, 1504–1518. [CrossRef]
17. Papadopoulos, E.; Jenni, S.; Kabha, E.; Takrouri, K.J.; Yi, T.; Salvi, N.; Luna, R.E.; Gavathiotis, E.; Mahalingam, P.; Arthanari, H.; et al. Structure of the eukaryotic translation initiation factor eIF4E in complex with 4EGI-1 reveals an allosteric mechanism for dissociating eIF4G. *Proc. Natl. Acad. Sci. USA* **2014**, *111*, 3187–3195. [CrossRef]
18. Li, W.F.; He, Z.; Li, S.F.; Huang, Y.K.; Zhang, Z.X.; Jiang, D.M.; Wang, X.Y.; Luo, Z.M. Molecular characterization of a new strain of *Sugarcane streak mosaic virus* (SCSMV). *Arch. Virol.* **2011**, *156*, 2101–2104. [CrossRef]
19. Li, X.J.; Li, C.J.; Wu, Z.D.; Tian, C.Y.; Hu, X.; Qiu, L.H.; Wu, J.M. Expression characteristic and gene diversity analysis of *ScHTD2* in sugarcane. *Acta Agron. Sin.* **2022**, *48*, 1601–1613. (In Chinese)

20. Wang, X.Y.; Li, W.F.; Huang, Y.K.; Zhang, R.Y.; Shan, H.L.; Yin, J.; Luo, Z.M. Molecular detection and phylogenetic analysis of viruses causing mosaic symptoms in new sugarcane varieties in China. *Eur. J. Plant Pathol.* **2017**, *148*, 931–940. [CrossRef]
21. Adams, M.J.; Antoniw, J.F.; Fauquet, C.M. Molecular criteria for genus and species discrimination within the family *Potyviridae*. *Arch. Virol.* **2005**, *150*, 459–479. [CrossRef]
22. Zhu, M.; Chen, Y.T.; Ding, S.X.; Webb, S.L.; Zhou, T.; Nelson, R.S.; Fan, Z.F. Maize elonginc interacts with the viral genome-linked protein, VPg, of *Sugarcane mosaic virus* and facilitates virus infection. *New Phytol.* **2014**, *203*, 1291–1304. [CrossRef]
23. Zhang, R.Y.; Li, W.F.; Huang, Y.K.; Pu, C.H.; Wang, X.Y.; Shan, H.L.; Cang, X.Y.; Luo, Z.M.; Yin, J. Genetic diversity and population structure of *Sugarcane streak mosaic virus* in Yunnan province, China. *Trop. Plant Pathol.* **2018**, *43*, 514–519. [CrossRef]
24. Zhu, M. Studies of the Molecular Interactions between Maize Elongin C and Vpg of *Sugarcane mosaic virus*. Ph.D. Thesis, China Agricultural University, Beijing, China, 2014. (In Chinese).
25. Moury, B.; Charron, C.; Janzac, B.; Simon, V.; Gallois, L.J. Evolution of plant eukaryotic initiation factor 4E (eIF4E) and potyvirus genome-linked protein (VPg): A game of mirrors impacting resistance spectrum and durability. *Infect. Genet. Evol.* **2014**, *27*, 472–480. [CrossRef] [PubMed]
26. Kim, J.; Kang, W.H.; Hwang, J.; Yang, H.B.; Dosun, K.; Oh, C.S.; Kang, B.C. Transgenic Brassica rapa plants over-expressing eIF(iso)4E variants show broad-spectrum *Turnip mosaic virus* (TuMV) resistance. *Mol. Plant Pathol.* **2014**, *15*, 615–626. [CrossRef] [PubMed]
27. Li, H.; Shirako, Y. Association of VPg and eIF4E in the host tropism at the cellular level of *Barley yellow mosaic virus* and *Wheat yellow mosaic virus* in the genus *Bymovirus*. *Virology* **2015**, *476*, 159–167. [CrossRef] [PubMed]
28. Geng, G.W.; Gu, K.; Yu, C.M.; Li, X.D.; Tian, Y.P.; Yuan, X.F. Sequence analysis of translation initiation factor eIF4E from two varieties of wheat with differential resistance to *Wheat yellow mosaic virus*. *Acta Phytopathol. Sin.* **2017**, *47*, 568–572. (In Chinese) [CrossRef]
29. Huang, Z.X.; Zhou, F.; Wang, Q.N.; Jin, Y.F.; Fu, C.; Hu, H.X.; Zhang, C.M.; Chang, H.L.; Ji, J.L.; Wu, Q.W.; et al. Genetic diversity assessment of *Saccharum spontaneum* L. native of domestic and overseas with phenotype agronomic traits. *J. Plant Genet. Resour.* **2012**, *13*, 825–829. (In Chinese) [CrossRef]
30. Li, W.F.; Wang, X.Y.; Huang, Y.K.; Shan, H.L.; Luo, Z.M.; Ying, X.M.; Zhang, R.Y.; Shen, K.; Yin, J. Screening sugarcane germplasm resistant to *Sorghum mosaic virus*. *Crop Prot.* **2013**, *43*, 27–30. [CrossRef]
31. Li, Y.H.; Li, J.; Qin, W.; Wang, X.Y.; Zhang, R.Y.; Zhao, J.; Shan, H.L.; Li, W.F.; Huang, Y.K. Diseases investigation and resistance analysis of new sugarcane varieties in the regional test demonstration of the national sugar system. *Sugar Crops China* **2022**, *44*, 58–63. (In Chinese) [CrossRef]
32. Zhang, J.; Arro, J.; Chen, Y.Q.; Ming, R. Haplotype analysis of sucrose synthase gene family in three *Saccharum* species. *BMC Genom.* **2013**, *14*, 314. [CrossRef]
33. Wang, X.W. Germplasm Identification and Evaluation of Resistance to TuMV and eIF(iso)4E Gene Cloning and Sequence Analysis in *Brassica campestris* L. ssp. Master's Thesis, Chinese Academy of Agricultural Sciences, Beijing, China, 2015. (In Chinese).
34. Zhu, Y.L.; Song, Q.J.; Hyten, D.L.; Van, C.P.; Matukumalli, L.K.; Grimm, D.R.; Hyatt, S.M.; Fickus, E.W.; Young, N.D.; Cregan, P.B. Single-nucleotide polymorphisms in Soybean. *Genetics* **2003**, *163*, 1123–1134. [CrossRef]
35. Rao, Q. SNP markers and soybean genome mapping. *Prog. Mod. Biomed.* **2008**, *8*, 2184–2186. [CrossRef]
36. Nielsen, R. Molecular signatures of natural selection. *Annu. Rev. Genet.* **2005**, *39*, 197–218. [CrossRef]
37. Ruffel, S.; Dussault, M.H.; Palloix, A.; Moury, B.; Bendahmane, A.; Robaglia, C.; Caranta, C. A natural recessive resistance gene against potato virus Y in pepper corresponds to the eukaryotic initiation factor 4E (eIF4E). *Plant J.* **2002**, *32*, 1067–1075. [CrossRef] [PubMed]
38. Ruffel, S.; Gallois, J.L.; Moury, B.; Robaglia, C.; Palloix, A.; Caranta, C. Simultaneous mutations in translation initiation factors eIF4E and eIF(iso)4E are required to prevent *Pepper veinal mottle virus* infection of pepper. *J. Gen. Virol.* **2006**, *87*, 2089–2098. [CrossRef] [PubMed]
39. Yeam, I.; Cavatorta, J.R.; Ripoll, D.R.; Kang, B.C. Functional dissection of naturally occurring amino acid substitutions in eIF4E that confers recessive potyvirus resistance in Plants. *Plant Cell* **2007**, *19*, 2913–2928. [CrossRef] [PubMed]
40. Stein, N.; Perovic, D.; Kumlehn, J.; Pellio, B.; Stracke, S.; Streng, S.; Ordon, F.; Graner, A. The eukaryotic translation initiation factor, 4E, confers multiallelic recessive Bymovirus resistance in *Hordeum vulgare* (L.). *Plant J.* **2005**, *42*, 912–922. [CrossRef]
41. Miyoshi, H.; Suehiro, N.; Tomoo, K.; Muto, S.; Takahashi, T.; Tsukamoto, T.; Ohmori, T.; Natsuaki, T. Binding analysis for the interaction between plant virus genome-linked protein (VPg) and plant translational initiation factors. *Biochimie* **2006**, *88*, 329–340. [CrossRef]
42. Li, W.F.; Shan, H.L.; Zhang, R.Y.; Wang, X.Y.; Yang, K.; Luo, Z.M.; Yin, J.; Cang, X.Y.; Li, J.; Huang, Y.K. Identification of resistance to *Sugarcane streak mosaic virus* (SCSMV) and *Sorghum mosaic virus* (SrMV) in new elite sugarcane varieties/clones in China. *Crop Prot.* **2018**, *110*, 77–82. [CrossRef]
43. Wilhelm, J.; Pingoud, A. Real-time polymerase chain reaction. *ChemBioChem* **2003**, *4*, 1120–1128. [CrossRef]
44. Tamura, K.; Stecher, G.; Peterson, D.; Filipski, A.; Kumar, S. MEGA6: Molecular evolutionary genetics analysis version 6.0. *Mol. Biol. Evol.* **2013**, *30*, 2725–2729. [CrossRef]

plants

MDPI

Article

Transcriptomic Profiling of Sugarcane White Leaf (SCWL) Canes during Maturation Phase

Karan Lohmaneeratana [1], Kantinan Leetanasaksakul [2] and Arinthip Thamchaipenet [1,3,*]

1 Department of Genetics, Faculty of Science, Kasetsart University, Bangkok 10900, Thailand; karan.l@ku.th
2 National Center for Genetic Engineering and Biotechnology, National Science and Technology Development Agency, Pathumthani 12120, Thailand; kantinan.lee@biotec.or.th
3 Omics Center for Agriculture, Bioresources, Food and Health, Kasetsart University (OmiKU), Bangkok 10900, Thailand
* Correspondence: arinthip.t@ku.ac.th

Abstract: Sugarcane white leaf (SCWL) disease, caused by *Candidatus* Phytoplasma sacchari, results in the most damage to sugarcane plantations. Some SCWL canes can grow unnoticed through the maturation phase, subsequently resulting in an overall low sugar yield, or they can be used accidentally as seed canes. In this work, 12-month-old SCWL and asymptomatic canes growing in the same field were investigated. An abundance of phytoplasma in SCWL canes affected growth and sugar content as well as alterations of transcriptomic profiles corresponding to several pathways that responded to the infection. Suppression of photosynthesis, porphyrin and chlorophyll metabolism, coupled with an increase in the expression of chlorophyllase, contributed to the reduction in chlorophyll levels and photosynthesis. Blockage of sucrose transport plausibly occurred due to the expression of sugar transporters in leaves but suppression in stalks, resulting in low sugar content in canes. Increased expression of genes associated with MAPK cascades, plant hormone signaling transduction, callose plug formation, the phenylpropanoid pathway, and calcium cascades positively promoted defense mechanisms against phytoplasma colonization by an accumulation of lignin and calcium in response to plant immunity. Significant downregulation of *CPK* plausibly results in a reduction in antioxidant enzymes and likely facilitates pathogen invasion, while expression of sesquiterpene biosynthesis possibly attracts the insect vectors for transmission, thereby enabling the spread of phytoplasma. Moreover, downregulation of flavonoid biosynthesis potentially intensifies the symptoms of SCWL upon challenge by phytoplasma. These SCWL sugarcane transcriptomic profiles describe the first comprehensive sugarcane–phytoplasma interaction during the harvesting stage. Understanding molecular mechanisms will allow for sustainable management and the prevention of SCWL disease—a crucial benefit to the sugar industry.

Keywords: sugarcane white leaf; phytoplasma; transcriptome; plant defense response; biotic stress

Citation: Lohmaneeratana, K.; Leetanasaksakul, K.; Thamchaipenet, A. Transcriptomic Profiling of Sugarcane White Leaf (SCWL) Canes during Maturation Phase. *Plants* **2024**, *13*, 1551. https://doi.org/10.3390/plants13111551

Academic Editors: Nataliya V. Melnikova and San-Ji Gao

Received: 9 April 2024
Revised: 24 May 2024
Accepted: 31 May 2024
Published: 4 June 2024

1. Introduction

Sugarcane (*Saccharum* sp.), a grass monocot plant in the Family *Poaceae*, is one of the most important economic crops for the sugar industry as well as the production of ethanol, molasses, and animal feed. Thailand is the second-largest sugar exporter after Brazil and is ranked fourth for sugar production worldwide (US Department of Agriculture, 2023/2024). However, sugar production from cane has been seriously affected by several sugarcane diseases. One of the most destructive effects on sugar yield from canes in Thailand and other regions is sugarcane white leaf (SCWL) disease. The economic loss caused by SCWL disease on income from sugarcane cultivation was estimated to be around USD 30–40 million annually [1].

SCWL disease is caused by an obligate pathogenic bacterium that lacks a cell wall, *Candidatus* Phytoplasma sacchari, which so far has been impracticable to culture in laboratory conditions [2]. SCWL phytoplasma is transmitted via insect vectors, *Matsumuratettix hiroglyphicus* Matsumura [3] and *Yamatotettix flavovittatus* Matsumura [4]. After transmission, the bacterium generally localizes in phloem and uses metabolic pathways from the sugarcane host [5]. Based on 16S rRNA classification, SCWL phytoplasma is closely related to sugarcane grassy shoot (SCGS) and rice yellow dwarf (RYD) [6,7].

Symptoms of SCWL disease can be observed in every growth stage of sugarcane from seed canes or cane setts to the mature stage. In general, new emerging buds and shoots from the cane setts emerge with bushy slim white leaves of a soft texture that grow from slender chlorotic shoots [7]. Degrees of leaf chlorosis start from pale green to white with considerable tiller proliferation and stunting [8]. Variation in symptoms depends on the stage of sugarcane growth and development of disease. Although SCWL sugarcane is recognized and eliminated while it is young, less symptomatic canes can be missed visually and remain in the field as SCWL carriers. At the maturation phase (12 months old), when sugarcanes contain the highest sugar yield [9], such SCWL canes are harvested together with the healthy ones and result in a low yield of sugar production. Moreover, they could be unintentionally prepared as seed canes for the next propagation cycle that will quickly spread SCWL disease to other sugarcane plantations.

In Thailand, Khon Kaen 3 (KK3) is one of the most popular sugarcane cultivars for the sugar industry due to its high sugar productivity and vigorous growth in various soil conditions [10], but KK3 is not tolerant to SCWL disease. In this study, 12-month-old SCWL KK3, growing together with asymptomatic KK3 in the same field, were harvested and investigated. To determine the sugarcane genes that play important roles in SCWL pathogenesis, transcriptomic profiles of SCWL sugarcane leaves and stalks were undertaken as both organs perform the distinct metabolic functions of photosynthesis and sugar storage, respectively. Molecular mechanisms and related pathways of the sugarcane host affected by SCWL phytoplasma have been identified.

2. Results

2.1. Growth Characteristics and Phytoplasma Detection of SCWL Sugarcanes

Twelve-month-old SCWL sugarcanes showed symptoms of chlorosis leaf, 0.65 times smaller stalk diameter, 0.65 times shorter height, and 0.69 times lower sugar content than those of asymptomatic sugarcanes collected from the same plantation (Table 1, Figure S1A–C). To evaluate the numbers of active phytoplasma in leaves and stalks of sugarcane samples, the gene expression ratio of 16S/18S rRNA of phytoplasma/sugarcane was calculated. SCWL sugarcanes harbored 10^4 times more phytoplasma than those of asymptomatic canes (Table 1). After propagation of the SCWL seed canes under greenhouse conditions for 2 months, newly emerged buds produced tillers and stunting with slim and narrow white leaves (Figure S1D).

Table 1. Growth parameters, sugar content, and phytoplasma ratio of asymptomatic and symptomatic sugarcanes.

Sugarcane	Height (cm)	Diameter (cm)	Sugar Content (°Brix)	16S/18S rRNA Gene Ratio	
				Leaves	Stalks
Asymptom	302.20 ± 12.77	3.34 ± 0.13	22.84 ± 0.25	$2.35 \times 10^{-5} \pm 0.82 \times 10^{-5}$	$8.96 \times 10^{-5} \pm 3.00 \times 10^{-5}$
Symptom	199.20 ± 16.76 *	2.18 ± 0.16 *	15.86 ± 1.14 *	0.33 ± 0.10 *	0.57 ± 0.14 *

Asterisk (*) indicates significant differences by unpaired *t*-test analysis.

2.2. RNA-Sequencing and Assembly

Total RNA-seq data of leaves and stalks of SCWL and asymptomatic sugarcanes were analyzed in triplicate (BioProject number PRJNA719388). More than 671 million raw reads (sequencing depth of 53.8–59.8 million of 150 bp paired-end read per library) were generated. About 647 million clean reads were filtered and 69.48% to 80.66% of the reads were mapped to the monoploid sugarcane genome reference (https://sugarcane-genome.cirad.fr; accessed on 14 February 2020) [11] (Table S1). Principal component analysis (PCA) of gene expression showed that SCWL and asymptomatic sugarcanes clustered separately in both leaves and stalks (Figure 1) and the volcano plot showed a reasonable distribution of gene expression (Figure S2). In leaves, 4799 and 1485 differentially expressed genes (DEGs; p-value < 0.05, $\log_2$ FC > 1) were upregulated and downregulated, respectively, while in stalks, 994 and 743 DEGs were upregulated and downregulated, respectively (Table 2). Co-expression of DEGs between leaves and stalks contained 330 upregulated and 92 downregulated DEGs, respectively (Table 2).

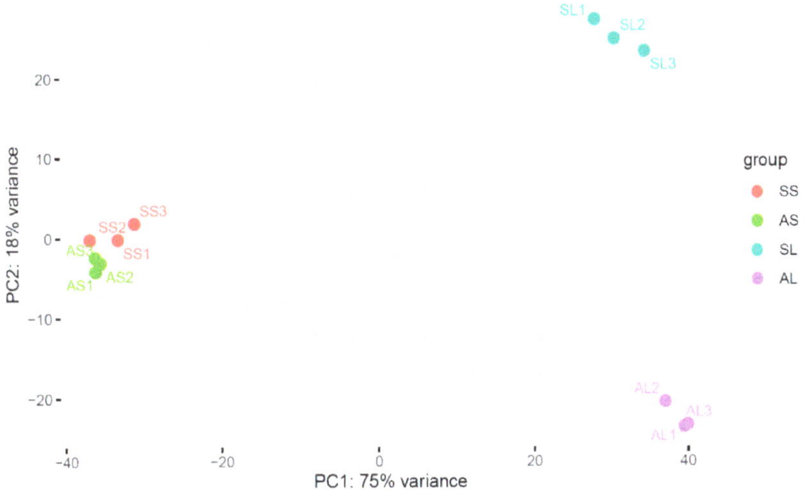

Figure 1. Principal component analysis (PCA) of gene expression from RNA-seq data of asymptomatic and symptomatic sugarcanes. AS, asymptomatic stalks; SS, symptomatic stalks; AL: asymptomatic leaves; SL, symptomatic leaves. The two principal components determine 93% of the total variance.

Table 2. Number of DEGs in gene ontology (GO) and KEGG pathway in SCWL leaves and stalks.

Category	Number of DEGs	
	Leaves	Stalks
Significantly expressed gene (p-value < 0.05)	6612	2173
|$\log_2$Fold Change| > 1.0	6284	1737
- Upregulated genes	4799	994
- Downregulate genes	1485	743
Gene ontology	3711 (59.06%)	987 (56.82%)
- Molecular function	3350	901
- Cellular component	884	186
- Biological process	2523	675
KEGG pathway	2248 (35.79%)	531 (30.57%)
- Upregulated DEGs	1701	294
- Downregulated DEGs	547	237

2.3. Gene Ontology (GO) Annotation and KEGG Pathway-Enrichment Analysis

The significant upregulated and downregulated DEGs of leaves and stalks were annotated with GO terms (Table 2). In leaves, DEGs were categorized and enriched in biological processes (e.g., photosynthesis, carbohydrate metabolic processes, generation of precursor metabolites and energy, cellular amino acid and protein metabolic processes), cellular component (membrane and thylakoid), and molecular function (e.g., transporter activity, catalytic activity, kinase/transferase activity, carbohydrate binding, and transcription factor activity) (Figure S3A), while in stalks, significant DEGs were categorized in biological processes (e.g., regulation of metabolic process, macromolecule modification, regulation of gene expression, protein modification processes, and regulation of biological processes) and molecular function (e.g., transcription factor activity, transferase activity, kinase activity, and transporter activity) (Figure S3B).

As a result of the KEGG analysis, upregulated and downregulated DEGs of leaves and stalks were annotated for functionality (Table 2). The most frequent pathways were metabolism, genetic information processing, environmental information processing, cellular processes, and organismal systems (Figure 2). To attain pathway enrichment, the 'piano in R' package was used. The enriched DEGs in leaves were associated with various functions including photosynthesis, signal transduction, flavonoid biosynthesis, metabolism of terpenoids and polyketides (Figure 3, Table S2). The most enriched upregulated pathways included mitogen-activated protein kinase (MAPK) signaling pathways, monoterpenoid biosynthesis, and plant–pathogen interactions, while the most enriched downregulated pathways were photosynthesis, photosynthesis antenna proteins, porphyrin and chlorophyll metabolism, signal transduction, carotenoid biosynthesis, and carbon fixation in photosynthetic organisms (Table S2). In stalks, the enriched DEGs included plant–pathogen interactions and protein processing in the endoplasmic reticulum pathway (Figure 3, Table S3). The most enriched upregulated pathways encompassed brassinosteroid biosynthesis, sesquiterpenoid and triterpenoid biosynthesis, while the most enriched downregulated pathways were starch and sucrose metabolism, glycolysis/gluconeogenesis, diterpenoid biosynthesis, flavonoid biosynthesis and carbon fixation in photosynthetic organisms (Table S3).

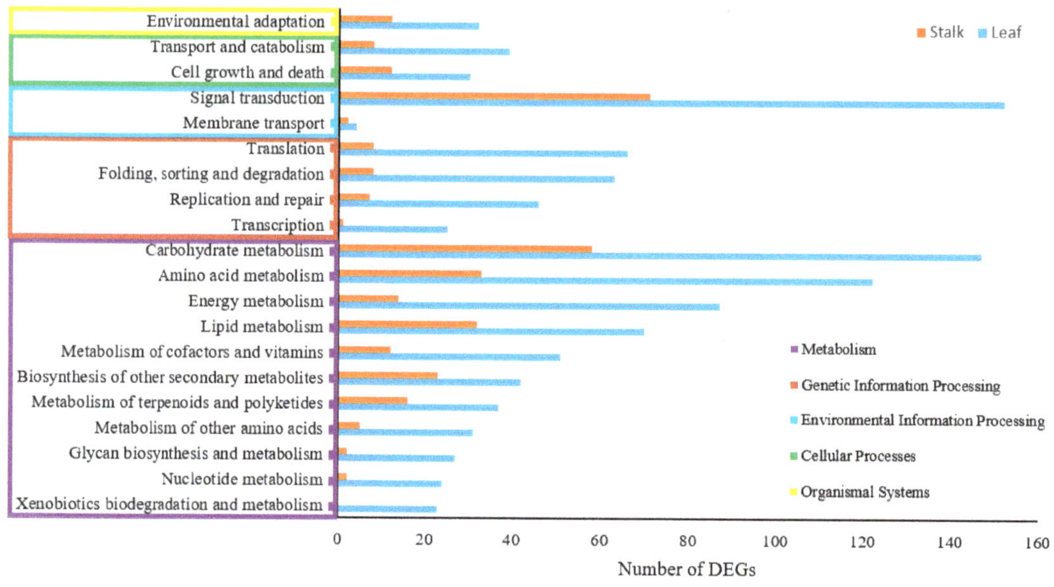

Figure 2. Number of DEGs in each category of KEGG pathways in SCWL leaves and stalks.

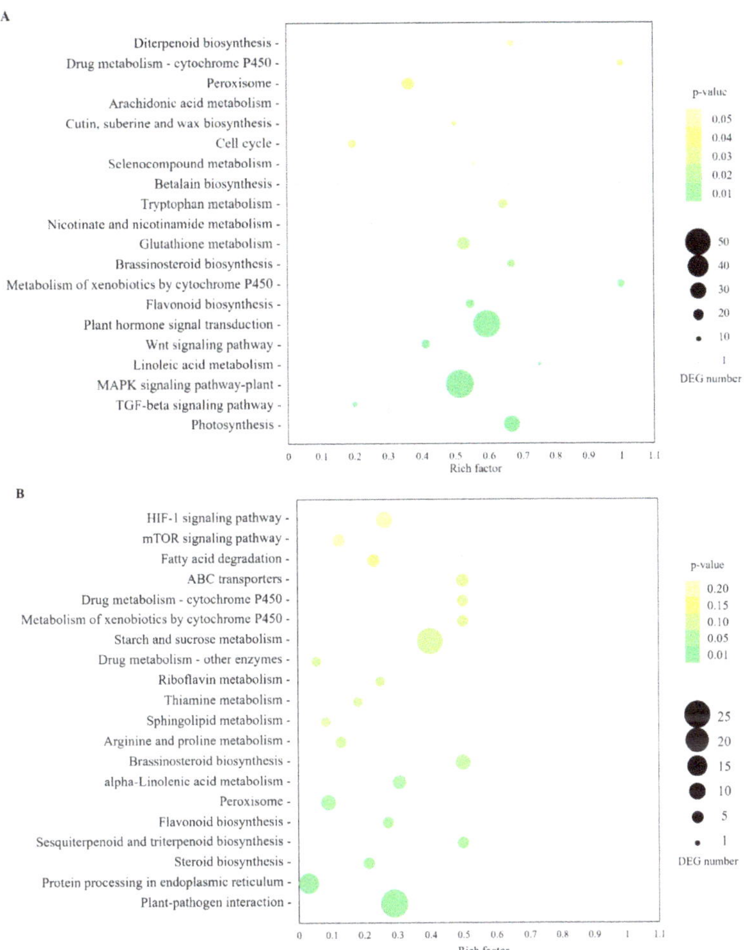

Figure 3. Top 20 KEGG pathway enrichment of DEGs between SCWL and asymptomatic sugarcanes in leaves (**A**) and stalks (**B**). The rich factor is a ratio of the number of DEGs and the total genes. The size and color of the bubble represent the number of DEGs and the *p*-value, respectively.

2.4. Phytoplasma-Affected Metabolic Processes of Sugarcane

From GO and KEGG pathway analyses, infection by phytoplasma evidently influenced the expression of many genes related to metabolic pathways in leaves and stalks including chlorophyll metabolism and photosynthesis, sucrose accumulation, plant–pathogen interactions, plant hormone signaling transduction, and secondary metabolites.

2.4.1. Chlorophyll Metabolism and Photosynthesis

In SCWL leaves, phytoplasma significantly suppressed the photosynthetic pathway and chlorophyll content, resulting in an alteration of the downstream pathways. Most DEGs of porphyrin and chlorophyll metabolism were downregulated including protoporphyrin IX-related genes; chlorophyllide b reductase (*NYC1*, K13606), geranylgeranyl diphosphate (*chlP*, K10960), 7-hydroxymethyl chlorophyll a reductase (*HCAR*, K18010), magnesium chelatases (*chlH*, K03404; K03405), magnesium-protoporphyrin IX monomethyl ester cyclase (*chlE*, K04035), protochlorophyllide reductase (*por*, K00218); and heme-related gene clusters [*hemA* (K02492), *hemC* (K01749), *hemE* (K01599), *hemL* (K01845), and *hemY* (K00231)]; while chlorophyllase (K08099) and cytochrome c oxidase assembly protein subunit 15 (*COX15*, K02259) were upregulated (Figure 4 and Table S4).

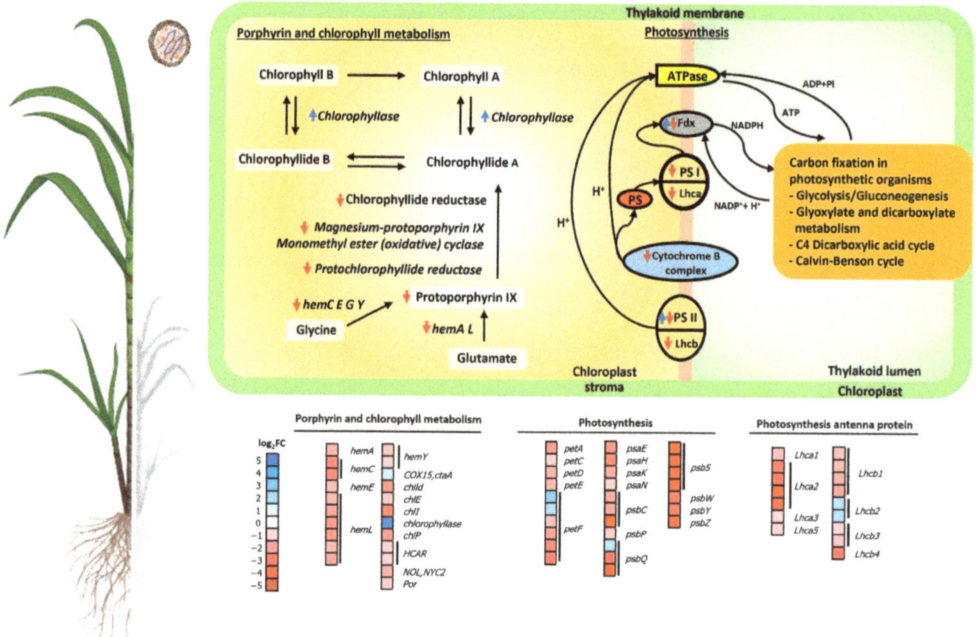

Figure 4. Effect of phytoplasma on photosynthesis and chlorophyll metabolism in SCWL sugarcane. Heatmaps of upregulated and downregulated DEGs of porphyrin and chlorophyll metabolism, photosynthesis, and photosynthesis antenna proteins are shown. Blue arrow, upregulation; red arrow, downregulation.

Meanwhile, most DEGs of photosynthesis were downregulated including cytochrome b6/f complex subunits [*PetA*, K02634; *PetC*, *PetD*, (K02636–7)], photosystem I subunits (*PsaE*, K02693; *PsaH*, K02695; *PsaK*, K02698; *PsaN*, K02701), photosystem II Psb proteins [*PsbC*, K02705; *PsbP*, K02717; *PsbS*, K03542; *PsbW*, K02721; *PsbY*, *PsbZ*, (K02723–4)], and photosynthetic electron transport, i.e., plastocyanin and ferredoxin [*PetE*, *PetF* (K02638–9)]. Moreover, the DEGs of photosynthesis antenna protein were also downregulated including the light-harvesting chlorophyll protein complex [*Lhca1*, *Lhca2*, *Lhca3*, (K08907–9); *Lhca5*, *Lhcb1*, *Lhcb2*, *Lhcb3*, *Lhcb4* (K08911–5) (Figure 4 and Table S4).

2.4.2. Regulation of Genes Related to Sucrose Accumulation

The accumulation of sucrose was negatively affected in sugarcanes infected by phytoplasma. In leaves, transcripts related to sucrose accumulation were altered by most DEGs of carbon fixation in photosynthetic organisms including the upregulation of aspartate aminotransferase (*GOT1*, K14454), phosphoenolpyruvate carboxykinase (*pecK*, K01610), malate dehydrogenase (*maeB*, K00029); and downregulation of fructose-1,6-biphosphatase I (*FBP*, K03841), pyruvate phosphate dikinase (*ppdK*, K01006), and ribulose-bisphosphate carboxylases (*rubisco*, K01601–2) (Figure 5 and Table S4). Forty-three DEGs of starch and sucrose metabolism were upregulated including alpha-amylase (*amy*, K01176), beta-amylase (*bmy*, K01177), beta-glucosidase (*bglX*, K05349), invertase (*sacA*, K01193), starch synthase (*glgA*, K00703), sucrose phosphate synthase (*SPS*, K00696), trehalose 6-phosphate phosphatase (*otsB*, K01087), and trehalose 6-phosphate synthase (*TPS*, K16055). Additionally, SWEET transporters (*SWEET*, *SWEET17*; K15382) were highly upregulated but the H$^+$/sugar co-transporter (SUT) (*SLC*, K15378) was downregulated (Figure 5 and Table S4).

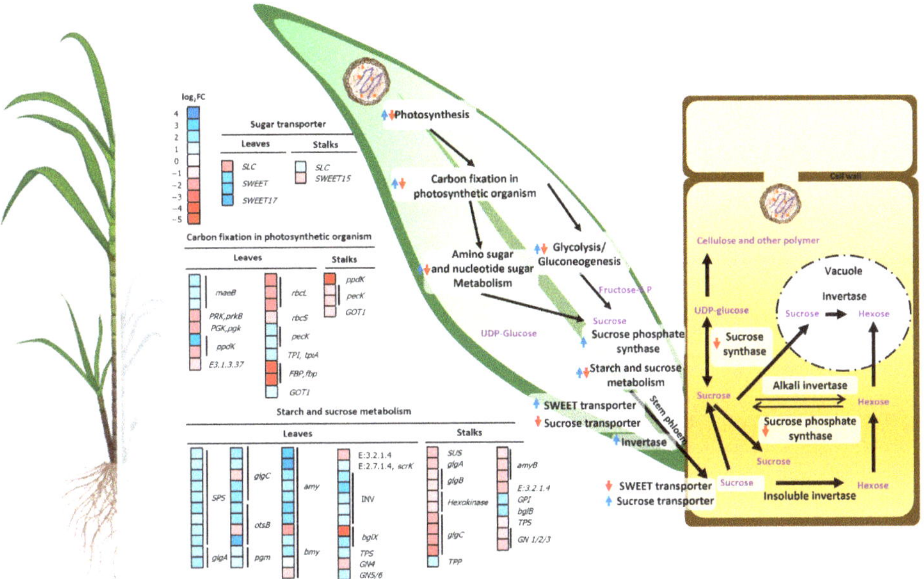

Figure 5. Effect of phytoplasma on sucrose accumulation in SCWL leaves and stalks. Heatmaps of up-regulated and downregulated DEGs of sugar transporters, carbon fixation in photosynthetic organisms, starch and sucrose metabolism are shown. Blue arrow, upregulation; red arrow, downregulation.

In stalks, carbon fixation in photosynthetic organisms was downregulated including *GOT1*, *pecK*, and *ppdK*. Seventeen DEGs of starch and sucrose metabolism were also suppressed including *bmy*, *glgA*, *glgC*, sucrose synthase (*SUS*, K00695), and *TPS*, whereas beta-glucosidase (K05350), glucose-6-phosphate isomerase (K01810), and *otsB* were upregulated (Figure 5 and Table S5). On the contrary to leaves, SWEET transporter, *SWEET15* (K15382), was downregulated but *SLC* was upregulated (Figure 5 and Table S5).

2.4.3. Regulation of Genes Related to Plant–Pathogen Interactions

Most DEGs in plant–pathogen interaction pathways were upregulated (Figure 6 and Table S4) including calmodulin (*CALM*, K02183), cathepsin F (*CTSF*, K01373), cyclic nucleotide-gated channel (*CNGC*, K05391), disease resistance proteins [*RAR1* (K13458), *RPM1* (K13457), and *RPS2* (K13459)], glycerol kinase (*GK*, K00864), mitogen-activated protein kinase kinase (*MAP2K1*, K04368; *MKK4/5*, K13413), MAP kinase substrate 1 (*MKS1*, K20725), and pathogenesis-related genes including transcriptional activator pattern-triggered immunity (*PTI5*, K13433), pathogenesis-related protein (*PR1*, K13449), and WRKY transcription factors (*WRKY2*, K18835; *WRKY22*, K13425; *WRKY33*, K13424). Downregulated DEGs included calcium-dependent protein kinase (*CPK*, K13412), elongation factor Tu (*tuf*, K02358), and 3-ketoacyl-CoA synthase 11 (*KCS*, K15397). In addition, one of the plant defense-related genes, invertase (K01193), was accelerated during sugarcane–phytoplasma interactions along with callose synthases (K11000) (Table S4). In stalks, upregulated DEGs included *RPM1*, *KCS*, *WRKY22*, *WRKY24*, and *WRKY33*, while *CALM*, *CNGC*, *CPK10*, and *PR1* were downregulated (Table S5).

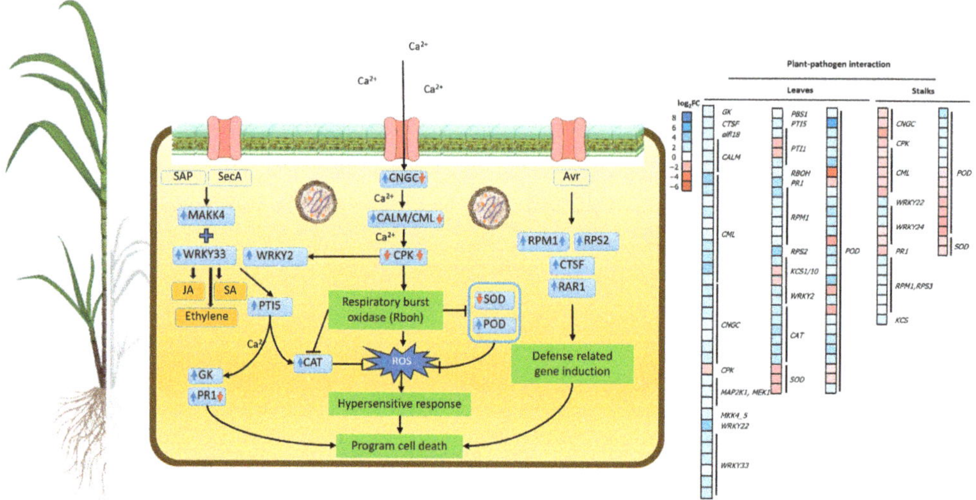

Figure 6. Effect of phytoplasma on plant–pathogen interaction in SCWL sugarcane leaves and stalks. Heatmaps of upregulated and downregulated DEGs of the plant–pathogen interaction are shown. Blue arrow, upregulation; red arrow, downregulation.

KEGG pathway analysis of SCWL leaves revealed 113 DEGs encoding transcription factors (TFs) which can be categorized into six types (Table S6). The largest category was helix-turn-helix proteins including MYB (*MYBP*, K09422), and MYB-related TF LHY (*LHY*, K12133). MYB TFs were upregulated along with other TFs including TGA (*TGA*, K14431), phytochrome-interacting factor 4 (*PIF4*, K16189), teosinte branched1/cycloidea/proliferating cell factors (*TCP*, K16221), and WRKYs (Table S6). In stalks, TFs such as *MYB*, *TGA*, *WRKY22*, and *WRKY33* were upregulated, while *PIF4* was downregulated (Table S7).

2.4.4. Regulation of Genes Related to Signal Transduction by Plant Hormones

In SCWL leaves, most DEGs of the plant hormone signal transduction pathway were altered (Table S4), including upregulated DEGs of abscisic acid (ABA) [e.g., ABA receptor PYR/PYL family (*PYL*, K14496), ABA responsive element binding factor (*ABF*, K14432), catalase (*CAT*, K03781), and MAPK kinase kinase (*MAPKKK17/18*, K20716)], ethylene (ETH) signaling pathway [e.g., ethylene receptor (*ETR*, K14509), ETH-responsive TF 1 (*ERF1*, K14516), endochitinase B (*ChiB*, K20547), and MAPK kinase (*MKK4/5*, K13413; *MKK9*, K20604)], auxin (AUX) (*AUX1*, K13946; *AUX/IAA*, K14484; *GH3*, K14487; *SAUR*, K14488), cytokinin (CK) (*A-ARR*, K14492; *AHP*, K14490; *HK*, K14489), gibberellin (GA) (*RGA*, K14494 and *PIF4*), jasmonic acid (JA) (*COI1*, K13463; *JAZ*, K13464), salicylic acid (SA) [*PR1*, K13449 and TGA TF (K14431)], and brassinosteroid (BR) (*TCH4*, K14504). On the contrary, *A-ARR* of CK, *PIF4* of GA, and *PR1* of SA in stalks were suppressed, whereas *JAR1* and *JAZ* of JA and *NPR1* of SA were upregulated (Table S5).

2.4.5. Regulation of Genes Related to Secondary Metabolites

In leaves, key genes in flavonoid biosynthesis were downregulated (Table S4), including chalcone synthase (*CHS*, K00660), caffeoyl-CoA O-methyltransferase (*CCoAOMT*, K00588), chalcone isomerases (*CHIL*, K01859), flavonoid 3,5-hydroxylase (*CYP75A*, K13083), and flavonoid 3-monooxygenase (*CYP75B1*, K05280), while some were also downregulated in stalks (Table S5). Genes in phenylpropanoid biosynthesis were upregulated in SCWL leaves (Table S4), including cinnamyl-alcohol dehydrogenase (*CAD*, K00083), cinnamoyl-CoA reductase (*CCR*, K09753), coniferyl-aldehyde dehydrogenase (*REF1*, K12355), peroxidase (*POD*, K00430), phenylalanine ammonia-lyase (*PAL*, K10755), and shikimate O-hydroxycinnamoyltransferase (*HCT*, K13065), while some were found to be upregulated in stalks (Table S5).

In leaves, the DEGs of BR biosynthesis were upregulated (Table S4), including typhasterol/6-deoxotyphasterol 2-alpha-hydroxylase (*CYP92A6*, K20623), PHYB activation-tagged suppressor 1 (*CYP734A1*, K15639), steroid 22-alpha-hydroxylase (*DWF4*, K09587), and steroid 5-alpha-reductase (*DET2*, K09591), while some were also upregulated in stalks (Table S5). In SCWL leaves, the DEGs of diterpenoid biosynthesis were upregulated (Table S4), including ent-kaurene synthase (*GA2*, K04121), ent-kaurenoic acid monooxygenase (*KAO*, K04123), GA 2-beta-dioxygenase (*GA2ox*, K04125), and GA-44 dioxygenase (*GA20ox*, K05282), whereas some from stalks were downregulated including ent-kaurene oxidase (*GA3*, K04122), *GA2ox* and *GA20ox* (Table S5). Monoterpenoid, sesquiterpenoid, and triterpenoid biosynthetic genes including germacrene D synthase (K15803) and NAD+-dependent farnesol dehydrogenase (*FLDH*, K15891) were upregulated in both leaves and stalks (Tables S4 and S5).

In leaves, the DEGs of carotenoid biosynthesis including abscisic-aldehyde oxidase (*AAO*, K09842), abscisic acid 8′-hydroxylase 2 (*CYP707A*, K09843), β-carotene 3-hydroxylase (*crtZ*, K15746), carotenoid epsilon hydroxylase (*LUT1*, K09837), 9-cis-epoxycarotenoid dioxygenase (*NCED*, K09840), and xanthoxin dehydrogenase (K09841) were downregulated but abscisic-aldehyde oxidase was upregulated (Table S4). In stalks, *AAO* was upregulated, while β-carotene isomerase and *NCED* were downregulated (Table S5).

2.5. Verification of DEGs by Real-Time PCR

To validate RNA-seq data, a total of twenty randomly selected candidate genes from leaves and stalks with upregulation and downregulation (Table S8) were verified by real-time PCR. A comparison of RNA-seq with real-time PCR showed a correlation coefficient (R^2) of 0.795925. The results confirmed the expression trends of all candidate genes in accordance with the analysis of the transcriptome data (Figure 7).

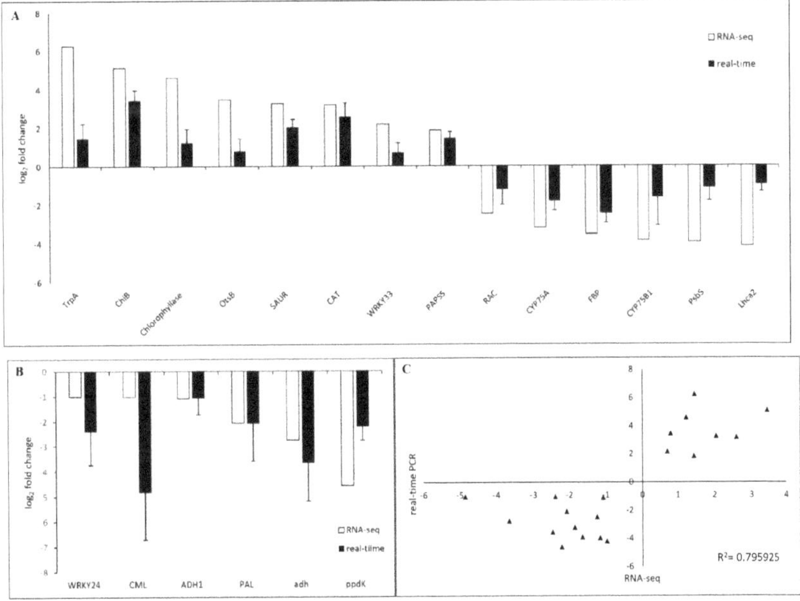

Figure 7. Validation and comparison of randomly selected DEGs of RNA-seq data (white bar) and real-time PCR quantification (black bar) of leaves (**A**), stalks (**B**) and correlation coefficients (R^2) between RNA-seq and real-time PCR results for differential gene expression in leaves and stalks of asymptomatic and symptomatic sugarcane (**C**).

3. Discussion

The pathogenicity of SCWL phytoplasma for sugarcane is still unclear. Since phytoplasmas are difficult to culture in laboratory conditions and in vitro infection is exceptionally impracticable [2], 12-month-old SCWL and asymptomatic *Saccharum* hybrid cv. KK3 were collected from the same field and used for investigation to simulate the natural condition. The maturation phase is an important stage for harvesting since it gives the highest sugar yield with an economically acceptable marginal return for the sugar industry [9]. The phenotypic characteristics of SCWL sugarcanes including height, stalk diameter, and sugar content were significantly lower, which corresponded to a 10^4 times higher phytoplasma expression ratio detected than those of asymptomatic canes. Furthermore, new buds emerging from such SCWL seed canes generated SCWL symptoms of a white grassy shoot phenotype with chlorosis leaves [6].

3.1. SCWL Phytoplasma Affects Chlorophyll Metabolism and Photosynthesis

In SCWL leaves, *NYC1* and *HCAR* that, respectively, reduce chlorophyll *b* (Chl *b*) and chlorophyll *a* (Chl *a*) in chlorophyll metabolism [12,13] were repressed similarly to that reported in grapevine leaves infected with Flavescence dorée (FD) phytoplasma [14]. Moreover, *hemA*, *hemC*, *hemE*, *hemL*, and *hemY* involved in the formation of protoporphyrin IX, a key precursor of chlorophyll biosynthesis, were downregulated along with *chlH*, *chlE* and *por* (Figure 4) that modify Mg^{2+} branching and form an isocyclic pentanone ring of chlorophyll [13,15]. High expression of chlorophyllase additionally degrades chlorophyll by cleaving the phytol tail and removing magnesium from the porphyrin ring [16,17]. Such downregulated mechanisms of porphyrin and chlorophyll gene expression together with the enhancement of chlorophyllase activity in leaves of SCWL sugarcane evidently explain the symptoms of leaf chlorosis and low chlorophyll content caused by SCWL phytoplasma. The findings agree with the growth limitation of phytoplasma-infected sesame plants where photosynthetic rates were reduced due to the depletion of chlorophyll content, photosystem II photoprotection, and photosynthetic capacity [18,19].

In this work, photosynthesis-related genes were mostly downregulated in SCWL leaves, which agrees with the observation for phytoplasma-infected Chinese jujube leaves in which the *Lhcb* gene family and photosystem II were downregulated [20,21]. Such suppression was also found in genes associated with photosystem I and II, including the cytochrome b6/f complex and ATP synthase in phytoplasma-infected grapevine [22,23], periwinkle plant [24], and coconut palm [25]. As part of defense mechanisms in response to biotic stress, photosynthesis genes are globally downregulated [26], which decrease phloem loading and carbohydrate accumulation of the infected leaves as a consequence [23] (Figures 4 and 5).

The other downregulated pathway affecting photosynthesis, photoprotection, and signaling in SCWL sugarcanes was the carotenoid pathway. A reduction in carotenoid content was found in phytoplasma-infected apple, Chinese jujube, cranberry, grapevine, and Napier grass [14,20,27–29]. Low levels of the antioxidant carotenoids in phytoplasma-infected plants have a profound effect on plant physiology and response to stress since carotenoids play important roles in protecting cell damage caused by ROS [30], and protect chlorophyll and the photosynthetic apparatus from photodamage [31]. Moreover, carotenoids are precursors for plant hormone ABA; they are involved in the stress response [32] and work together with chlorophyll in the process of light harvesting during photosynthesis [33]. Therefore, downregulation of carotenoid-related genes in SCWL sugarcanes must have a downstream effect on ABA levels, which impacts the ability of the plant to respond to the phytoplasma and reduces the efficiency of photosynthesis.

3.2. SCWL Phytoplasma Alters the Gene Expression of Sucrose Accumulation

In SCWL leaves, several genes related to carbon fixation in photosynthetic organisms, such as *GOT1*, *maeB*, and *pecK*, were upregulated (Figure 5) similarly to those in leaves of cranberry infected with phytoplasmas [29], *Paulownia fortunei* [34], pepper, and *Ziziphus jujuba* Mill [35]. Moreover, *ppdK* that converts pyruvate to phosphoenolpyruvate (PEP) in the C4 photosynthesis pathway [36] was negatively regulated in SCWL leaves similar to that of phytoplasma-infected Paulownia [37]. In addition, *rubisco* was also downregulated like that of phytoplasma-infected Chinese jujube, indicating the limitation of photosynthesis imposed by the carboxylation [20].

In this study, the accumulation of sugar was negatively affected by SCWL phytoplasma by the upregulation of genes involved in starch and sucrose metabolism of leaves (source organ) such as alpha-amylase, beta-amylase, invertase, starch synthase, *SPS* and *TPS* (Figure 5), similarly to those of phytoplasma-infected Chinese jujube [21], grapevine [22], and tomato [38]. Such high expression of invertase, *SPS*, and *TPS* in SCWL leaves, but downregulation in stalks, suggests that sucrose transport is blocked and leads to a high sucrose content in leaves but low in stalks (the sink organ) [38]. Furthermore, such a block would affect sugar signaling and auxin crosstalk that play important roles in plant growth and development [39] and result in thinner stalks compared to the asymptomatic sugarcanes. Likewise, a sugar transporter, of the SWEET family, that maintains the equilibrium between membranes [40] was upregulated in SCWL leaves but downregulated in stalks (Figure 5), which would lead to unbalanced carbohydrate content between source and sink organs, similarly to those reported in coconut palm [41], periwinkle, tobacco [42], and tomato infected by phytoplasmas [43]. On the contrary, SUT that transports sucrose from the photosynthetic organ to the sieve element–companion cell complex was downregulated in SCWL leaves (Figure 5), which is similar to phytoplasma-infected tomato and plants expressing *SUT* antisense RNA, in which the rate of sucrose exudation was lower in the infected plants [43]. Moreover, phytoplasma possibly acts as an additional 'sink' to accumulate carbohydrates from photosynthesis and then blocks sugar transport from the leaf source to stalk sink [41]. Altogether, the loading of sugar from leaves to stalks was disrupted, which resulted in lower sugar content in SCWL stalks than that of asymptomatic sugarcanes.

3.3. SCWL Phytoplasma Alters the Expression of Genes Related to Plant–Pathogen Interactions

Sugars are not only the primary component furnishing energy but also a structural material for defense responses in plants [44]. Pathogens utilize sugar for their survival, while plants adapt their sugar production to stimulate plant defense responses. Invertases are one of the sucrose hydrolytic enzymes for carbohydrate degradation [45], and a key regulator for sucrose accumulation in sugarcane stalks [46]. *SUS* that was upregulated in SCWL stalks is localized in both companion cells and sieve elements of phloem and supplies UDP glucose to induce callose plugs in the sieve pores [47]. Such obstruction of the sieve tubes by callose deposition was previously reported in phytoplasma-infected *C. roseus*, *Euphorbia pulcherrina* [48], and *Vitis vinifera* L. cv. Chardonnay [23]. Moreover, invertase and callose synthases of SCWL sugarcanes were elevated in leaves, which may lead to sieve-tube occlusion, which is required to prevent the colonization of SCWL phytoplasmas. The results of this work are in accordance with those of phytoplasma-infected apple [49] and grapevine [23].

Calcium-binding proteins such as calmodulin (CAM) and calcium-dependent protein kinase (CPK) are involved in the signaling network in plant innate immunity [50]. CPK-CNGCs are cation transport channels regulated by CAM [51]. Those genes were upregulated in phytoplasma-infected American cranberry [29], *P. fortunei* [52], and Mexican lime [53], similarly to the results for SCWL sugarcanes. Upregulation of calcium signaling cascades sensed by a high expression of *CALM* in SCWL sugarcanes leads to Ca^{2+} accumulation. This accumulation should induce the expression of *CPK*, followed by *rboh*, which is a main source for ROS and plays an important role in the plant disease response [54]. Upregulation of *rboh* in SCWL sugarcanes implies that the pathogenic development of SCWL phytoplasma possibly required host Rboh to induce ROS and cell death, which resulted in the anchoring of SCWL disease. However, *CPK* expression in SCWL canes at the maturation phase was unexpectedly downregulated. The result would lead to a decrease in the activity of SOD, POD, and CAT, correlating with the silencing *CPK* in powdery mildew-infected wheat [55] and *CPK* mutants of Arabidopsis that exhibited vigorous pathogen growth and development of disease symptoms [56]. Therefore, downregulation of *CPK* may be a mechanism whereby SCWL phytoplasma triggers sugarcane to allow for more invasions of the pathogen.

MAPK cascades play important roles in the regulation of innate immune responses in plants [57] and directly control gene expression by phosphorylating TFs [58]. These genes were highly expressed in SCWL sugarcanes, indicating that infection by phytoplasma accelerated the expression of the MAPK signaling pathway and boosted the stress response in sugarcane. Furthermore, plant disease resistance genes involved in the innate immune response against pathogen invasion such as *RAR1*, *RPS2*, and *PTI5* [59] were upregulated in SCWL leaves, similarly to those found in witches' broom disease (WBD) in *Nerium indicum* Mill [60]. PTI5, a transcriptional activator of ETH-responsive element binding protein, was enriched in SCWL sugarcanes, which potentially induces the expression of *CAT* [61]. In concordance with our results, *PR1* encoding a toxic protein against pathogen invasion [62] was upregulated in periwinkle leaf yellowing (PLY) phytoplasma-infected periwinkle [63], *Ca.* P. solani-infected tomato [30], *Ca.* P. asteris-infected Arabidopsis [64], *Ca.* P. mali-infected apple [65], and *Ca.* P. solani-infected grapevine [66]. Acceleration of *GK* in SCWL sugarcanes agrees with that of WBD in *Nerium indicum L.*, which is involved in phytoplasma resistance [67]. In this work, WRKY22 that modulates ETH, JA, and SA signaling [68] was upregulated in SCWL sugarcanes. Upregulation of WRKY33 in SCWL sugarcanes is similar to that of *Xanthomonas albilineans*-infected sugarcanes [69]. Moreover, upregulation of *PTI5* together with *WRKYs* in SCWL sugarcanes may activate the defense-related genes involved in phytoalexin production, which plays a crucial role in plant defense against pathogens and contributes to the overall resistance of the plant [70]. Notably, some DEGs were upregulated in leaves but downregulated in stalks due to the fact that leaves are frequently exposed to pathogens and promptly activate defense mechanisms to prevent the invasion [71], whereas stalks primarily serve as structural and nutrient transport rather than having a defense function [72].

3.4. SCWL Phytoplasma Accelerates Plant Hormone Signaling Transduction

Genes responsible for plant hormone signaling pathways were mostly upregulated in SCWL sugarcanes including ABA, AUX, BR, CK, ETH, GA, JA, and SA, which are involved in plant defense responses, particularly host–pathogen interactions [73]. Similar to these results, ABA signaling pathway-related genes such as *PYL*, *ABF*, and *NCED* of phytoplasma-infected apple, Mexican lime, mulberry, and *Paulownia fortunei* were upregulated [53]. Genes involved in the AUX biosynthesis of phytoplasma-infected Mexican lime [53] and sesame [32] were upregulated along with CK and GA-related genes, which are similar to those of SCWL sugarcanes in this work. Furthermore, SCWL phytoplasma caused the upregulation of genes involved in the ETH signaling pathway such as *ERF* and *ETR*, similar to those of *Ca.* P. solani-infected grapevine [74], *Ca.* P. aurantifolia-infected Mexican lime [53], and *Paulownia* witches' broom phytoplasma-infected *P. fortunei* [75]. Upregulation of ETH genes is associated with the activation of genes involved in the production of ROS, which are important components of the plant defense response [22].

Phytoplasma infection controls cell balance and plant development by secretion of the SAP11 effector that targets TCP TF, which, in turn, controls cell proliferation, cell maturation, plant development, and senescence by binding to the lipoxygenase (*LOX*) promoter in *Ca.* P. asteris AY-WB-infected Arabidopsis [76] and mediates the production of oxylipin, a precursor of JA biosynthesis [77]. TCP5 binds directly to the promoter of the *PIF4* gene to increase its expression level and interacts with the PIF4 protein [78]. In tomato and grapevine infected with *Ca.* P. solani, *LOX* was upregulated [30,79]. In this work, TCPs were upregulated in SCWL leaves and resulted in an enhancement of the expression of *LOX*, which may lead to increased levels of JA in sugarcanes. Accumulation of JA can in turn trigger the expression of the downstream JA-responsive genes involved in the plant defense against SCWL phytoplasma. Upregulation of such genes in '*Ca.* P. asteris' strain AY-WB-infected Arabidopsis was similarly observed [76].

The expression of several SA-related genes such as *PR1* was upregulated in SCWL sugarcane, similarly to those of phytoplasma-infected apple, Arabidopsis, grapevine, Madagascar periwinkle, and tomato [30,64,65,80,81]. The PR1 regulates gene expression by interacting with TGA TF that interacts with the promoter of *PR* itself in the presence of SA [82]. Thus, the induction of *PR1* expression would promote the biosynthesis of the plant SA [83]. Upregulation of *TCH4* in SCWL sugarcane is similar to that of phytoplasma-infected paulownia [84], which is likely to be involved in the biosynthesis of hemicellulose, a complex carbohydrate component of plant cell walls [85]. Together with the downregulation of *BRI1*, these results suggest a potential plant defense mechanism by sugarcanes against SCWL phytoplasma.

3.5. SCWL Phytoplasma Alter the Expression of Genes Involved in Secondary Metabolites

Phenylpropanoid precursors including benzenoids, coumarins, flavonoids, hydroxycinnamates, and lignin are involved in plant development and plant–pathogen interactions relating to environmental stress and disease tolerance [86]. In this study, *CAD*, *CCR*, *4CL*, *PAL*, and *POD* (which play roles in the phenylpropanoid pathway) were upregulated in SCWL sugarcanes, leading to an increase in lignin content [87] that would be related to strengthening the cell walls of sugarcane in response to an attack by phytoplasma. Furthermore, several upregulated MYB TFs that are responsible for the regulation of biosynthetic genes for phenylpropanoid and lignin were detected in SCWL sugarcanes like those reported in Arabidopsis [88]. Therefore, the contents of various phenylpropanoid compounds including lignin, flavonoids, and coumarins, which play crucial roles in the resistance against phytoplasma infection [89], were potentially increased in SCWL sugarcane, acting as antioxidants to protect sugarcane from phytoplasma infection.

Flavonoids and isoflavonoids act as antioxidants to protect plants from pathogens by boosting growth and development [90]. The downregulation of *CHS* and *CYP75B1*, which are key genes of flavonoid biosynthesis, in SCWL sugarcane suggests a reduction in flavonoid accumulation, which, in turn, could contribute to an escalation in disease symptoms when the plant is exposed to a pathogenic challenge [91]. Genes related to diterpenoid biosynthesis in phytoplasma-infected lime were upregulated including *GA2*, *GA3*, *KAO*, *GA2ox*, and *GA20ox* [53] like those found in SCWL sugarcanes in this study. Diterpenoids exhibit a species-specific chemical defense against pathogens [92]. A common precursor of diterpenoids, geranylgeranyl diphosphate, also contributes to GA biosynthesis [93]. Moreover, upregulation of sesquiterpenoid and triterpenoid biosynthesis in SCWL sugarcane may increase β-caryophyllene, δ-elemene, and germacrene D, similarly to those phytochemicals found in phytoplasma-infected *Hypericum perforatum* L. [94]. The increase in sesquiterpene biosynthesis in SCWL sugarcanes may attract the insect vector to disperse SCWL phytoplasma as suggested for AP phytoplasma-infected apple and tobacco [95]. Hence, changes in the biosynthesis of flavonoids, terpenoids, and sesquiterpenes in SCWL sugarcane may constitute a component of the pathogen's strategy for disease dissemination.

4. Conclusions

SCWL sugarcanes at the maturation phase harbor 10^4 times more phytoplasma than asymptomatic plants growing in the same field and have smaller stalk diameters, shorter height, and lower sugar content. Transcriptomic analysis revealed alterations in gene expression for several pathways in response to an infection by phytoplasma including photosynthesis, porphyrin and chlorophyll metabolism, starch and sucrose metabolism, plant hormone signaling transduction, flavonoid biosynthesis, and plant–pathogen interactions. Suppression of porphyrin and chlorophyll metabolism coupled with an increase in chlorophyllase expression led to a reduction in chlorophyll levels and in photosynthesis that corresponded to the white leaf phenotype. Enhancement of sugar transporters in leaves, but suppression in stalks, suggests that sucrose transport is blocked and results in lower sucrose levels in stalks. Upregulation of *SUS* leads to a UDP-glucose-induced callose plug formation in sieve pores that serves as a defense mechanism against colonization by phytoplasma. The expression of genes associated with calcium cascades led to Ca^{2+} accumulation, which contributed to plant immunity. However, downregulation of *CPK* would result in a reduction in SOD, POD, and CAT activities, consequently facilitating an invasion by the pathogen. Furthermore, enrichment of signaling transduction pathways (e.g., *MAPK, CAT, PR1, GK*, and transcription factors) and plant hormone signaling transduction (e.g., ABA, AUX, CK, GA, JA, SA, and ETH-responsive genes) positively promotes the defense response of sugarcane against phytoplasma. It is speculated that phytoplasma secretion effectors might target TCPs and cause an expression of *LOX* and subsequently increase the JA level. Upregulation of the phenylpropanoid pathway led to lignin accumulation in response to an attack by phytoplasma, while the expression of sesquiterpene biosynthesis possibly attracts the insect vectors for transmission, thereby facilitating the spread of phytoplasma to other sugarcanes. However, downregulation of flavonoid biosynthesis would potentially intensify the SCWL symptoms upon challenge by phytoplasma. These SCWL sugarcane transcriptomic profiles represent the first comprehensive view of genes and pathways involved in the response of sugarcane towards infection by phytoplasma, which will facilitate in finding specific corresponding marker genes for further sustainable development of disease prevention, protection, and management.

5. Materials and Methods

5.1. Sugarcane Samples and Plant Growth Parameters

Saccharum hybrid cv. KK3 was grown in a field of the MitrPhol sugarcane plantation, Nong Kung Si district, Kalasin province, Thailand (16°42′56.0″ N 103°22′34.2″ E). Treatments of organic fertilizers in January and chemical fertilizers in February [N-P-K = 21-7-18, 312.5 kg/ha April (N-P-K = 46-0-0, 156.25 kg/ha and 21-7-18, 156.25 kg/ha), and July (N-P-K = 21-7-18, 312.5 kg/ha)] were applied. Five of the 12-month-old symptomatic and asymptomatic SCWL sugarcanes were randomly harvested from the same location. Phenotypic characteristics of SCWL sugarcanes were determined visually based on the basic paler leaves compared to the asymptomatic ones. Sugarcane stalks of the 14th–16th internode from the root and 1st–3rd leaves from shoots were immediately sampled between 11 a.m. and 4 p.m. and kept in an RNA storage reagent (Tiangen, Beijing, China) at −20 °C until use. Plant growth parameters including height and stalk diameter were measured. Sugar content was analyzed using a master refractometer (ATAGO, Tokyo, Japan) and recorded as a Brix percentage.

5.2. RNA Extraction

Both stalk and leaf samples were cut into small pieces and immediately frozen with liquid nitrogen and ground to a fine powder. Total RNA was extracted using a GF-1 total RNA extraction kit (Vivantis, Shah Alam, Malaysia), as described by the manufacturer. RNA quality and quantity were analyzed using a Nanodrop ND-100 (Thermo Scientific, Waltham, MA, USA).

5.3. Detection of Phytoplasma in Sugarcane Using Real-Time PCR

DNA samples of SCWL sugarcane were prepared using the CTAB method [96]. The phytoplasma 16S rRNA gene was amplified by nested PCR using primers described previously [48]. The PCR master mix contained 100 ng DNA template, 1× Phusion HF buffer, 200 µM dTNPs, 0.5 µM each primer, and Phusion Hot Start II DNA polymerase (0.02 U/µL) (Thermo Scientific, Waltham, MA, USA). Sugarcane 18S rRNA gene was amplified using 18S rRNA-F and 18S rRNA-R primers [97] following the PCR reaction above. The purified PCR products were sent for sequencing at Macrogen (Seoul, Republic of Korea). The numbers of phytoplasma 16S rRNA and sugarcane 18S rRNA genes were calculated using the formula of Staroscik [98] to generate standard curves.

To determine the number of active phytoplasma in sugarcanes, total RNA was converted to cDNA using random hexamers and RevertAid First strand cDNA synthesis (Thermo Scientific, Waltham, MA, USA). Real-time PCR was performed using a Master Cycler Realplex 4 (Eppendorf, Hauppauge, NY, USA) with a KAPA SYBR FAST qPCR master mix (Kapa Biosystems, Seoul, Republic of Korea) with reaction conditions: 95 °C for 3 min, 40 cycles at 95 °C for 10 s, and 62 °C for 1 min. Quantification of 16S rRNA genes of phytoplasma and 18S rRNA genes of sugarcane was calculated based on the standard curve using comparative cycle threshold (Ct) values [99]. The content of phytoplasma in each sample was normalized using the corresponding sugarcane 18S rRNA gene.

5.4. RNA Sequencing and Transcriptome Analysis

Total RNA samples of 3 biological replicates of symptomatic SCWL leaves (SL), asymptomatic leaves (AL), symptomatic SCWL stalks (SS), and asymptomatic stalks (AS) were sent for RNA sequencing at NovogeneAIT Genomics Singapore Pte (Singapore). Briefly, RNA samples with an RNA integrity number (RIN) of more than 6.5 were subjected to rRNA depletion, mRNA random fragmentation, cDNA construction, and terminal end ligation with poly-A and sequencing adaptors. Then, cDNA libraries were size selected and quantified before sequencing using an Illumina Hiseq X instrument (2×150 bp) (Illumina, San Diego, CA, USA).

Raw reads were uploaded to Galaxy Project (usegalaxy.org) and initially were quality checked using FastQC [100]. Low-quality sequences at Q20 and adapter contamination were filtered out using Trimmomatic [101]. The clean reads were mapped to the monoploid sugarcane genome reference (https://sugarcane-genome.cirad.fr; accessed on 14 February 2020) using Bowtie2/TopHat [102]. The mapped reads were assembled using Cufflink, merged using Cuffmerge, and performed DEGs using Cuffdiff [103]. Principal component analysis (PCA) was performed using DESeq2 [104].

DEGs were cut off at $\log_2$ FC $\geq$ 1 with a false discovery rate (FDR) of <0.05. Gene function was categorized using gene id and protein products from the gene ontology (GO) [11] and general feature format (GFF) annotation file provided by the reference genome sequence [11]. GO terms of all DEGs were subjected to singular enrichment analysis (SEA) in agriGO [105]. The GO term enrichment was calculated by comparing the frequency of gene sets with different GO terms relative to reference gene sets of Plant GO slim using Fisher as statistical test method, Benjamini–Yekutieli as a multi-test adjustment method, and 5 as a minimum number of mapping entries [106]. Then, biological pathways and molecular interaction networks of DEGs were classified by mapping amino acid sequences using GhostKOALA and KEGG mapper [107,108]. Pathway enrichment analysis was performed using the piano in R package based on geneSetStat "reporter" [109]. Pathways with p values of <0.05 were considered as significantly enriched pathways.

5.5. Quantification of Candidate Genes by Real-Time PCR

One microgram of total RNA was used for cDNA synthesis with oligodT primer using RevertAid First strand cDNA synthesis (Thermo Scientific, Waltham, MA, USA). The candidate genes were randomly selected from the significantly enriched pathways and primers were accordingly designed using Primer BLAST [110] (Table S8). Real-time PCR was performed in 3 biological and 3 technical replicates using a Master Cycler Realplex 4 (Eppendorf, Hauppauge, NY, USA) with KAPA SYBR FAST qPCR master mix (Kapa Biosystems, Seoul, Republic of Korea) using reaction conditions: 95 °C for 3 min, 40 cycles at 95 °C for 10 s, and 59 °C for 1 min. Relative levels of the expression of DEGs of SCWL and asymptomatic sugarcanes were determined using the $\Delta\Delta C_T$ method [111] with the *GAPDH* gene used as an internal control for data normalization.

5.6. Statistical Analysis

Plant growth parameters were analyzed by *t*-test to determine statistical significance between groups at *p*-values of <0.05 by Microsoft Excel data analysis.

Supplementary Materials: The following supporting information can be downloaded at: https://www.mdpi.com/article/10.3390/plants13111551/s1. Figure S1: Effect of phytoplasma on sugarcane growth and physiology; Figure S2: Volcano plots of upregulated and downregulated DEGs in comparison between SCWL and asymptomatic sugarcane leaves and stalks; Figure S3: Gene ontology enrichment analysis of DEGs under phytoplasma infection in sugarcane leaves and stalks; Table S1: Quality assessment of RNA-seq data; Table S2: List of enrichment pathways in leaves; Table S3: List of enrichment pathways in stalks; Table S4: List of differentially expressed genes (DEGs) within the targeted pathway in leaves; Table S5: List of differentially expressed genes (DEGs) within the targeted pathway in stalks; Table S6: List of transcription factors in leaves; Table S7: List of transcription factors in stalks; Table S8: List of candidate genes and primers used for real-time PCR in this study.

Author Contributions: Conceptualization, supervision, funding acquisition, A.T.; methodology, investigation, K.L. (Karan Lohmaneeratana) and A.T.; data curation, visualization, K.L. (Karan Lohmaneeratana) and K.L. (Kantinan Leetanasaksakul); writing—original draft preparation, K.L. (Karan Lohmaneeratana) and A.T.; writing—review and editing, A.T. All authors have read and agreed to the published version of the manuscript.

Funding: K.L. (Karan Lohmaneeratana) has been awarded a PhD scholarship under Research and Researcher Fund for Industry (RRi), National Research Council of Thailand (NRCT5-RRI63002-P02) and Mitr Phol Innovation & Research Center. This work was supported by Kasetsart University Research and Development Institute [FF(KU)5.64] and Bioinformatics Academic Association of Thailand (BAT).

Data Availability Statement: RNA sequences are deposited to GenBank database as a sequence read archive (SRA) accession number SRR14169813-24 in the BioProject number PRJNA719388.

Acknowledgments: We especially thank Mitr Phol Innovation & Research Center for providing sugarcane samples for the experiment.

Conflicts of Interest: The authors declare no conflicts of interest.

References

1. Hanboonsong, Y.; Wangkeeree, J.; Kobori, Y. Integrated management of the vectors of sugarcane white leaf disease in Thailand: An update. *Int. Sugar J.* **2017**, *119*, 220–223.
2. Wongkaew, P.; Fletcher, J. Sugarcane white leaf phytoplasma in tissue culture: Long-term maintenance, transmission, and oxytetracycline remission. *Plant Cell Rep.* **2004**, *23*, 426–434. [CrossRef] [PubMed]
3. Matsumoto, T.; Lee, C.; Teng, W. Studies on sugarcane white leaf disease of Taiwan, with special reference to the transmission by a leafhopper, *Epitettix hiroglyphicus* Mats. *Jpn. J. Phytopathol.* **1969**, *35*, 251–259. [CrossRef]
4. Hanboonsong, Y.; Ritthison, W.; Choosai, C.; Sirithorn, P. Transmission of sugarcane white leaf phytoplasma by *Yamatotettix flavovittatus*, a new leafhopper vector. *J. Econ. Entomol.* **2006**, *99*, 1531–1537. [CrossRef]
5. Marcone, C.; Neimark, H.; Ragozzino, A.; Lauer, U.; Seemuller, E. Chromosome sizes of phytoplasmas composing major phylogenetic groups and subgroups. *Phytopathology* **1999**, *89*, 805–810. [CrossRef] [PubMed]
6. Wongkaew, P.; Hanboonsong, Y.; Sirithorn, P.; Choosai, C.; Boonkrong, S.; Tinnangwattana, T.; Kitchareonpanya, R.; Damak, S. Differentiation of phytoplasmas associated with sugarcane and gramineous weed white leaf disease and sugarcane grassy shoot disease by RFLP and sequencing. *Theor. Appl. Genet.* **1997**, *95*, 660–663. [CrossRef]
7. Marcone, C. Phytoplasma diseases of sugarcane. *Sugar Tech.* **2002**, *4*, 79–85. [CrossRef]
8. Nakashima, K.; Murata, N. Destructive plant diseases caused by mycoplasma-like organisms in Asia. *Outlook Agric.* **1993**, *22*, 53–58. [CrossRef]
9. Hagos, H.; Mengistu, L.; Mequanint, Y. Determining optimum harvest age of sugarcane varieties on the newly establishing sugar project in the tropical areas of Tendaho, Ethiopia. *Adv. Crop Sci. Technol.* **2014**, *2*, 156. [CrossRef]
10. Kamwilaisak, K.; Jutakridsada, P.; Iamamornphanth, W.; Saengprachatanarug, K.; Kasemsiri, P.; Konyai, S.; Posom, J.; Chindaprasirt, P. Estimation of sugar content in sugarcane (*Saccharum* spp.) variety Lumpang 92-11 (LK 92-11) and Khon Kaen 3 (KK 3) by near infrared spectroscopy. *Eng. J.* **2021**, *25*, 69–83. [CrossRef]
11. Garsmeur, O.; Droc, G.; Antonise, R.; Grimwood, J.; Potier, B.; Aitken, K.; Jenkins, J.; Martin, G.; Charron, C.; Hervouet, C.; et al. A mosaic monoploid reference sequence for the highly complex genome of sugarcane. *Nat. Commun.* **2018**, *9*, 2638. [CrossRef] [PubMed]
12. Matile, P.; Hortensteiner, S.; Thomas, H. Chlorophyll degradation. *Annu. Rev. Plant Physiol. Plant Mol. Biol.* **1999**, *50*, 67–95. [CrossRef] [PubMed]
13. Fiedor, L.; Zbyradowski, M.; Pilch, M. Tetrapyrrole pigments of photosynthetic antennae and reaction centers of higher plants: Structures, biophysics, functions, biochemistry, mechanisms of regulation, applications. In *Advances in Botanical Research*; Grimm, B., Ed.; Academic Press: Cambridge, MA, USA, 2019; Volume 90, pp. 1–33.

14. Teixeira, A.; Martins, V.; Frusciante, S.; Cruz, T.; Noronha, H.; Diretto, G.; Geros, H. Flavescence dorée-derived leaf yellowing in grapevine (*Vitis vinifera* L.) is associated to a general repression of isoprenoid biosynthetic pathways. *Front. Plant Sci.* **2020**, *11*, 896. [CrossRef] [PubMed]

15. Wu, Y.; Jin, X.; Liao, W.; Hu, L.; Dawuda, M.M.; Zhao, X.; Tang, Z.; Gong, T.; Yu, J. 5-Aminolevulinic acid (ALA) alleviated salinity stress in cucumber seedlings by enhancing chlorophyll synthesis pathway. *Front. Plant Sci.* **2018**, *9*, 635. [CrossRef] [PubMed]

16. Srivastava, L.M. Vegetative storage protein, tuberization, senescence, and abscission. In *Plant Growth and Development*; Srivastava, L.M., Ed.; Academic Press: San Diego, CA, USA, 2002; pp. 473–502. [CrossRef]

17. Bertamini, M.; Nedunchezhian, N. Effects of phytoplasma [stolbur-subgroup (Bois noir-BN)] on photosynthetic pigments, saccharides, ribulose 1, 5-bisphosphate carboxylase, nitrate and nitrite reductases, and photosynthetic activities in field-grown grapevine (*Vitis vinifera* L. cv. Chardonnay) leaves. *Photosynthetica* **2001**, *39*, 119–122. [CrossRef]

18. Ji, X.; Gai, Y.; Zheng, C.; Mu, Z. Comparative proteomic analysis provides new insights into mulberry dwarf responses in mulberry (*Morus alba* L.). *Proteomics* **2009**, *9*, 5328–5339. [CrossRef] [PubMed]

19. Ahmed, E.A.; Farrag, A.A.; Kheder, A.A.; Shaaban, A. Effect of phytoplasma associated with sesame phyllody on ultrastructural modification, physio-biochemical traits, productivity and oil quality. *Plants* **2022**, *11*, 477. [CrossRef] [PubMed]

20. Liu, Z.; Zhao, J.; Liu, M. Photosynthetic responses to phytoplasma infection in chinese jujube. *Plant Physiol. Biochem.* **2016**, *105*, 12–20. [CrossRef] [PubMed]

21. Xue, C.; Liu, Z.; Dai, L.; Bu, J.; Liu, M.; Zhao, Z.; Jiang, Z.; Gao, W.; Zhao, J. Changing host photosynthetic, carbohydrate, and energy metabolisms play important roles in phytoplasma infection. *Phytopathology* **2018**, *108*, 1067–1077. [CrossRef] [PubMed]

22. Hren, M.; Nikolic, P.; Rotter, A.; Blejec, A.; Terrier, N.; Ravnikar, M.; Dermastia, M.; Gruden, K. 'Bois noir' phytoplasma induces significant reprogramming of the leaf transcriptome in the field grown grapevine. *BMC Genom.* **2009**, *10*, 460. [CrossRef] [PubMed]

23. Santi, S.; De Marco, F.; Polizzotto, R.; Grisan, S.; Musetti, R. Recovery from stolbur disease in grapevine involves changes in sugar transport and metabolism. *Front. Plant Sci.* **2013**, *4*, 171. [CrossRef] [PubMed]

24. Jagoueix-Eveillard, S.; Tarendeau, F.; Guolter, K.; Danet, J.L.; Bove, J.M.; Garnier, M. *Catharanthus roseus* genes regulated differentially by mollicute infections. *Mol. Plant Microbe Interact.* **2001**, *14*, 225–233. [CrossRef] [PubMed]

25. Nejat, N.; Cahill, D.M.; Vadamalai, G.; Ziemann, M.; Rookes, J.; Naderali, N. Transcriptomics-based analysis using RNA-Seq of the coconut (*Cocos nucifera*) leaf in response to yellow decline phytoplasma infection. *Mol. Genet. Genom.* **2015**, *290*, 1899–1910. [CrossRef] [PubMed]

26. Bilgin, D.D.; Zavala, J.A.; Zhu, J.; Clough, S.J.; Ort, D.R.; DeLucia, E.H. Biotic stress globally downregulates photosynthesis genes. *Plant Cell Environ.* **2010**, *33*, 1597–1613. [CrossRef] [PubMed]

27. Bertamini, M.; Grando, M.S.; Nedunchezhian, N. Effects of phytoplasma infection on pigments, chlorophyll-protein complex and photosynthetic activities in field grown apple leaves. *Biol. Plant* **2003**, *46*, 237–242. [CrossRef]

28. Asudi, G.O.; Omenge, K.M.; Paulmann, M.K.; Reichelt, M.; Grabe, V.; Mithofer, A.; Oelmuller, R.; Furch, A.C.U. The physiological and biochemical effects on napier grass plants following napier grass stunt phytoplasma infection. *Phytopathology* **2021**, *111*, 703–712. [CrossRef] [PubMed]

29. Pradit, N.; Rodriguez-Saona, C.; Kawash, J.; Polashock, J. Phytoplasma infection influences gene expression in American cranberry. *Front. Ecol. Evol.* **2019**, *7*, 178. [CrossRef]

30. Ahmad, J.N.; Renaudin, J.; Eveillard, S. Expression of defence genes in stolbur phytoplasma infected tomatoes, and effect of defence stimulators on disease development. *Eur. J. Plant Pathol.* **2013**, *139*, 39–51. [CrossRef]

31. Young, A.; Frank, H. Energy transfer reactions involving carotenoids: Quenching of chlorophyll fluorescence. *J. Photochem. Photobiol. B Biol.* **1996**, *36*, 3–15. [CrossRef] [PubMed]

32. Youssef, S.; Safwat, G.; Baset, A.; Shalaby, A.; El-Beltagi, H. Effect of phytoplasma infection on plant hormones, enzymes and their role in infected sesame. *Fresenius Environ. Bull.* **2018**, *27*, 5727–5735.

33. Polívka, T.; Frank, H.A. Molecular factors controlling photosynthetic light harvesting by carotenoids. *Acc. Chem. Res.* **2010**, *43*, 1125–1134. [CrossRef] [PubMed]

34. Wei, Z.; Wang, Z.; Li, X.; Zhao, Z.; Deng, M.; Dong, Y.; Cao, X.; Fan, G. Comparative analysis of *Paulownia fortunei* response to phytoplasma infection with dimethyl sulfate treatment. *Int. J. Genom.* **2017**, *2017*, 6542075. [CrossRef] [PubMed]

35. Ye, X.; Wang, H.; Chen, P.; Fu, B.; Zhang, M.; Li, J.; Zheng, X.; Tan, B.; Feng, J. Combination of iTRAQ proteomics and RNA-seq transcriptomics reveals multiple levels of regulation in phytoplasma-infected *Ziziphus jujuba* Mill. *Hortic. Res.* **2017**, *4*, 17080. [CrossRef] [PubMed]

36. Minges, A.; Groth, G. Small-molecule inhibition of pyruvate phosphate dikinase targeting the nucleotide binding site. *PLoS ONE* **2017**, *12*, e0181139. [CrossRef] [PubMed]

37. Cao, Y.; Fan, G.; Wang, Z.; Gu, Z. Phytoplasma-induced changes in the acetylome and succinylome of *Paulownia tomentosa* provide evidence for involvement of acetylated proteins in witches' broom disease. *Mol. Cell Proteom.* **2019**, *18*, 1210–1226. [CrossRef] [PubMed]

38. Wei, W.; Inaba, J.; Zhao, Y.; Mowery, J.D.; Hammond, R. Phytoplasma infection blocks starch breakdown and triggers chloroplast degradation, leading to premature leaf senescence, sucrose reallocation, and spatiotemporal redistribution of phytohormones. *Int. J. Mol. Sci.* **2022**, *23*, 1810. [CrossRef] [PubMed]

39. Goren, S.; Lugassi, N.; Stein, O.; Yeselson, Y.; Schaffer, A.A.; David-Schwartz, R.; Granot, D. Suppression of sucrose synthase affects auxin signaling and leaf morphology in tomato. *PLoS ONE* **2017**, *12*, e0182334. [CrossRef] [PubMed]
40. Eom, J.S.; Chen, L.Q.; Sosso, D.; Julius, B.T.; Lin, I.W.; Qu, X.Q.; Braun, D.M.; Frommer, W.B. SWEETs, transporters for intracellular and intercellular sugar translocation. *Curr. Opin. Plant Biol.* **2015**, *25*, 53–62. [CrossRef] [PubMed]
41. Maust, B.E.; Espadas, F.; Talavera, C.; Aguilar, M.; Santamaria, J.M.; Oropeza, C. Changes in carbohydrate metabolism in coconut palms infected with the lethal yellowing phytoplasma. *Phytopathology* **2003**, *93*, 976–981. [CrossRef] [PubMed]
42. Lepka, P.; Stitt, M.; Moll, E.; Seemüller, E. Effect of phytoplasmal infection on concentration and translocation of carbohydrates and amino acids in periwinkle and tobacco. *Physiol. Mol. Plant Pathol.* **1999**, *55*, 59–68. [CrossRef]
43. Marco, F.; Batailler, B.; Thorpe, M.R.; Razan, F.; Le Hir, R.; Vilaine, F.; Bouchereau, A.; Martin-Magniette, M.L.; Eveillard, S.; Dinant, S. Involvement of SUT1 and SUT2 sugar transporters in the impairment of sugar transport and changes in phloem exudate contents in phytoplasma-infected plants. *Int. J. Mol. Sci.* **2021**, *22*, 745. [CrossRef] [PubMed]
44. Morkunas, I.; Ratajczak, L. The role of sugar signaling in plant defense responses against fungal pathogens. *Acta Physiol. Plant* **2014**, *36*, 1607–1619. [CrossRef]
45. Tauzin, A.S.; Giardina, T. Sucrose and invertases, a part of the plant defense response to the biotic stresses. *Front. Plant Sci.* **2014**, *5*, 293. [CrossRef] [PubMed]
46. Gayler, K.; Glasziou, K. Physiological functions of acid and neutral invertases in growth and sugar storage in sugar cane. *Physiol. Plant.* **1972**, *27*, 25–31. [CrossRef]
47. Koch, K. Sucrose metabolism: Regulatory mechanisms and pivotal roles in sugar sensing and plant development. *Curr. Opin. Plant Biol.* **2004**, *7*, 235–246. [CrossRef] [PubMed]
48. Christensen, N.M.; Nicolaisen, M.; Hansen, M.; Schulz, A. Distribution of phytoplasmas in infected plants as revealed by real-time PCR and bioimaging. *Mol. Plant Microbe Interact.* **2004**, *17*, 1175–1184. [CrossRef] [PubMed]
49. Musetti, R.; Paolacci, A.; Ciaffi, M.; Tanzarella, O.A.; Polizzotto, R.; Tubaro, F.; Mizzau, M.; Ermacora, P.; Badiani, M.; Osler, R. Phloem cytochemical modification and gene expression following the recovery of apple plants from apple proliferation disease. *Phytopathology* **2010**, *100*, 390–399. [CrossRef] [PubMed]
50. Tena, G.; Boudsocq, M.; Sheen, J. Protein kinase signaling networks in plant innate immunity. *Curr. Opin. Plant Biol.* **2011**, *14*, 519–529. [CrossRef] [PubMed]
51. Wang, L.; Li, M.; Liu, Z.; Dai, L.; Zhang, M.; Wang, L.; Zhao, J.; Liu, M. Genome-wide identification of CNGC genes in Chinese jujube (*Ziziphus jujuba* Mill.) and ZjCNGC2 mediated signalling cascades in response to cold stress. *BMC Genom.* **2020**, *21*, 191. [CrossRef] [PubMed]
52. Yan, L.; Fan, G.; Li, X. Genome-wide analysis of three histone marks and gene expression in *Paulownia fortunei* with phytoplasma infection. *BMC Genom.* **2019**, *20*, 234. [CrossRef] [PubMed]
53. Mardi, M.; Farsad, L.K.; Gharechahi, J.; Salekdeh, G.H. In-depth transcriptome sequencing of Mexican lime trees infected with *Candidatus* Phytoplasma aurantifolia. *PLoS ONE* **2015**, *10*, e0130425. [CrossRef] [PubMed]
54. Ranjan, A.; Jayaraman, D.; Grau, C.; Hill, J.H.; Whitham, S.A.; Ane, J.M.; Smith, D.L.; Kabbage, M. The pathogenic development of *Sclerotinia sclerotiorum* in soybean requires specific host NADPH oxidases. *Mol. Plant Pathol.* **2018**, *19*, 700–714. [CrossRef] [PubMed]
55. Yue, J.-Y.; Jiao, J.-L.; Wang, W.-W.; Jie, X.-R.; Wang, H.-Z. Silencing of the calcium-dependent protein kinase TaCDPK27 improves wheat resistance to powdery mildew. *BMC Plant Biol.* **2023**, *23*, 134. [CrossRef] [PubMed]
56. Lu, Y.-J.; Li, P.; Shimono, M.; Corrion, A.; Higaki, T.; He, S.Y.; Day, B. Arabidopsis calcium-dependent protein kinase 3 regulates actin cytoskeleton organization and immunity. *Nat. Commun.* **2020**, *11*, 6234. [CrossRef] [PubMed]
57. Liu, Z.; Zhao, Z.; Xue, C.; Wang, L.; Wang, L.; Feng, C.; Zhang, L.; Yu, Z.; Zhao, J.; Liu, M. Three main genes in the MAPK cascade involved in the Chinese jujube-phytoplasma interaction. *Forests* **2019**, *10*, 392. [CrossRef]
58. Yang, S.H.; Sharrocks, A.D.; Whitmarsh, A.J. Transcriptional regulation by the MAP kinase signaling cascades. *Gene* **2003**, *320*, 3–21. [CrossRef] [PubMed]
59. Azevedo, C.; Sadanandom, A.; Kitagawa, K.; Freialdenhoven, A.; Shirasu, K.; Schulze-Lefert, P. The RAR1 interactor SGT1, an essential component of R gene-triggered disease resistance. *Science* **2002**, *295*, 2073–2076. [CrossRef] [PubMed]
60. Wang, S.; Wang, S.; Li, M.; Su, Y.; Sun, Z.; Ma, H. Combined transcriptome and metabolome analysis of *Nerium indicum* L. elaborates the key pathways that are activated in response to witches' broom disease. *BMC Plant Biol.* **2022**, *22*, 291. [CrossRef] [PubMed]
61. He, P.; Warren, R.F.; Zhao, T.; Shan, L.; Zhu, L.; Tang, X.; Zhou, J.M. Overexpression of Pti5 in tomato potentiates pathogen-induced defense gene expression and enhances disease resistance to *Pseudomonas syringae* pv. tomato. *Mol. Plant Microbe Interact.* **2001**, *14*, 1453–1457. [CrossRef] [PubMed]
62. Agrios, G. How plants defend themselves against pathogens. In *Plant Pathology*; Elsevier: Amsterdam, The Netherlands, 2005; pp. 207–248. [CrossRef]
63. Tai, C.F.; Lin, C.P.; Sung, Y.C.; Chen, J.C. Auxin influences symptom expression and phytoplasma colonisation in periwinkle infected with periwinkle leaf yellowing phytoplasma. *Ann. Appl. Biol.* **2013**, *163*, 420–429. [CrossRef]
64. Lu, Y.T.; Li, M.Y.; Cheng, K.T.; Tan, C.M.; Su, L.W.; Lin, W.Y.; Shih, H.T.; Chiou, T.J.; Yang, J.Y. Transgenic plants that express the phytoplasma effector SAP11 show altered phosphate starvation and defense responses. *Plant Physiol.* **2014**, *164*, 1456–1469. [CrossRef] [PubMed]

65. Musetti, R.; Farhan, K.; De Marco, F.; Polizzotto, R.; Paolacci, A.; Ciaffi, M.; Ermacora, P.; Grisan, S.; Santi, S.; Osler, R. Differentially-regulated defence genes in *Malus domestica* during phytoplasma infection and recovery. *Eur. J. Plant Pathol.* **2013**, *136*, 13–19. [CrossRef]

66. Paolacci, A.R.; Catarcione, G.; Ederli, L.; Zadra, C.; Pasqualini, S.; Badiani, M.; Musetti, R.; Santi, S.; Ciaffi, M. Jasmonate-mediated defence responses, unlike salicylate-mediated responses, are involved in the recovery of grapevine from bois noir disease. *BMC Plant Biol.* **2017**, *17*, 118. [CrossRef] [PubMed]

67. Yang, Y.; Zhao, J.; Liu, P.; Xing, H.; Li, C.; Wei, G.; Kang, Z. Glycerol-3-phosphate metabolism in wheat contributes to systemic acquired resistance against *Puccinia striiformis* f. sp. tritici. *PLoS ONE* **2013**, *8*, e81756. [CrossRef] [PubMed]

68. Hsu, F.C.; Chou, M.Y.; Chou, S.J.; Li, Y.R.; Peng, H.P.; Shih, M.C. Submergence confers immunity mediated by the WRKY22 transcription factor in arabidopsis. *Plant Cell* **2013**, *25*, 2699–2713. [CrossRef] [PubMed]

69. Ntambo, M.S.; Meng, J.Y.; Rott, P.C.; Henry, R.J.; Zhang, H.L.; Gao, S.J. Comparative transcriptome profiling of resistant and susceptible sugarcane cultivars in response to infection by *Xanthomonas albilineans*. *Int. J. Mol. Sci.* **2019**, *20*, 6138. [CrossRef] [PubMed]

70. Zhou, J.; Tang, X.; Martin, G.B. The Pto kinase conferring resistance to tomato bacterial speck disease interacts with proteins that bind a cis-element of pathogenesis-related genes. *EMBO J.* **1997**, *16*, 3207–3218. [CrossRef] [PubMed]

71. Meddya, S.; Meshram, S.; Sarkar, D.; S, R.; Datta, R.; Singh, S.; Avinash, G.; Kumar Kondeti, A.; Savani, A.K.; Thulasinathan, T. Plant stomata: An unrealized possibility in plant defense against invading pathogens and stress tolerance. *Plants* **2023**, *12*, 3380. [CrossRef] [PubMed]

72. Ana, L.-M.; Rogelio, S.; Xose Carlos, S.; Rosa Ana, M. Cell wall composition impacts structural characteristics of the stems and thereby the biomass yield. *J. Agric. Food Chem.* **2022**, *70*, 3136–3141. [CrossRef] [PubMed]

73. De León, I.P.; Montesano, M. Activation of defense mechanisms against pathogens in mosses and flowering plants. *Int. J. Mol. Sci.* **2013**, *14*, 3178–3200. [CrossRef] [PubMed]

74. Hren, M.; Ravnikar, M.; Brzin, J.; Ermacora, P.; Carraro, L.; Bianco, P.; Casati, P.; Borgo, M.; Angelini, E.; Rotter, A. Induced expression of sucrose synthase and alcohol dehydrogenase I genes in phytoplasma-infected grapevine plants grown in the field. *Plant Pathol.* **2009**, *58*, 170–180. [CrossRef]

75. Fan, G.; Xu, E.; Deng, M.; Zhao, Z.; Niu, S. Phenylpropanoid metabolism, hormone biosynthesis and signal transduction-related genes play crucial roles in the resistance of *Paulownia fortunei* to paulownia witches' broom phytoplasma infection. *Genes. Genom.* **2015**, *37*, 913–929. [CrossRef]

76. Sugio, A.; Kingdom, H.N.; MacLean, A.M.; Grieve, V.M.; Hogenhout, S.A. Phytoplasma protein effector SAP11 enhances insect vector reproduction by manipulating plant development and defense hormone biosynthesis. *Proc. Natl. Acad. Sci. USA* **2011**, *108*, E1254–E1263. [CrossRef] [PubMed]

77. Schommer, C.; Palatnik, J.F.; Aggarwal, P.; Chetelat, A.; Cubas, P.; Farmer, E.E.; Nath, U.; Weigel, D. Control of jasmonate biosynthesis and senescence by miR319 targets. *PLoS Biol.* **2008**, *6*, e230. [CrossRef] [PubMed]

78. Han, X.; Yu, H.; Yuan, R.; Yang, Y.; An, F.; Qin, G. Arabidopsis transcription factor TCP5 controls plant thermomorphogenesis by positively regulating PIF4 activity. *iScience* **2019**, *15*, 611–622. [CrossRef] [PubMed]

79. Dermastia, M.; Nikolic, P.; Chersicola, M.; Gruden, K. Transcriptional profiling in infected and recovered grapevine plant responses to 'Candidatus Phytoplasma solani'. *Phytopathog. Mollicutes* **2015**, *5*, S123–S124. [CrossRef]

80. Giorno, F.; Guerriero, G.; Biagetti, M.; Ciccotti, A.M.; Baric, S. Gene expression and biochemical changes of carbohydrate metabolism in in vitro micro-propagated apple plantlets infected by 'Candidatus Phytoplasma mali'. *Plant Physiol. Biochem.* **2013**, *70*, 311–317. [CrossRef] [PubMed]

81. Liu, L.Y.; Tseng, H.I.; Lin, C.P.; Lin, Y.Y.; Huang, Y.H.; Huang, C.K.; Chang, T.H.; Lin, S.S. High-throughput transcriptome analysis of the leafy flower transition of *Catharanthus roseus* induced by peanut witches'-broom phytoplasma infection. *Plant Cell Physiol.* **2014**, *55*, 942–957. [CrossRef] [PubMed]

82. Fu, Z.Q.; Dong, X. Systemic acquired resistance: Turning local infection into global defense. *Annu. Rev. Plant Biol.* **2013**, *64*, 839–863. [CrossRef] [PubMed]

83. Robert-Seilaniantz, A.; Grant, M.; Jones, J.D. Hormone crosstalk in plant disease and defense: More than just jasmonate-salicylate antagonism. *Annu. Rev. Phytopathol.* **2011**, *49*, 317–343. [CrossRef] [PubMed]

84. Mou, H.Q.; Lu, J.; Zhu, S.F.; Lin, C.L.; Tian, G.Z.; Xu, X.; Zhao, W.J. Transcriptomic analysis of paulownia infected by paulownia witches'-broom phytoplasma. *PLoS ONE* **2013**, *8*, e77217. [CrossRef] [PubMed]

85. Stratilova, B.; Kozmon, S.; Stratilova, E.; Hrmova, M. Plant xyloglucan xyloglucosyl transferases and the cell wall structure: Subtle but significant. *Molecules* **2020**, *25*, 5619. [CrossRef] [PubMed]

86. Geng, D.; Shen, X.; Xie, Y.; Yang, Y.; Bian, R.; Gao, Y.; Li, P.; Sun, L.; Feng, H.; Ma, F.; et al. Regulation of phenylpropanoid biosynthesis by MdMYB88 and MdMYB124 contributes to pathogen and drought resistance in apple. *Hortic. Res.* **2020**, *7*, 102. [CrossRef] [PubMed]

87. Dixon, R.A.; Achnine, L.; Kota, P.; Liu, C.J.; Reddy, M.S.; Wang, L. The phenylpropanoid pathway and plant defence-a genomics perspective. *Mol. Plant Pathol.* **2002**, *3*, 371–390. [CrossRef] [PubMed]

88. Zhong, R.; Ye, Z.H. MYB46 and MYB83 bind to the SMRE sites and directly activate a suite of transcription factors and secondary wall biosynthetic genes. *Plant Cell Physiol.* **2012**, *53*, 368–380. [CrossRef] [PubMed]

89. Xie, M.; Zhang, J.; Tschaplinski, T.J.; Tuskan, G.A.; Chen, J.G.; Muchero, W. Regulation of lignin biosynthesis and its role in growth-defense tradeoffs. *Front. Plant Sci.* **2018**, *9*, 1427. [CrossRef] [PubMed]

90. Simmonds, M.S.; Stevenson, P.C. Effects of isoflavonoids from *Cicer* on larvae of *Heliocoverpa armigera*. *J. Chem. Ecol.* **2001**, *27*, 965–977. [CrossRef] [PubMed]

91. Karre, S.; Kumar, A.; Yogendra, K.; Kage, U.; Kushalappa, A.; Charron, J.-B. HvWRKY23 regulates flavonoid glycoside and hydroxycinnamic acid amide biosynthetic genes in barley to combat Fusarium head blight. *Plant Mol. Biol.* **2019**, *100*, 591–605. [CrossRef] [PubMed]

92. Murphy, K.M.; Zerbe, P. Specialized diterpenoid metabolism in monocot crops: Biosynthesis and chemical diversity. *Phytochemistry* **2020**, *172*, 112289. [CrossRef] [PubMed]

93. Olszewski, N.; Sun, T.P.; Gubler, F. Gibberellin signaling: Biosynthesis, catabolism, and response pathways. *Plant Cell* **2002**, *14*, S61–S80. [CrossRef] [PubMed]

94. Bruni, R.; Pellati, F.; Bellardi, M.G.; Benvenuti, S.; Paltrinieri, S.; Bertaccini, A.; Bianchi, A. Herbal drug quality and phytochemical composition of *Hypericum perforatum* L. affected by ash yellows phytoplasma infection. *J. Agric. Food Chem.* **2005**, *53*, 964–968. [CrossRef] [PubMed]

95. Rid, M.; Mesca, C.; Ayasse, M.; Gross, J. Apple proliferation phytoplasma influences the pattern of plant volatiles emitted depending on pathogen virulence. *Front. Ecol. Evol.* **2016**, *3*, 152. [CrossRef]

96. Honeycutt, R.J.; Sobral, B.W.S.; Keim, P.; Irvine, J.E. A rapid DNA extraction method for sugarcane and its relatives. *Plant Mol. Biol. Rep.* **1992**, *10*, 66–72. [CrossRef]

97. Gai, Y.P.; Yuan, S.S.; Liu, Z.Y.; Zhao, H.N.; Liu, Q.; Qin, R.L.; Fang, L.J.; Ji, X.L. Integrated phloem sap mRNA and protein expression analysis reveals phytoplasma-infection responses in mulberry. *Mol. Cell Proteom.* **2018**, *17*, 1702–1719. [CrossRef] [PubMed]

98. Staroscik, A. Calculator for Determining the Number of Copies of a Template. Available online: https://cels.uri.edu/gsc/cndna.html (accessed on 25 December 2019).

99. Ikten, C.; Ustun, R.; Catal, M.; Yol, E.; Uzun, B. Multiplex real-time qPCR assay for simultaneous and sensitive detection of phytoplasmas in sesame plants and insect vectors. *PLoS ONE* **2016**, *11*, e0155891. [CrossRef] [PubMed]

100. Andrews, S. FastQC: A Quality Control Tool for High Throughput Sequence Data. Available online: http://www.bioinformatics.babraham.ac.uk/projects/fastqc/ (accessed on 3 October 2019).

101. Bolger, A.M.; Lohse, M.; Usadel, B. Trimmomatic: A flexible trimmer for Illumina sequence data. *Bioinformatics* **2014**, *30*, 2114–2120. [CrossRef] [PubMed]

102. Trapnell, C.; Pachter, L.; Salzberg, S.L. TopHat: Discovering splice junctions with RNA-Seq. *Bioinformatics* **2009**, *25*, 1105–1111. [CrossRef] [PubMed]

103. Trapnell, C.; Williams, B.A.; Pertea, G.; Mortazavi, A.; Kwan, G.; van Baren, M.J.; Salzberg, S.L.; Wold, B.J.; Pachter, L. Transcript assembly and quantification by RNA-Seq reveals unannotated transcripts and isoform switching during cell differentiation. *Nat. Biotechnol.* **2010**, *28*, 511–515. [CrossRef] [PubMed]

104. Love, M.I.; Huber, W.; Anders, S. Moderated estimation of fold change and dispersion for RNA-seq data with DESeq2. *Genome Biol.* **2014**, *15*, 550. [CrossRef] [PubMed]

105. Tian, T.; Liu, Y.; Yan, H.; You, Q.; Yi, X.; Du, Z.; Xu, W.; Su, Z. agriGO v2.0: A GO analysis toolkit for the agricultural community, 2017 update. *Nucleic Acids Res.* **2017**, *45*, W122–W129. [CrossRef] [PubMed]

106. Benjamini, Y.; Yekutieli, D. The control of the false discovery rate in multiple testing under dependency. *Ann. Stat.* **2001**, *29*, 1165–1188. [CrossRef]

107. Kanehisa, M.; Sato, Y. KEGG Mapper for inferring cellular functions from protein sequences. *Protein Sci.* **2020**, *29*, 28–35. [CrossRef] [PubMed]

108. Kanehisa, M.; Sato, Y.; Morishima, K. BlastKOALA and GhostKOALA: KEGG tools for functional characterization of genome and metagenome sequences. *J. Mol. Biol.* **2016**, *428*, 726–731. [CrossRef] [PubMed]

109. Varemo, L.; Nielsen, J.; Nookaew, I. Enriching the gene set analysis of genome-wide data by incorporating directionality of gene expression and combining statistical hypotheses and methods. *Nucleic Acids Res.* **2013**, *41*, 4378–4391. [CrossRef] [PubMed]

110. Ye, J.; Coulouris, G.; Zaretskaya, I.; Cutcutache, I.; Rozen, S.; Madden, T.L. Primer-BLAST: A tool to design target-specific primers for polymerase chain reaction. *BMC Bioinform.* **2012**, *13*, 134. [CrossRef] [PubMed]

111. Schmittgen, T.D.; Livak, K.J. Analyzing real-time PCR data by the comparative C(T) method. *Nat. Protoc.* **2008**, *3*, 1101–1108. [CrossRef] [PubMed]

Article

Screening of Sugarcane Proteins Associated with Defense against *Leifsonia xyli* subsp. *xyli*, Agent of Ratoon Stunting Disease

Xiao-Qiu Zhang [1,†], Yong-Jian Liang [2,†], Bao-Qing Zhang [1], Mei-Xin Yan [1], Ze-Ping Wang [1], Dong-Mei Huang [1], Yu-Xin Huang [1], Jing-Chao Lei [1], Xiu-Peng Song [1,*] and Dong-Liang Huang [1,*]

1   Key Laboratory of Sugarcane Biotechnology and Genetic Improvement (Guangxi), Ministry of Agriculture and Rural Affairs/Guangxi Key Laboratory of Sugarcane Genetic Improvement/Sugarcane Research Institute, Guangxi Academy of Agricultural Sciences, Nanning 530007, China; zhangxiaoqiuxhd@163.com (X.-Q.Z.); zbqsxau@126.com (B.-Q.Z.); yanmeixin@gxaas.net (M.-X.Y.); yaheng830619@163.com (Z.-P.W.); huang18260951004@163.com (D.-M.H.); huangyuxin13@163.com (Y.-X.H.); jchlei1130@163.com (J.-C.L.)
2   Guangxi South Subtropical Agricultural Science Research Institute, Chongzuo 532415, China; yongjianliang_605@163.com
*   Correspondence: xiupengsong@163.com (X.-P.S.); hdl666@163.com (D.-L.H.)
†   These authors contributed equally to this work.

Abstract: Sugarcane is the most important sugar crop and one of the leading energy-producing crops in the world. Ratoon stunting disease (RSD), caused by the bacterium *Leifsonia xyli* subsp. *xyli*, poses a huge threat to ratoon crops, causing a significant yield loss in sugarcane. Breeding resistant varieties is considered the most effective and fundamental approach to control RSD in sugarcane. The exploration of resistance genes forms the foundation for breeding resistant varieties through molecular technology. The *pglA* gene is a pathogenicity gene in *L. xyli* subsp. *xyli*, encoding an endopolygalacturonase. In this study, the pglA gene from *L. xyli* subsp. *xyli* and related microorganisms was analyzed. Then, a non-toxic, non-autoactivating pglA bait was successfully expressed in yeast cells. Simultaneously the yeast two-hybrid library was generated using RNA from the *L. xyli* subsp. *xyli*-infected sugarcane. Screening the library with the pglA bait uncovered proteins that interacted with pglA, primarily associated with ABA pathways and the plant immune system, suggesting that sugarcane employs these pathways to respond to *L. xyli* subsp. *xyli*, triggering pathogenicity or resistance. The expression of genes encoding these proteins was also investigated in *L. xyli* subsp. *xyli*-infected sugarcane, suggesting multiple layers of regulatory mechanisms in the interaction between sugarcane and *L. xyli* subsp. *xyli*. This work promotes the understanding of plant–pathogen interaction and provides target proteins/genes for molecular breeding to improve sugarcane resistance to *L. xyli* subsp. *xyli*.

Keywords: sugarcane; endopolygalacturonase; molecular breeding

Citation: Zhang, X.-Q.; Liang, Y.-J.; Zhang, B.-Q.; Yan, M.-X.; Wang, Z.-P.; Huang, D.-M.; Huang, Y.-X.; Lei, J.-C.; Song, X.-P.; Huang, D.-L. Screening of Sugarcane Proteins Associated with Defense against *Leifsonia xyli* subsp. *xyli*, Agent of Ratoon Stunting Disease. *Plants* **2024**, *13*, 448. https://doi.org/10.3390/plants13030448

Academic Editor: Assunta Bertaccini

Received: 10 December 2023
Revised: 29 January 2024
Accepted: 1 February 2024
Published: 3 February 2024

1. Introduction

Sugarcane is the most important sugar crop and one of the leading energy-producing crops in the world. Cane sugar accounts for about 87% of the total sugar production in China [1]. *Leifsonia xyli* subsp. *xyli*, the agent of RSD, is a Gram-positive bacterium. *L. xyli* subsp. *xyli* colonizes the vascular bundles of the sugarcane stem, mesophyll cells, apical tissue, and the area around the leaf sheath [2,3]. *L. xyli* subsp. *xyli* adheres to the inside and outside of cell walls, making the cell wall of cane stem and leaves dissolve or break [4,5]. Due to its highly contagious nature and the absence of symptoms, ratoon stunting disease (RSD) caused by *L. xyli* subsp. *xyli* is a significant concern for sugarcane cultivation [6]. RSD poses a significant threat to ratoon crops, causing reduced yields, stalk abnormalities, and delayed harvest [7]. In China, RSD is also a serious disease widely spread in sugarcane planting regions, with an incidence between 48.9 and 100% [8]. To manage RSD, a combination of practices, such as using disease-free seed cane and adopting resistant varieties, are crucial. Hot water treatment can control RSD to some extent, but it comes with high costs and cannot eliminate the pathogen completely [7,9].

Breeding of varieties with resistance to RSD is considered the most effective and fundamental approach to controlling RSD on sugarcane. However, due to the fact that sugarcane is a genetically complex allopolyploid plant, it is difficult to combine various favorable traits together through conventional hybrid breeding. Therefore, developing a sugarcane variety resistant to RSD through a conventional breeding program is a challenging task. Molecular breeding is an important method for developing RSD-resistant sugarcane varieties, as it has been successful in improving various traits in multiple crops [10–12].

The proteins directly interacting with pathogenic proteins are the ones most likely to participate in defense against the pathogen or be involved in pathogenicity. Endo-PGs, encoded by the *pglA* gene, are essential factors for plant pathogens to colonize the host and degrade pectin in the plant cell walls, resulting in the maceration of host tissues [13,14]. The *pglA* gene has been proved to be a pathogenicity gene in *Pseudomonas solanacearum* [15], *Xylella fastidiosa* [16], and *Xanthomonas oryzae* pv. *oryzae* [17]. Even the endo-PG protein encoded by *pglA* can cause disease. Necrotic lesions are formed after inoculating the proteins SsPG3 and SsPG6 from *Sclerotinia sclerotiorum* on the leaves of *Arabidopsis thaliana* [18]. The leaves of cotton appear chlorotic, and the plants exhibit dwarfism after inoculation with the proteins of VDPG1 and FOVPG1 derived from *Verticillium dahliae* and *Fusarium oxysporum* f. sp. *vasinfectum* [19].

The genome of *L. xyli* subsp. *xyli* has been sequenced in our previous work [4] (https://www.ncbi.nlm.nih.gov/nuccore/LFYU00000000 accessed on 26 May 2023) and by Monteiro-Vitorello et al. [20], as well as by Wang et al. [21]. In the genome, a *pglA* (https://www.ncbi.nlm.nih.gov/gene/2939326#reference-sequences accessed on 26 May 2023), annotated to be an endopolygalacturonase (endo-PG; EC 3.2.1.15), was proposed to be a pathogenicity gene of *L. xyli* subsp. *xyli* by Monteiro-Vitorello et al. [20]. Therefore, the proteins directly interacting with the pglA gene of *L. xyli* subsp. *xyli* will be the best candidate factors against this pathogen or involved in its pathogenicity. The yeast two-hybrid (Y2H) method is the most suitable solution to screen these interacting proteins, since Y2H has been extensively utilized to unravel protein–protein interactions [22] and has helped to reveal the molecular mechanisms underlying the pathogenic processes [23,24]. Thus, in this work, the Y2H system was used to identify the proteins directly interacting with the pglA protein. This approach aims to accelerate the understanding of the mechanisms related to pglA-triggered pathogenicity and to provide target proteins/genes for improving sugarcane resistance to *L. xyli* subsp. *xyli* by molecular breeding.

2. Results

2.1. Characteristics of pglA Gene and Protein

The open reading frame (ORF) of the *pglA* gene encompasses conserved domains, notably the glycosyl hydrolases family 28 (spanning positions 577 to 1320), and the transcription termination factor Rho (encompassing positions 477-1196 and 951-1460). Furthermore, the pglA protein is anticipated to possess a signal peptide of Sec/SPI. This signal peptide is projected to be cleaved between positions 23 and 24 (Figure 1A,B). The pglA protein shows high similarity with those in other two *L. xyli* subsp. *xyli* strains (Figure 1C). Subsequently, this signal peptide was excised to construct the pglA bait vector.

2.2. Toxicity, Autoactivation and Protein Expression of pglA Bait

Colonies were able to grow on the SD/-Trp and SD/-Trp/X media, similar to those on the positive control pGBKT7 medium. However, no colony developed on the SD/-Trp/X/A medium (Figure 2A). This observation suggests that there was no toxicity or autoactivation exhibited by the pglA bait toward yeast cells.

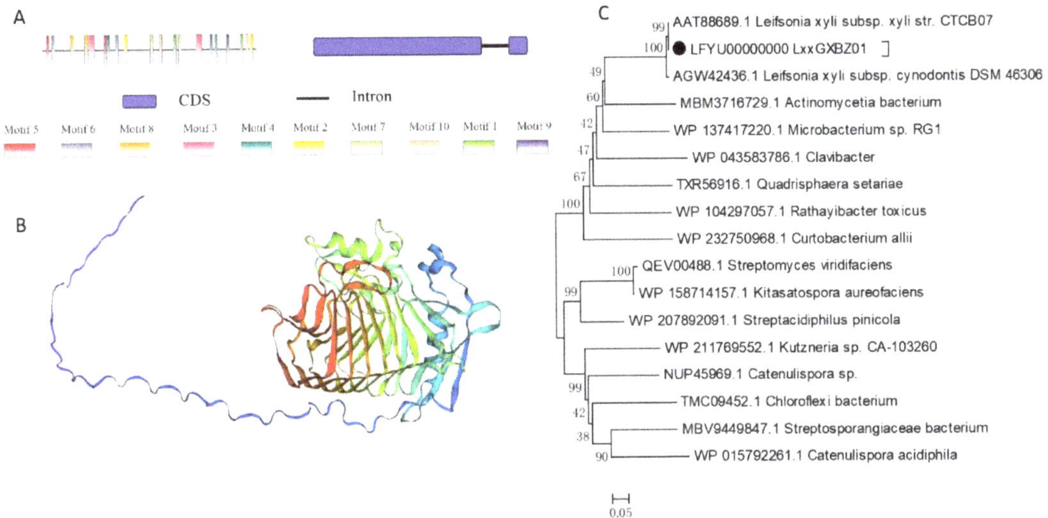

Figure 1. Characteristics of *pglA* gene and protein. (**A**) Schematic structure of gene and protein generated by TBtools-II software. (**B**) Tertiary structure of pglA. The colors illustrate the sequence transitioning from the N-terminus to the C-terminus, progressing from blue to red. (**C**) Phylogenetic analysis of pglA in bacterium.

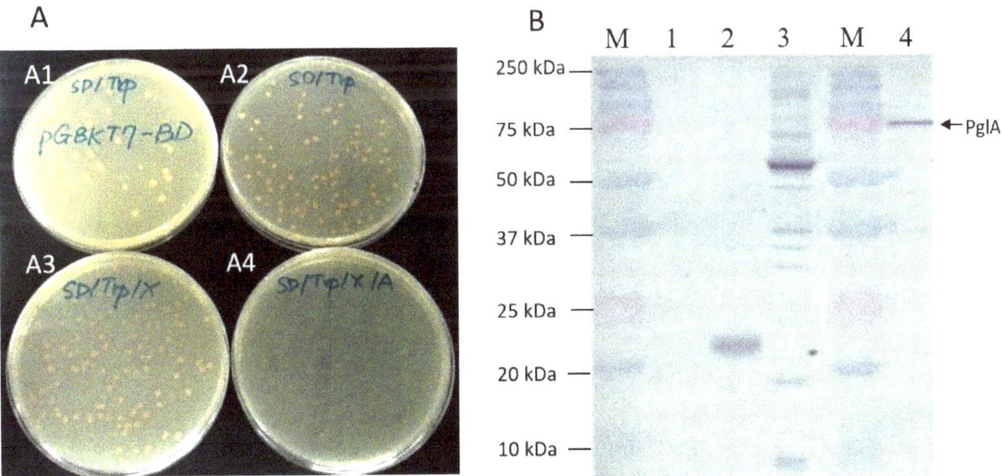

Figure 2. Testing for toxicity, autoactivation and protein expression of pglA bait. (**A**) The protein of pglA bait was expressed in Y2HGold. The strain of Y2HGold, transformed with the pGBKT7-BD vector, was spread on SD-Trp medium (**A1**). The transformed cells were spread on SD/-Trp (**A2**), SD/-Trp/X (**A3**), and SD/-Trp/X/A media (**A4**), respectively. (**B**) Protein expression of pglA bait. M = marker; 1 = Y2HGold without recombinant plasmid; 2 = Y2HGold was transformed with pGBKT7-BD plasmid; 3 = Y2HGold was transformed with pGBKT7-53 plasmid; 4 = Y2HGold was transformed with pGBKT7-pglA plasmid.

No target protein was expressed in the Y2HGold strain lacking recombinant plasmid. However, a target protein with a size of 22 kDa was expressed in the Y2HGold strain containing the pGBKT7-BD plasmid, and a target protein with a size of 57 kDa was expressed in the Y2HGold strain harboring the pGBKT7-53 plasmid. A target protein with a size of

74 kDa was detected in the Y2HGold carrying the pGBKT7-pglA plasmid. The outcomes affirm the successful expression of the pglA bait in the Y2HGold strain (Figure 2B).

2.3. Construction of Y2H Library

The total RNA extraction yielded distinct bands corresponding to the ribosomal 28 S and 18 S (Figure 3A), and the total RNA displayed an OD_{260}/OD_{280} ratio of 2.11, suggesting the high quality of the total RNA. Subsequently, cDNA was synthesized using SMART technology (Clontech) (Figure 3B), and purification using a Chroma Spin-1000 column eliminated smaller cDNA fragments (Figure 3C). The purified cDNA was then cloned into the pGADT7 vector to construct a Y2H library. Remarkably, the Y2H library consisted of over 1.0×10^6 primary clones in total, with a final library titer surpassing 4.0×10^7 cfu/mL. The inserted cDNA fragments ranged in size from 0.4 to 2.0 kb (Figure 3D). Impressively, the cDNA library showcased an approximate 100% recombination rate.

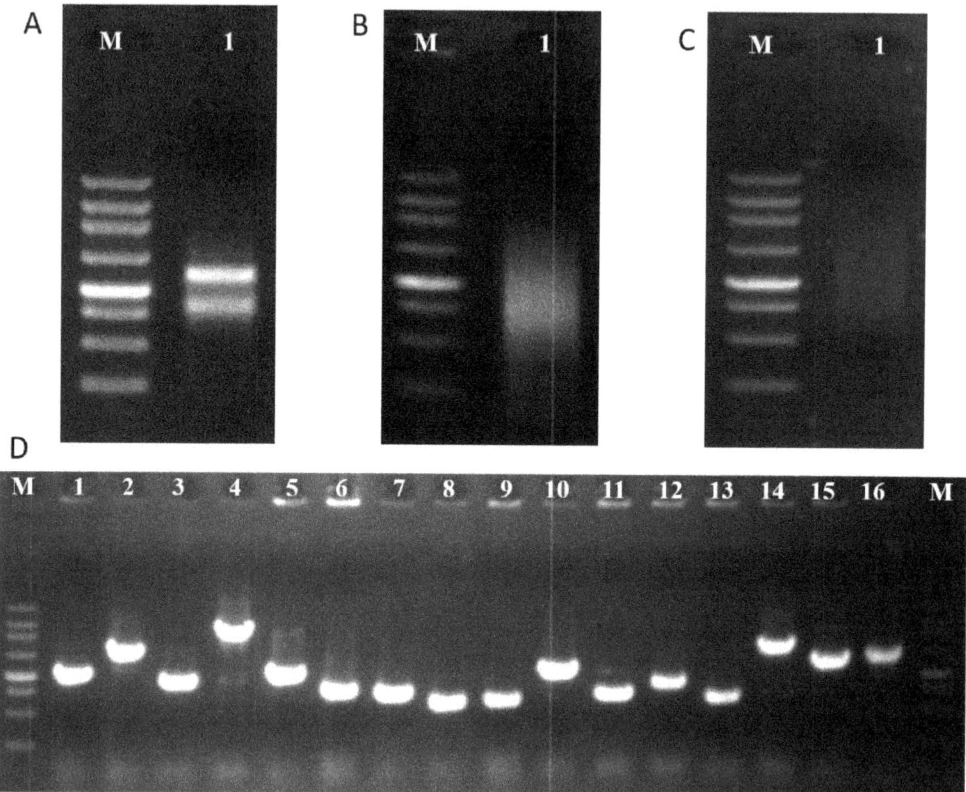

Figure 3. Construction of Y2H library. (**A**) Agarose gel electrophoresis of total RNA from *S. officinarum*. (**B**) cDNA after normalization was evaluated by 1% agarose gel electrophoresis. (**C**) Purified cDNA after eliminating small cDNA fragments using a Chroma Spin-1000 column. (**D**) PCR amplification for the inserted fragments of cDNA library. M = 250 bp DNA ladder (Takara, Dalian, China).

2.4. Screening of pglA-Interacting Proteins

Under a 1/10,000 dilution, a count of 560 clones emerged on SD/-Leu medium, while more than 2000 clones grew on SD/-Trp medium, and 84 clones formed on SD/-Leu/-Trp medium. This accumulation led to a total of 9.66×10^6 zygotes clones (Figure 4A–C). Upon re-inoculating the blue clones from the SD/-Leu/-Trp/X/A medium onto the SD/-Ade/-His/-Leu/-Trp/X-a-Gal/AbA medium, seven clones surfaced (Figure 4D). PCR amplifications of these clones exhibited a distinct primary band, affirming that the positive clones carried the exclusive AD plasmid (Figure 4D). The colonies of both the positive and negative controls exhibited normal growth, providing confidence in the reliability of our results (Figure 4F,G).

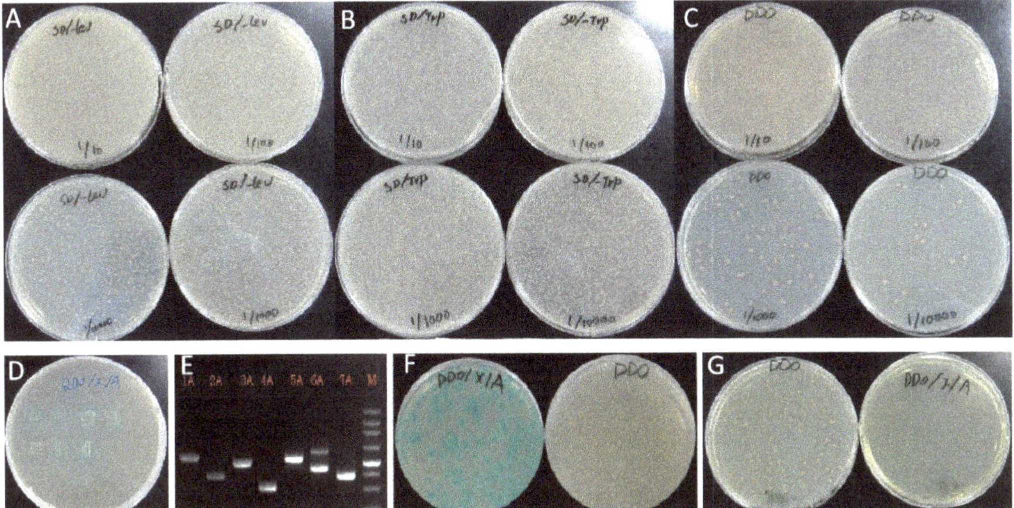

Figure 4. Screening of pglA-interacting proteins. Clones grew on the medium of SD/-Leu (**A**), SD/-Trp (**B**) and SD/-Leu/-Trp (**C**), respectively. Seven clones grew on an SD/-Ade/-His/-Leu/-Trp/X-a-Gal/AbA medium (**D**) and the PCR amplifications of seven clones showed that the main band was distinct (**E**). The colonies of positive and negative controls grew normally (**F,G**).

2.5. Identification of Proteins Interacting with pglA

The outcomes from both BLAST and UniProt analyses identified a total of six characterized proteins. They included like heterochromatin protein (LHP1), SNF1-related protein kinase regulatory subunit beta-1, histidine-rich calcium-binding protein, DNA-directed RNA polymerase III subunit RPC4, E3 ubiquitin-protein ligase RGLG2, and polyubiquitin-C. Furthermore, one protein was recognized as the yeast vector pDEST-GADT7 (Table 1).

Table 1. The information of the proteins interacting with pglA.

No.	Gene Name	UniProtKB Name	Protein Name	Homology Function in *Sorghum bicolor*
1A	AtLHP1	A0A1P8B9G6_ARATH	Like heterochromatin protein (LHP1)	Probable chromo domain protein LHP1
2A	KINB1	KINB1_ARATH	SNF1-related protein kinase regulatory subunit beta-1	SNF1-related protein kinase regulatory subunit beta-1
3A	Hrc	A0A8I5ZZ96	Histidine-rich calcium-binding protein	Sarcoplasmic reticulum histidine-rich calcium-binding protein
4A	/	/	/	Yeast vector pDEST-GADT7
5A	RPC4	RPC4_YEAST	DNA-directed RNA polymerase III subunit RPC4	DNA-directed RNA polymerase III subunit RPC4
6A	RGLG2	RGLG2_ARATH	E3 ubiquitin-protein ligase RGLG2	E3 ubiquitin-protein ligase RGLG2
7A	UBC	UBC_HUMAN	Polyubiquitin-C	Polyubiquitin

2.6. Re-Test of Proteins Interaction

To verify the protein interaction, one protein, 2A (SNF1-related protein kinase regulatory subunit beta-1), was selected for further analysis. The partial sequence of 2A with

619 bp was obtained by sequencing. Then RACE technology was used to obtain the full-length sequence of the 2A gene, SoSnRK1β1 (OP390183). By constructing bait vectors and performing Y2H analysis, the co-cultured strains grew normally with a blue appearance on DDO/X, DDO/X/A, QDO/X/A media, indicating that the pGADT7-SoSnRK1β1 plasmid interacted with the pGBKT7-pglA plasmid (Figure 5A). No false positives of the protein were expressed by the SoSnRK1β1-Y187 strains (Figure 5B), confirming the reliability of the proteins identified in this work.

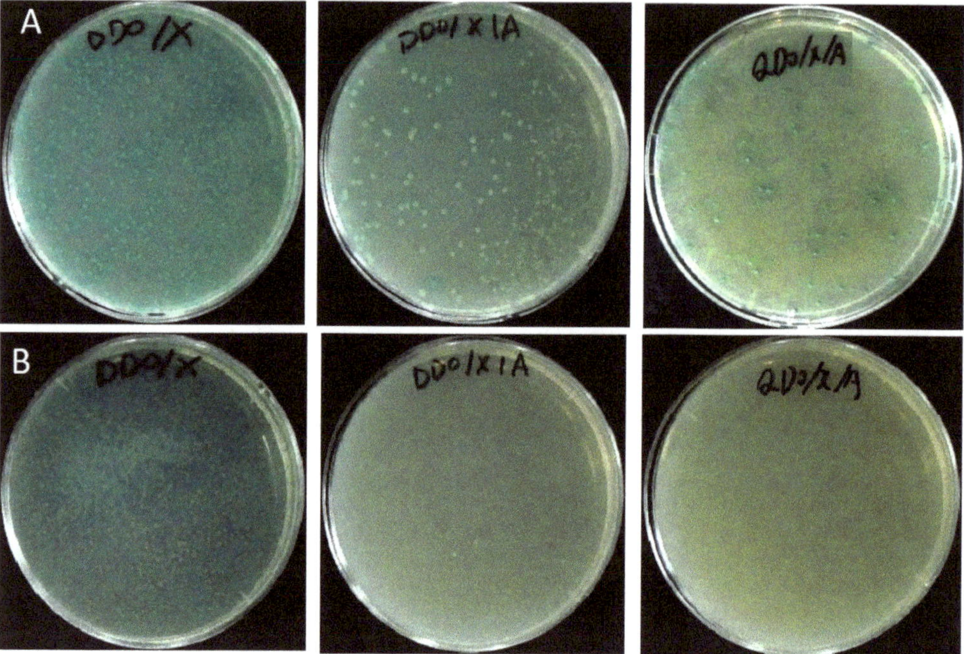

Figure 5. Verification of the interaction between SoSnRK1β1 and pglA. (**A**) The co-cultured strains grew normally with a blue appearance on DDO/X, DDO/X/A and QDO/X/A media. (**B**) The co-culture strains grew normally on DDO/X medium, and two blue colonies grew on DDO/X/A medium. No colonies grew on the QDO/X/A medium.

2.7. Gene Expression Analysis

The gene expression of these six proteins was analyzed based on the transcriptome database of sugarcane infected by *L. xyli* subsp. *xyli* [4]. After *L. xyli* subsp. *xyli* inoculation for 60 days, the gene expression levels of 1A and 2A in *L. xyli* subsp. *xyli*-infected plants were higher than those in the control plants. The gene expression level of 3A (c70487_g1) showed no difference between *L. xyli* subsp. *xyli*-infected and the control plants. However, the gene expression levels of 5A, 6A, and 7A were lower in *L. xyli* subsp. *xyli*-infected plants than those in the control plants. After *L. xyli* subsp. *xyli* inoculation for 90 days, the gene expression level of 3A (c70487_g1) did not show a significant difference between *L. xyli* subsp. *xyli*-infected and the control plants. However, the gene expression levels of 1A, 2A, 6A, and 7A in *L. xyli* subsp. *xyli*-infected plants were higher than those in the control plants, whereas the gene expression level of 5A in *L. xyli* subsp. *xyli*-infected plants was lower than that in the control (Figure 6). These results indicate that genes were positively, negatively, or not regulated by *L. xyli* subsp. *xyli* infection, but the interaction happens simultaneously at the protein level. These results also suggest that different levels of regulatory mechanisms participate in the interaction between sugarcane and *L. xyli* subsp. *xyli*.

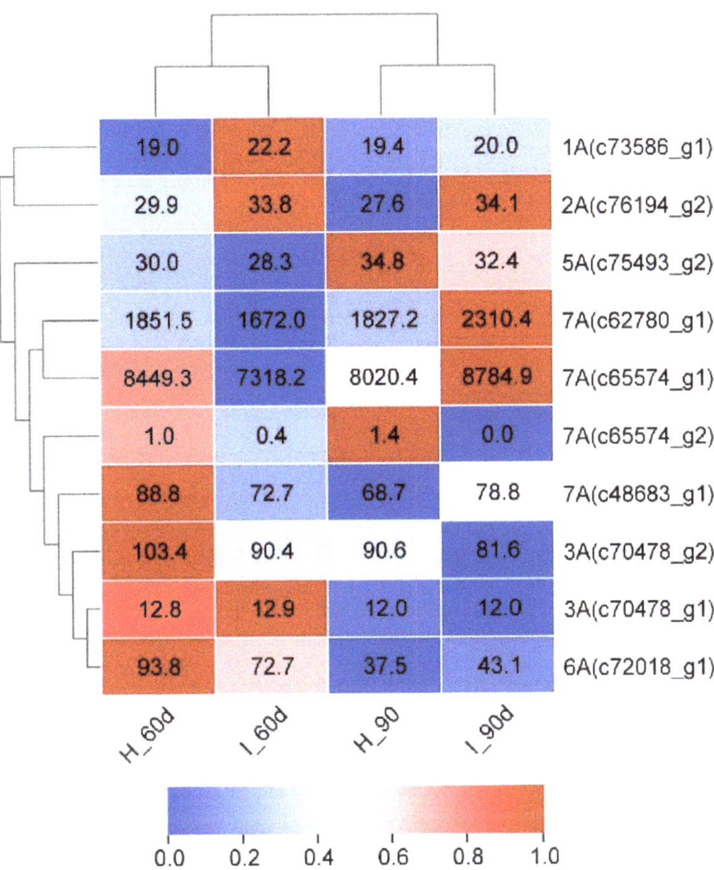

Figure 6. Expression analysis of the proteins encoded by genes identified in this work when infected by *L. xyli* subsp. *xyli*. Heat map drawing based on RPKM values using TBtools software. Different shades of color indicate different levels of gene expression. RPKM values are showed in the box. H_90d: plants inoculated with water after 90 days (control); I_90d: plants inoculated with *L. xyli* subsp. *xyli* after 90 days; H_60d: plants inoculated with water after 60 days (control); I_60d: plants inoculated with *L. xyli* subsp. *xyli* after 60 days.

3. Discussion

Sugarcane plays a pivotal role as a major source of sugar and renewable energy, contributing to the global economy. RSD profoundly impacts sugarcane by reducing yields and stalk length, hindering sugar production [25]. Developing sugarcane varieties resistant to RSD stands as the most economically effective approach to mitigate its impact on the sugarcane industry. Identifying RSD-resistant genes is a crucial endeavor in breeding sugarcane varieties resistant to RSD through molecular technology. Ento-PG has been proved to be a pathogenic enzyme in plants [18–20]. In *L. xyli* subsp. *xyli*, the agent of RSD in sugarcane, a *pglA* gene also encodes a pathogenic ento-PG. The identification of proteins that directly interact with ento-PG may accelerate the deciphering of the mechanism underlying RSD in sugarcane, providing target proteins/genes for RSD-resistant variety improvement by molecular breeding.

The yeast two-hybrid (Y2H) method is a widely used technique for identifying protein–protein interactions in various organisms, including plants [26]. A high-quality Y2H library is vital for dissecting protein functions and interactions. Factors such as library titer, recombination rate, and inserted fragments size gauge the cDNA library quality [27]. Ideally, the library titer should surpass 1.7×10^5 cfu/mL. In addition, the bait protein, which must be efficiently expressed in the yeast host without causing toxicity or autoactivation of reporter genes, is the focal point of a Y2H experiment [28]. Additionally, the signal peptide is usually removed from the bait protein to prevent improper localization or secretion in the yeast cells. By analyzing the *pglA* gene, it was observed that it had the signal peptide of Sec/SPI. Consequently, the signal peptide was deleted to construct a bait vector. Further work affirmed that the pglA bait protein could be expressed in yeast cells, and no toxicity or autoactivation was exhibited by the pglA bait toward yeast cells. These findings form the foundation for the further identification of pglA-interacting proteins through Y2H. Furthermore, the present study achieved a final library titer exceeding $>4.0 \times 10^7$ cfu/mL, with inserted cDNA fragments with lengths ranging from 0.4 to 2.0 kb. The recombination rate approached 100%, indicating that a superior-quality library was constructed for further research. Eventually, the presence of six proteins that directly interacted with pglA was successfully identified (Table 1). For instance, six proteins were screened to interact with TpLAP, the leucine aminopeptidase of *Taenia pisiformis* [29]. Similarly, the pore-forming toxin-like gene *PFT*, responsible for conferring resistance against *Fusarium* head blight disease (FHB) in wheat, was found to interact with 23 proteins [24].

Like heterochromatin protein (LHP1) has been found to mitigate abscisic acid (ABA) sensitivity by directly suppressing the expression of the ABA-responsive genes [30]. It also influences the isochorismate synthase 1 (ICS) gene within the salicylic acid (SA) biosynthesis pathway, up-regulating the inactivating enzyme salicylate/benzoate carboxyl methyltransferase (BSMT1), leading to an alteration in salicylic acid content [30]. This subsequently contributes to a decrease in the ABA level in *L. xyli* subsp. *xyli*-infected stalks [4]. Thus, pglA might interact with LHP1 to suppress ABA biosynthesis, potentially inducing stunting in sugarcane, a hallmark of RSD-infected plants [31]. LHP1 is also recognized for its role in regulating plant height, along with influencing leaf and shoot development [32].

SNF1-related protein kinase 1 (SnRK1), the plant ortholog of the yeast sucrose non-fermenting 1 (SNF1) and the mammalian AMP-activated protein kinase (AMPK), constitutes a heterotrimeric complex featuring an α catalytic subunit, a β regulatory subunit, and a γ regulatory subunit [33,34]. SnRK1-mediated signaling profoundly influences the ABA biosynthesis pathway and plant responses to viral, bacterial, fungal, and oomycete pathogens [35,36]. The SnRK1 is activated in response to ABA via SnRK2-containing complexes [36]. In addition, SnRK1 bolsters plant disease resistance through the phosphorylation and destabilization of the WRKY3 repressor [37]. In this work, sugarcane SnRK1 might enhance resistance to *L. xyli* subsp. *xyli* via interaction with pglA, thereby mediating the ABA pathway and immune response. Consequently, SnRK1 emerges as a potential putative primary target for enhancing *L. xyli* subsp. *xyli* tolerance through molecular breeding [35].

Histidine-rich calcium-binding protein (HRC) was initially identified within the QTL *Fhb1* region of wheat [38]. Plants possessing functional *HRC* genes exhibited susceptibility, while those with mutated *HRC* alleles displayed resistance to *Fusarium* head blight (FHB) in wheat [38]. Similarly, silencing the *HRC* gene enhanced multiple disease resistance in potatoes [39]. Additionally, the HRC protein was characterized in *Leymus chinensis* (*LcHRC*), where it modulates abscisic acid (ABA)-responsive gene expression through interaction with the histone deacetylation protein (*AtPWWP3*) [40]. Consequently, HRC negatively correlates with plant disease. Therefore, sugarcane HRC may participate in disease resistance or pathogen pathogenesis by interacting with the pathogenic protein ento-PG.

Ubiquitination governs a myriad of cellular functions in response to biotic and abiotic cues. E3 ubiquitin (Ub)-protein ligase is vital for post-translational modifications (PTMs) that intricately regulate various steps of plant immune signaling [41]. K48 polyubiquitination, a well-studied form, leads to targeted protein degradation through the ubiquitin-proteasome system (UPS) [42]. In *Arabidopsis*, Marino et al. [43] identified the E3 Ub-ligase MIEL1 interacting with the TF MYB30, enhancing resistance responses via down-regulating *MIEL1* to accumulate MYB30 after inoculation with bacteria. Yu et al. [44] demonstrated E3 Ub-ligase EIRP1 interacting with the nuclear TF VpWRKY11, promoting defense in Chinese wild grapevine (*Vitis pseudoreticulata*). *Arabidopsis'* E3 Ub-ligase RGLG1 and RGLG2 respond to drought stress, interacting with ERF53, with RGLG2 negatively regulating drought responses [45]. The discovery of RGLG2 interacting with pglA in *L. xyli* subsp. *xyli*-infected sugarcane suggests its involvement in bacterial stress response. Additionally, E3 Ub-ligases are known to regulate ABA signaling during abiotic stress [41].

The RNA polymerase III subunit RPC4, a member of DNA-dependent RNA polymerases (Pols) III, plays a role in the transcription of 5S rRNAs and tRNAs. This process is crucial for precise gene transcription and protein synthesis [46,47]. Pol III, the largest RNA polymerase, is highly conserved across eukaryotic organisms [48]. Nguyen et al. [49] discovered that weak or absent *RPC4* expression caused the loss-of-function alleles in rice *DGS1*-nivara^s and *DGS2-T65^s*, leading to hybrid incompatibility. In *N. benthamiana*, Nemchinov et al. [50] demonstrated that *RPC5L* silencing disrupted core Pol III transcripts and diverse cellular processes, including stress responses. Despite classic Pol III genes being considered house-keeping genes, their regulation remains understudied [51]. Likewise, Pol III regulation in sugarcane under *L. xyli* subsp. *xyli* infection, despite RPC4′s interaction with pglA, remains poorly investigated.

4. Materials and Methods

4.1. Sequence Characteristics of the pglA Gene

The sequence of *pglA* gene (https://www.ncbi.nlm.nih.gov/gene/2939326#reference-sequences accessed on 26 May 2023) was obtained from strain *Lxx*GXBZ01 (LFYU00000000), which was sequenced in our previous work [4] (https://www.ncbi.nlm.nih.gov/nuccore/LFYU00000000 accessed on 26 May 2023). The conserved domains of pglA were analyzed using the Conserved Domain Search Service (CD Search) on NCBI. The SignalP-5.0 (https://services.healthtech.dtu.dk/service.php?SignalP-5.0 accessed on 10 May 2023) was employed to predict the presence of signal peptide. The BLASTp tool (https://blast.ncbi.nlm.nih.gov/Blast.cgi?PROGRAM=blastp&PAGE_TYPE=BlastSearch&LINK_LOC=blasthome accessed on 10 May 2023) on NCBI was used to identify the homologous amino acid sequences from other pathogens.

4.2. Construction, Toxicity Testing, Autoactivation and Expression of pglA Bait

The sequence of *pglA*, excluding signal peptide sequence (1–81 bp), was obtained through direct gene synthesis. The resulting fragment was subsequently inserted into the pGBKT7 vector. The recombinant vector, pGBKT7-pglA, was then introduced into *Saccharomyces cerevisiae* strain Y2HGold using Yeastmaker^{TM} Yeast Transformation System 2 (Clontech, CA, USA). To evaluate the potential toxicity and autoactivation of the pglA bait, the transformed cells were plated onto three distinct media of SD/-Trp, SD/-Trp/X, and SD/-Trp/X/A, respectively. Following plating, the cells were cultured at 30 °C for 3–5 days. The strain of Y2HGold containing the pGBKT7-BD vector was spread on an SD-Trp medium, as a positive control.

To determine the expression of pglA bait, the transformed cells were cultured in an SD/-Trp broth medium at 30 °C under 200 rpm for 4–8 h, until the OD$_{600}$ value reached the range of 0.4–0.6. As positive control, the Y2HGold strain carrying the pGBKT7-53 vector and the Y2HGold strain carrying the pGBKT7-BD vector were also cultured in an SD/-Trp broth medium. For negative control, the wild type Y2HGold strain was cultured in a YPDA broth medium. The various yeast cells were harvested via centrifugalization, and the total protein was extracted using the Yeast Protein Extraction Reagent (Takara, Dalian, China).

4.3. Construction of Y2H Library

Due to the inability of *L. xyli* subsp. *xyli* to colonize resistant sugarcane varieties, the highly susceptible *Saccharum officinarum* L. cultivar Badila was selected to ensure successful infection and protein interaction. The plants were cultivated in a germplasm repository located at 108°22′, 22°48′ within the Sugarcane Research Institute at Guangxi Academy of Agricultural Sciences, in Nanning, Guangxi, China.

For library construction, the total RNA was extracted from the first internode of stalk above the ground, excluding the epidermis, using the MiniBEST plant RNA extraction kit (Takara, Dalian, China). Subsequently, cDNA synthesis was carried out using the SMART cDNA library construction kit and the Advantage 2 PCR kit (Clontech, CA, USA). The resulting cDNA was then normalized using the TRIMMER-DIRECT cDNA normalization kit (Evrogen, Moscow, Russia) according to the manufacturer's instructions. The normalized cDNA was further amplified using the cDNA normalization kit and Advantage 2 PCR kit (Takara, Dalian, China). Following amplification, the normalized cDNA was purified using the MiniBEST DNA fragment purification kit (Takara, Dalian, China). To eliminate low-molecular-weight cDNA fragments and small DNA contaminants, the cDNA was excised from a 1% agarose gel after *Sfi*I digestion and purified using CHROMA SPIN-1000 columns (Clontech, CA, USA). The purified cDNA was directionally cloned into the pGADT7-SfI vector (a library of prey proteins with the Gal4 activation domain; Clontech, CA, USA) at the *Sfi*I A (5′-GGCCATTACGGCC-3′) and *Sfi*I B (3′-CCGGCGGAGCCGG-5′) sites to establish the primary cDNA library.

The primary library was transformed into HST08 competent cells through electro-transformation under the condition of 1.8 KV, 200 Ω, 25 μF. The transformed mixture was spread onto LB media supplemented with ampicillin (Amp⁺), followed by overnight incubation at 37 °C until colonies became visible. Subsequently, the transformation efficiency and number of primary colonies were calculated. From the pool of primary colonies, 16 were randomly selected and subjected to PCR amplification using the primer pGADT7-F/R (GGAGTACCCATACGACGTACC/ TATCTACGATTCATCTGCAGC) to assess the insert sizes and the recombination rate of the Y2H library. To validate the normalization results, 96 colonies from the primary library were selected for sequencing. The primary library was the re-transformed into HST08 competent cells, spread across 10 plates with LB medium, and cultured at 37 °C overnight. The plasmids were harvested using the NucleoBond Xtra Midi EF kit (MN, NRW, Germany), and transformed into *S. cerevisiae* strain Y187 using the YeastmakerTM Yeast Transformation System 2 (Clontech, CA, USA). The transformed Y187 cells were then plated onto 100 SD/-Leu medium plates and cultured at 30 °C for 3 days. Colonies of transformed Y187 were collected using a freezing medium containing 25% glycerin and stored at −80 °C for further use within the Matchmaker TM gold Y2H system.

4.4. Screening of pglA-Interacting Proteins

The pglA-Y2HGold strains were cultured on a solid SD/-Trp medium for 3 days, yielding yeast colonies with a diameter of 2–3 mm. These colonies were co-cultured with the Y2H library strains using liquid SD/-Trp medium in a yeast-mating way, under shaking at 40 rpm and incubation at 30 °C for 24 h. To conduct gradient dilution, a small amount of the co-culture yeast cells was extracted for dilutions of 1/10, 1/100, 1/1000, and 1/10,000, respectively. For each gradient dilution, a 100 μL droplet was spread onto SD/-Leu, SD/-Trp, and SD/-Leu/-Trp media. The remaining dilution was plated on SD/-Leu/-Trp/ X-a-Gal/AbA medium across 50 plates. As a positive control, the pGBKT7-53 plasmid and the pGADT7-T plasmid were co-transformed into the Y2HGold strain. Correspondingly, the pGBKT7-Lam plasmid and the pGADT7-T plasmid were co-transformed into the Y2HGold strain to serve as the negative control. The strains of positive and negative controls were applied to SD/-Leu/-Trp and SD/-Leu/-Trp/ X-a-Gal/AbA media, respectively. The blue colonies were inoculated on an SD/-Ade/-His/-Leu/-Trp/X-a-Gal/AbA medium, and the Matchmaker Insert Check PCR Mix 2 kit was used to amplify the developed clones. The PCR product was purified, followed by sequencing. The sequences were then analyzed using the BLAST (https://blast.ncbi.nlm.nih.gov accessed on 20 July 2023) and Uniprot (https://www.uniprot.org/uniprot/ accessed on 20 July 2023) platforms to unveil protein names and functions. The homology function in *Sorghum bicolor* was determined using BLAST.

5. Conclusions

The proteins interacting with pglA are mainly involved in ABA pathways and the immune system, suggesting that sugarcane interacts with *L. xyli* subsp. *xyli* through these two pathways to either cause pathogenicity or develop resistance (Figure 7). This is the first report of protein interactions between sugarcane and pglA from *L. xyli* subsp. *xyli*. This study enhances the understanding of pglA's role in pathogenicity and provides potential target proteins/genes for molecular breeding to enhance sugarcane's resistance against *L. xyli* subsp. *xyli*. Future research could focus on validating the roles of these proteins in conferring resistance and elucidating the underlying mechanisms.

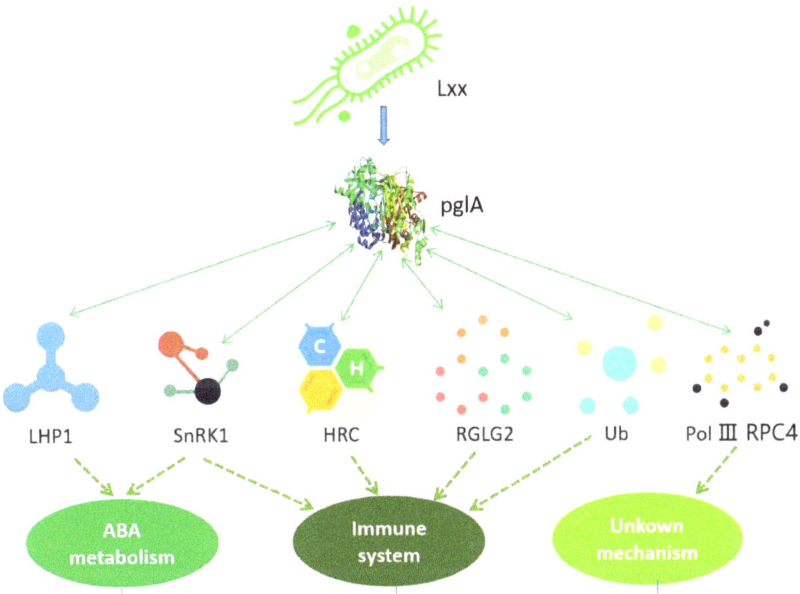

Figure 7. Putative mechanism of sugarcane response to *L. xyli* subsp. *xyli* pglA. Endo-PG, which is encoded by the pglA gene in *L. xyli* subsp. *xyli*, exhibits a direct interaction with sugarcane LHP1, leading to the suppression of the ABA pathway and consequent sugarcane stunting. Additionally, endo-PG also engages in interactions with sugarcane proteins, including SnRK1, HRC, RGLG2, and Ub (ubiquitin), which collectively confer resistance against *L. xyli* subsp. *xyli*. Notably, SnRK1 may also intersect with ABA biosynthesis in response to *L. xyli* subsp. *xyli* invasion. The dynamic interplay between endo-PG and these sugarcane proteins ultimately determines the pathogenicity or resistance of sugarcane.

Author Contributions: Conceptualization, X.-Q.Z., Y.-J.L., X.-P.S. and D.-L.H.; methodology, X.-Q.Z., Y.-J.L., B.-Q.Z., M.-X.Y., Z.-P.W., D.-M.H., Y.-X.H. and J.-C.L.; original draft preparation, X.-Q.Z., Y.-J.L. and X.-P.S.; review and editing, D.-L.H.; supervision, X.-P.S. and D.-L.H.; project administration, X.-P.S. and D.-L.H.; funding acquisition, X.-Q.Z. and Y.-J.L. All authors have read and agreed to the published version of the manuscript.

Plants **2024**, 13, 448

Funding: This research was funded by the National Natural Science Foundation of China (32001606), the Natural Science Foundation of Guangxi (2020GXNSFBA297040, 2023GXNSFAA026495), the Specific Research Project of Guangxi for Research Bases and Talents (GKAD17195100), the Guangxi Major Science and Technology Project (GuikeAA22117004), the Fund for Guangxi Innovation Teams of Modern Agriculture Technology (nycytxgxcxtd-2021-03), and the Fund of GXAAS (2021YT007).

Data Availability Statement: The data that support the findings of this study are available from the corresponding author upon reasonable request.

Conflicts of Interest: The authors declare no conflict of interest.

References

1. Qi, Y.W.; Gao, X.N.; Zeng, Q.Y.; Zheng, Z.; Wu, C.W.; Yang, R.Z.; Feng, X.M. Sugarcane breeding, germplasm development and related molecular research in China. *Sugar Tech* **2022**, *24*, 73–85. [CrossRef]
2. Quecine, M.C.; Silva, T.M.; Carvalho, G.; Saito, S.; Mondin, M.; Teixeira-Silva, N.S.; Camargo, L.E.A.; Monteiro-Vitorello, C.B. A stable *Leifsonia xyli* subsp. *xyli* GFP-tagged strain reveals a new colonization niche in sugarcane tissues. *Plant Pathol.* **2016**, *65*, 154–162. [CrossRef]
3. Marques, J.P.R.; Cia, M.C.; Batista-de-Andrade-Granato, A.; Muniz, L.F.; Appezzato-da-Glória, B.; Camargo, L.E.A. Histopathology of the shoot apex of sugarcane colonized by *Leifsonia xyli* subsp. *xyli*. *Phytopathology* **2022**, *12*, 2062–2071. [CrossRef] [PubMed]
4. Zhang, X.Q.; Chen, M.H.; Liang, Y.J.; Xing, Y.X.; Comstock, J.C.; Yang, L.T.; Li, Y.R. Morphological and physiological responses of sugarcane to *Leifsonia xyli* subsp. *xyli* infection. *Plant Dis.* **2016**, *100*, 2499–2506. [CrossRef] [PubMed]
5. Zhang, X.Q.; Liang, Y.J.; Song, X.P.; Wang, Z.P.; Zhang, B.Q.; Lei, J.C.; Yang, R.L.; Yan, M.X. Changes in gene expression levels and chloroplast anatomy induced by *Leifsonia xyli* subsp. *xyli* in sugarcane. *J. Plant Interact.* **2021**, *16*, 564–574. [CrossRef]
6. Zhu, K.; Yuan, D.; Zhang, X.Q.; Yang, L.T.; Li, Y.R. The physiological characteristics and associated gene expression of sugar cane inoculated with *Leifsonia xyli* subsp. *xyli*. *J. Phytopathol.* **2018**, *166*, 44–52. [CrossRef]
7. Chakraborty, M.; Ford, R.; Soda, N.; Strachan, S.; Ngo, C.N.; Bhuiyan, S.A.; Shiddiky, M. Ratoon Stunting Disease (RSD) of sugarcane: A review emphasizing detection strategies and challenges. *Phytopathology* **2023**. [CrossRef]
8. Li, W.F.; Shen, K.; Huang, Y.K.; Wang, X.Y.; Yin, J.; Luo, Z.M.; Zhang, R.Y.; Shan, H.L. Incidence of sugarcane ratoon stunting disease in the major cane-growing regions of China. *Crop Prot.* **2014**, *60*, 44–47. [CrossRef]
9. Carvalho, G.; Da Silva, T.G.E.R.; Munho, A.T.; Monteiro-Vitorello, C.B.; Azevedo, R.A.; Melotto, M.; Camargo, L.E.A. Development of a qPCR for *Leifsonia xyli* subsp. *xyli* and quantification of the effects of heat treatment of sugarcane cuttings on *Lxx*. *Crop Prot.* **2016**, *80*, 51–55. [CrossRef]
10. Oladosu, Y.; Rafii, M.Y.; Samuel, C.; Fatai, A.; Magaji, U.; Kareem, I.; Kamarudin, Z.S.; Muhammad, I.; Kolapo, K. Drought resistance in rice from conventional to molecular breeding: A review. *Int. J. Mol. Sci.* **2019**, *20*, 3519. [CrossRef]
11. Ahmar, S.; Gill, R.A.; Jung, K.H.; Faheem, A.; Qasim, M.U.; Mubeen, M.; Zhou, W. Conventional and molecular techniques from simple breeding to speed breeding in crop plants: Recent Advances and Future Outlook. *Int. J. Mol. Sci.* **2020**, *21*, 2590. [CrossRef]
12. Du, H.; Fang, C.; Li, Y.; Kong, F.; Liu, B. Understandings and future challenges in soybean functional genomics and molecular breeding. *J. Integr. Plant Biol.* **2023**, *5*, 468–495. [CrossRef]
13. Collmer, A.; Keen, N.T. The role of pectic enzymes in plant pathogenesis. *Annu. Rev. Phytopathol.* **1986**, *24*, 383–409. [CrossRef]
14. Roper, M.C.; Greve, L.C.; Warren, J.G.; Labavitch, J.M.; Kirkpatrick, B.C. *Xylella fastidiosa* requires polygalacturonase for colonization and pathogenicity in *Vitis vinifera* grapevines. *Mol. Plant-Microbe Interact.* **2007**, *20*, 411–419. [CrossRef]
15. Denny, T.P.; Carney, B.F.; Schell, M.A. Inactivation of multiple virulence genes reduces the ability of *Pseudomonas solanacearum* to cause wilt symptoms. *Mol. Plant-Microbe Interact.* **1990**, *3*, 293–300. [CrossRef]
16. Warren, J.G.; Lincoln, E.; Kirkpatrick, B.C. Insights into the activity and substrate binding of *Xylella fastidiosa* polygalacturonase by modification of a unique QMK amino acid motif using protein chimeras. *PLoS ONE* **2015**, *10*, e0142694. [CrossRef]
17. Tayi, L.; Maku, R.V.; Patel, H.K.; Sonti, R.V. Identification of pectin degrading enzymes secreted by *Xanthomonas oryzae* pv. *oryzae* and determination of their role in virulence on rice. *PLoS ONE* **2016**, *11*, e0166396. [CrossRef]
18. Bashi, Z.D.; Rimmer, S.R.; Khachatourians, G.G.; Hegedus, D.D. *Brassica napus* polygalacturonase inhibitor proteins inhibit *Sclerotinia sclerotiorum* polygalacturonase enzymatic and necrotizing activities and delay symptoms in transgenic plants. *Can. J. Microbiol.* **2013**, *59*, 79–86. [CrossRef]
19. Liu, N.N.; Ma, X.W.; Sun, Y.; Hou, Y.X.; Zhang, X.Y.; Li, F.G. Necrotizing activity of *Verticillium dahliae* and *Fusarium oxysporum* f. sp. *vasinfectum* endopolygalacturonases in cotton. *Plant Dis.* **2017**, *101*, 1128–1138. [CrossRef]
20. Monteiro-Vitorello, C.B.; Camargo, L.E.; Van-Sluys, M.-A.; Kitajima, J.P.; Truffi, D.; Do-Amaral, A.M.; Harakava, R.; de-Oliveira, J.C.F.; Wood, D.; de-Oliveira, M.C.; et al. The genome sequence of the gram-positive sugarcane pathogen *Leifsonia xyli* subsp. *xyli*. *Mol. Plant-Microbe Interact.* **2004**, *17*, 827–836. [CrossRef]
21. Wang, J.H.; Wang, L.; Cao, G.; Zhang, M.Q.; Guo, Y. Draft genome sequence of *Leifsonia xyli* subsp. *xyli* strain gdw1. *Genome Announc.* **2016**, *4*, e01128-16. [CrossRef]

22. Vidal, M.; Fields, S. The yeast two-hybrid assay: Still finding connections after 25 years. *Nat. Methods* **2014**, *11*, 1203–1206. [CrossRef] [PubMed]

23. Gnanasekaran, P.; Ponnusamy, K.; Chakraborty, S. A geminivirus betasatellite encoded βC1 protein interacts with PsbP and subverts PsbP-mediated antiviral defence in plants. *Mol. Plant Pathol.* **2019**, *20*, 943–960. [CrossRef]

24. He, Y.; Wu, L.; Liu, X.; Zhang, X.; Jiang, P.; Ma, H. Yeast two-hybrid screening for proteins that interact with PFT in wheat. *Sci. Rep.* **2019**, *9*, 15521. [CrossRef] [PubMed]

25. Zhu, K.; Yang, L.T.; Li, C.X.; Lakshmanan, P.; Xing, Y.X.; Li, Y.R. A transcriptomic analysis of sugarcane response to *Leifsonia xyli* subsp. *xyli* infection. *PLoS ONE* **2021**, *16*, e0245613. [CrossRef]

26. Stynen, B.; Tournu, H.; Tavernier, J.; Van-Dijck, P. Diversity in genetic in vivo methods for protein-protein interaction studies: From the yeast two-hybrid system to the mammalian split-luciferase system. *Microbiol. Mol. Biol. Rev.* **2012**, *76*, 331–382. [CrossRef]

27. Gao, J.; Jing, J.; Yu, C.J.; Chen, J. Construction of a high-quality yeast two-hybrid library and its application in identification of interacting proteins with brn1 in *Curvularia lunata. Plant Pathol. J.* **2015**, *31*, 108–114. [CrossRef]

28. Holcroft, J.; Ganss, B. Identification of amelotin-and ODAM-interacting enamel matrix proteins using the yeast two-hybrid system. *Eur. J. Oral Sci.* **2011**, *119*, 301–306. [CrossRef]

29. Zhang, S. Screening and verification for proteins that interact with leucine aminopeptidase of *Taenia pisiformis* using a yeast two-hybrid system. *Parasitol. Res.* **2019**, *118*, 3387–3398. [CrossRef]

30. Ramirez-Prado, J.S.; Latrasse, D.; Rodriguez-Granados, N.Y.; Huang, Y.; Manza-Mianza, D.; Brik-Chaouche, R.; Jaouannet, M. The Polycomb protein LHP 1 regulates *Arabidopsis thaliana* stress responses through the repression of the MYC2-dependent branch of immunity. *Plant J.* **2019**, *100*, 1118–1131. [CrossRef]

31. Comstock, J.C. Ratoon stunting disease. *Sugar Tech* **2002**, *4*, 1–6. [CrossRef]

32. Mansilla, N.; Ferrero, L.; Ariel, F.D.; Lucero, L.E. The Potential Use of the Epigenetic Remodeler LIKE HETEROCHROMATIN PROTEIN 1 (LHP1) as a Tool for Crop Improvement. *Horticulture* **2023**, *9*, 199. [CrossRef]

33. Ramon, M.; Ruelens, P.L.Y.; Sheen, J.; Geuten, K.; Rolland, F. The hybrid four-CBS-domain KIN βγ subunit functions as the canonical γ subunit of the plant energy sensor SnRK1. *Plant J.* **2013**, *75*, 11–25. [CrossRef]

34. Emanuelle, S.; Doblin, M.S.; Stapleton, D.I.; Bacic, A.; Gooley, P.R. Molecular insights into the enigmatic metabolic regulator, SnRK1. *Trends Plant Sci.* **2016**, *21*, 341–353. [CrossRef]

35. Hulsmans, S.; Rodriguez, M.; De Coninck, B.; Rolland, F. The SnRK1 energy sensor in plant biotic interactions. *Trends Plant Sci.* **2016**, *21*, 648–661. [CrossRef]

36. Belda-Palazon, B.; Adamo, M.; Valerio, C.; Ferreira, L.J.; Confraria, A.; Reis-Barata, D.; Rodrigues, A.; Meyer, C.; Rodriguez, P.L.; Baena-González, E. A dual function of SnRK2 kinases in the regulation of SnRK1 and plant growth. *Nat. Plants* **2020**, *6*, 345–1353. [CrossRef]

37. Han, X.; Zhang, L.; Zhao, L.; Xue, P.; Qi, T.; Zhang, C.; Yuan, H.; Zhou, L.; Wang, D.; Qiu, J.; et al. SnRK1 phosphorylates and destabilizes WRKY3 to enhance barley immunity to powdery mildew. *Plant Commun.* **2020**, *1*, 100083. [CrossRef]

38. Su, Z.; Bernardo, A.; Tian, B.; Chen, H.; Wang, S.; Ma, H.; Cai, S.; Liu, D.; Zhang, D.; Li, T.; et al. A deletion mutation in TaHRC confers Fhb1 resistance to Fusarium head blight in wheat. *Nat. Genet.* **2019**, *51*, 1099–1105. [CrossRef]

39. Kushalappa, A.C.; Hegde, N.G.; Gunnaiah, R.; Sathe, A.; Yogendra, K.N.; Ajjamada, L. Apoptotic-like PCD inducing HRC gene when silenced enhances multiple disease resistance in plants. *Sci. Rep.* **2022**, *12*, 20402. [CrossRef]

40. Yang, J.; Zhang, T.; Mao, H.; Jin, H.; Sun, Y.; Qi, Z. A *Leymus chinensis* histidine-rich Ca^{2+}-binding protein binds Ca^{2+}/Zn^{2+} and suppresses abscisic acid signaling in *Arabidopsis. J. Plant Physiol.* **2020**, *252*, 153209. [CrossRef]

41. Serrano, I.; Campos, L.; Rivas, S. Roles of E3 ubiquitin-ligases in nuclear protein homeostasis during plant stress responses. *Front. Plant Sci.* **2018**, *9*, 139. [CrossRef] [PubMed]

42. Sadanandom, A.; Bailey, M.; Ewan, R.; Lee, J.; Nelis, S. The ubiquitin-proteasome system: Central modifier of plant signalling. *New Phytol.* **2012**, *196*, 13–28. [CrossRef] [PubMed]

43. Marino, D.; Froidure, S.; Canonne, J.; Ben Khaled, S.; Khafif, M.; Pouzet, C.; Jauneau, A.; Roby, D.; Rivas, S. Arabidopsis ubiquitin ligase MIEL1 mediates degradation of the transcription factor MYB30 weakening plant defense. *Nat. Commun.* **2013**, *4*, 1476. [CrossRef] [PubMed]

44. Yu, Y.; Xu, W.; Wang, J.; Wang, L.; Yao, W.; Yang, Y.; Xu, Y.; Ma, F.; Du, Y.; Wang, Y. The Chinese wild grapevine (*Vitis pseudoreticulata*) E3 ubiquitin ligase *Erysiphe necator*-induced RING finger protein 1 (EIRP1) activates plant defense responses by inducing proteolysis of the VpWRKY11 transcription factor. *New Phytol.* **2013**, *200*, 834–846. [CrossRef] [PubMed]

45. Cheng, M.C.; Hsieh, E.J.; Chen, J.H.; Chen, H.Y.; Lin, T.P. Arabidopsis RGLG2, functioning as a RING E3 ligase, interacts with AtERF53 and negatively regulates the plant drought stress response. *Plant Physiol.* **2012**, *158*, 363–375. [CrossRef]

46. Dieci, G.; Bosio, M.C.; Fermi, B.; Ferrari, R. Transcription reinitiation by RNA polymerase III. *Biochim. Biophys. Acta.* **2013**, *1829*, 331–341. [CrossRef]

47. Arimbasseri, A.G.; Maraia, R.J.T. RNA polymerase III advances: Structural and tRNA functional views. *Trends Biochem. Sci.* **2016**, *41*, 546–559. [CrossRef]

48. Wang, Q.; Daiß, J.L.; Xu, Y.; Engel, C. Snapshots of RNA polymerase III in action-A mini review. *Gene* **2022**, *821*, 146282. [CrossRef]

49. Nguyen, G.N.; Yamagata, Y.; Shigematsu, Y.; Watanabe, M.; Miyazaki, Y.; Doi, K.; Tashiro, K.; Kuhara, S.; Kanamori, H.; Wu, J.; et al. Duplication and loss of function of genes encoding RNA polymerase III subunit C4 causes hybrid incompatibility in rice. *G3-Genes Genomes Genet.* **2017**, *7*, 2565–2575. [CrossRef]

50. Nemchinov, L.G.; Boutanaev, A.M.; Postnikova, O.A. Virus-induced gene silencing of the RPC5-like subunit of RNA polymerase III caused pleiotropic effects in *Nicotiana benthamiana*. *Sci. Rep.* **2016**, *6*, 27785. [CrossRef]

51. Park, J.L.; Lee, Y.S.; Kunkeaw, N.; Kim, S.Y.; Kim, I.H.; Lee, Y.S. Epigenetic regulation of noncoding RNA transcription by mammalian RNA polymerase III. *Epigenomics* **2017**, *9*, 171–187. [CrossRef] [PubMed]

plants

Article

Insights into Reactive Oxygen Species Production-Scavenging System Involved in Sugarcane Response to *Xanthomonas albilineans* Infection under Drought Stress

Yao-Sheng Wei [1], Jian-Ying Zhao [1], Talha Javed [2], Ahmad Ali [1], Mei-Ting Huang [1], Hua-Ying Fu [1], Hui-Li Zhang [1] and San-Ji Gao [1,*]

[1] National Engineering Research Center for Sugarcane, Fujian Agriculture and Forestry University, Fuzhou 350002, China; dell990218@163.com (Y.-S.W.); zhaojyfafu@126.com (J.-Y.Z.); Ahmad03348454473@yahoo.com (A.A.); hmt159379@163.com (M.-T.H.); mddzyfhy@163.com (H.-Y.F.); hlzhang@fafu.edu.cn (H.-L.Z.)

[2] Institute of Tropical Bioscience and Biotechnology, Chinese Academy of Tropical Agricultural Sciences, Haikou 571101, China; talhajaved@itbb.org.cn

* Correspondence: gaosanji@fafu.edu.cn

Abstract: Plants must adapt to the complex effects of several stressors brought on by global warming, which may result in interaction and superposition effects between diverse stressors. Few reports are available on how drought stress affects *Xanthomonas albilineans* (*Xa*) infection in sugarcane (*Saccharum* spp. hybrids). Drought and leaf scald resistance were identified on 16 sugarcane cultivars using *Xa* inoculation and soil drought treatments, respectively. Subsequently, four cultivars contrasting to drought and leaf scald resistance were used to explore the mechanisms of drought affecting *Xa*–sugarcane interaction. Drought stress significantly increased the occurrence of leaf scald and *Xa* populations in susceptible cultivars but had no obvious effect on resistant cultivars. The ROS bursting and scavenging system was significantly activated in sugarcane in the process of *Xa* infection, particularly in the resistant cultivars. Compared with *Xa* infection alone, defense response via the ROS generating and scavenging system was obviously weakened in sugarcane (especially in susceptible cultivars) under *Xa* infection plus drought stress. Collectively, ROS might play a crucial role involving sugarcane defense against combined effects of *Xa* infection and drought stress.

Keywords: *Saccharum* spp. hybrids; reactive oxygen species; leaf scald; drought; defense response

Citation: Wei, Y.-S.; Zhao, J.-Y.; Javed, T.; Ali, A.; Huang, M.-T.; Fu, H.-Y.; Zhang, H.-L.; Gao, S.-J. Insights into Reactive Oxygen Species Production-Scavenging System Involved in Sugarcane Response to *Xanthomonas albilineans* Infection under Drought Stress. *Plants* **2024**, *13*, 862. https://doi.org/10.3390/plants13060862

Academic Editor: Babar Shahzad

Received: 6 February 2024
Revised: 11 March 2024
Accepted: 15 March 2024
Published: 17 March 2024

1. Introduction

The number and frequency of extreme weather events have increased significantly due to global climate change and pose a serious impact on the sustainable development of agriculture [1,2]. Recent research has shown that extreme environmental events have a significant impact on the pathogenicity and transmissibility of pathogenic microorganisms [3], which will impair plant disease resistance, alter defensive signaling pathways, and ultimately cause a dramatic decline in plant growth and survival [4,5]. Thus, it is crucial to comprehend how abiotic stresses affect biotic stresses in order to engineer plants for climate adaptation and to ensure the long-term viability of agriculture [3].

Crosstalk and trade-offs exist in plants in response to abiotic and biotic stresses [3]. Drought directly weakens the fitness and survival of a wide range of root and leaf pathogens that require moisture [6]. Furthermore, combined drought and pathogen stress alters physio-morphological traits such as photosynthesis, stomatal conductance, and transpiration rate along with plant growth and root morphology [7]. However, some cases reported that plants which are subjected to drought stress increase plant susceptibility by destroying the woody structure, resulting in the acceleration of some diseases, such as charcoal stalk rot in sorghum and smut in cereals [3].

Reactive oxygen species (ROS) burst is an effective approach for plants to fend off stressors in the early stages of unfavorable environmental conditions. Later, ROS act as signaling molecules that set off the organism's defensive systems [8,9]. However, the plant loses its ability to respond to osmotic stress when the ROS concentrations surpass the threshold value; in extreme circumstances, this may result in wilting or even death [10,11]. ROS homeostasis in plants mainly includes enzymatic and non-enzymatic scavenging systems [12]. The enzymatic scavenging system mainly includes superoxide dismutase (SOD), catalase (CAT), peroxidase (POD), ascorbate peroxidase (APX), and so on [12]. Among them, SOD is the first and most important line of defense in the enzymatic scavenging system due its ability to remove superoxide (O_2^-) and generate the disproportionation product H_2O_2 [10]. In response to pathogen infection, plants trigger hypersensitivity reactions and apoptosis through ROS burst, thereby limiting the proliferation of pathogens [13,14]. In addition, drought confers resistance to plants against subsequent pathogen infections by elevated levels of ROS generation and higher activity of the ROS scavenging system [7].

Sugarcane (*Saccharum* spp. hybrids), a classic C_4 crop with the highest photosynthetic rates among other crops, plays a crucial role in sugar and biofuel production, thereby accounting for 80% of sugar production in the world and approximately 90% in China. China's sugarcane-producing areas are mainly located in the arid slope land of southern and southwestern regions such as Guangxi and Yunnan provinces, where sugarcane yield loss often occurs due to poor irrigation facilities and soil water retention capacity coupled with uneven natural precipitation. Extreme drought can lead to a reduction of up to 60% in sugarcane yields [15]. The bacterial pathogen *Xanthomonas albilineans* (*Xa*) is a causal agent of leaf scald in sugarcane, distributed across the majority of sugarcane-planting counties worldwide [16]. However, the mechanism of sugarcane infection by *Xa* under drought stress remains unclear. This study aims to assess some newly released sugarcane cultivars resistant to drought stress and *Xa* infection. Additionally, the ROS production-scavenging system participating sugarcane response to both stressors has been illustrated.

2. Results

2.1. Identification of Sugarcane Cultivars Tolerant to Drought

The malondialdehyde (MDA) contents in the leaves of 16 sugarcane cultivars were measured under drought stress. The MDA contents of all sugarcane cultivars increased significantly after drought stress. Notably, the MDA contents of ROC22, GT16-1253, and LC09-15 were significantly lower than that of other cultivars. The leaf relative water contents and maximum quantum yield of PSII photochemistry (Fv/Fm) of all sugarcane cultivars decreased after drought stress. The non-photochemical quenching (qN) of FN04-3504, GT29, YG59, ZZ14, and ZZ13 cultivars was significantly decreased after drought stress. Following drought stress, the PSII real photosynthetic efficiency (Y(II)) suffered a significant drop in FN04-3504, GT29, YG59, YZ15-505, and ZZ13 (Figure 1).

The membership function values and comprehensive evaluation values (D values) of 16 sugarcane cultivars were calculated by using the membership function method of fuzzy mathematics (Table 1). As shown in Table 1, the D value was positively correlated with drought resistance. Among all the tested cultivars, LC09-15 had the highest D value, while GT29 had the smallest. Finally, the drought resistance of the 16 sugarcane cultivars at the seedling stage was graded from strong to weak (Table 1).

Table 1. Membership function values and ranking of comprehensive evaluation values (D values) of 16 sugarcane cultivars.

Variety	Membership Function			D Value	Rank
	μ1	μ2	μ3		
LC09-15	0.9487	0.4645	0.4755	0.7764	1
GT16-1253	1.0000	0.3525	0.3186	0.7611	2
ROC22	0.9633	0.2351	0.5164	0.7498	3

Table 1. *Cont.*

Variety	Membership Function			D Value	Rank
	μ1	μ2	μ3		
GT13-334	0.6118	0.6142	0.5090	0.5945	4
LC15-39	0.5540	0.7773	0.4478	0.5775	5
FN14-1854	0.4478	0.8729	0.6594	0.5639	6
GNY14-6210	0.4030	0.6747	1.0000	0.5569	7
ZZ8	0.4036	1.0000	0.4619	0.5253	8
YZ15-505	0.4472	0.9969	0.2409	0.5145	9
ZT1501	0.5105	0.7291	0.2766	0.5110	10
LC05-136	0.3308	0.6219	0.4996	0.4144	11
YG59	0.1545	0.6518	0.5294	0.3123	12
ZZ13	0.1805	0.1863	0.4778	0.2329	13
ZZ14	0.0000	0.6090	0.5375	0.2068	14
FN04-3504	0.0847	0.4907	0.0000	0.1460	15
GT29	0.0173	0.0000	0.5373	0.1038	16

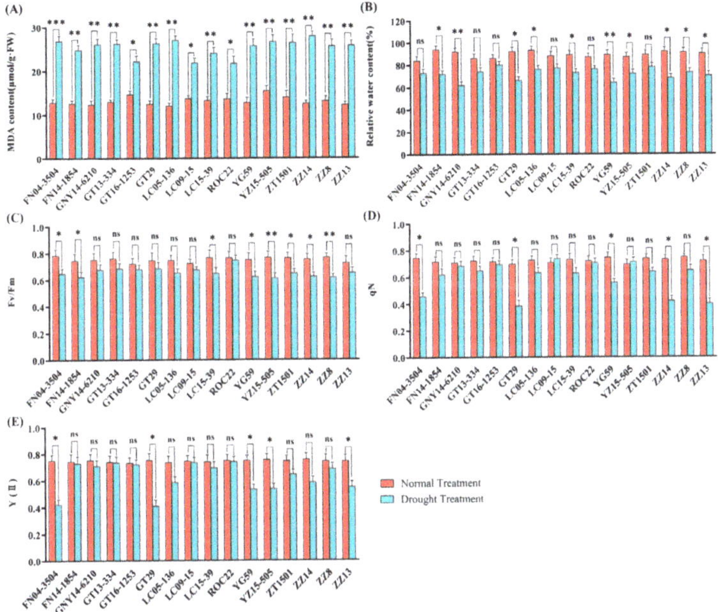

Figure 1. Changes in different physiological and biochemical indexes of 16 sugarcane cultivar leaves under normal and drought condition. (**A**) MDA contents, (**B**) relative water contents, (**C**) maximum quantum yield of PSII photochemistry (Fv/Fm), (**D**) non-photochemical quenching (qN), (**E**) PSII actual photosynthetic efficiency (Y(II)). Vertical bar values represent means ± SE. For each variety, significant differences between normal and drought treatments at $p < 0.05, 0.01$, and 0.001 (Student's *t*-test) are indicated by one, two, and three asterisks, respectively. ns, not significant.

2.2. Identification of Sugarcane Cultivars Resistant to Leaf Scald

The disease index was calculated according to the severity classification scale of leaf scald among 16 sugarcane cultivars. After Xa-FJ1 inoculation, the disease index varied for each variety of sugarcane but rose as the inoculation period was extended cultivars. Among them, ZZ13 had the lowest disease index (11.8%) after inoculation with the Xa-FJ1 strain, while ZZ14 had the highest disease index (58.8%). These findings demonstrated that the levels of resistance to the leaf scald varied significantly in these tested cultivars (Figure 2).

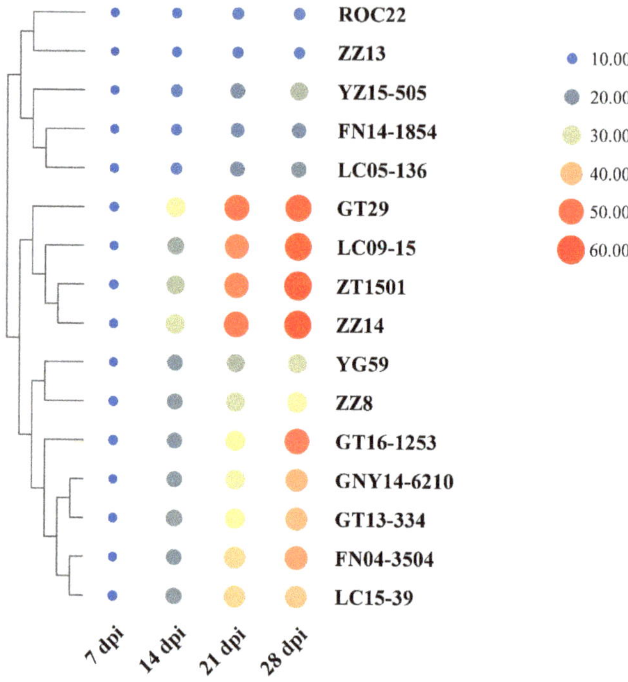

Figure 2. Disease index of 16 sugarcane cultivars under the infection by *X. albilineans* strain Xa-FJ1 at 7–28 days post-inoculation (dpi).

Based on the disease index 28 days post-inoculation (dpi) with the Xa-FJ1 strain, the resistance of 16 sugarcane cultivars to leaf scald disease was divided into four grades, as shown in Table 2. The disease index of the resistant group, which comprised ROC22 and ZZ13, varied between 11.8% and 12.6%. The disease index of the medium resistant group, including FN14-1854, LC05-136, and YZ15-505, ranged from 15.1% to 26.3%. The susceptible group consisted of seven cultivars (i.e., YG59, ZZ8, etc.), and the disease index ranged from 29.4% to 48.4%. There were four cultivars (GT29, ZZ14, LC09-15, and ZT1501) in the high susceptible group, and the disease index ranged from 52.8% to 58.8%.

Table 2. Different resistance grades of 16 sugarcane cultivars against *X. albilineans* strain Xa-FJ1 infection.

Grade	Variety (Line) Name	Number of Variety	Range of Disease Index (%)	Mean of Disease Index (%) [a]
Resistant	ROC22, ZZ13	2	11.8–12.6	12.2 A
Medium resistant	FN14-1854, LC05-136, YZ15-505	3	18.1–26.3	21.7 B
Susceptible	YG59, ZZ8, FN04-3504, LC15-39, GNY14-6210, GT13-334, GT16-1253	7	29.4–48.4	38.9 C
High susceptible	GT29, ZZ14, LC09-15, ZT1501	4	52.8–58.8	55.8 D

[a] Different letter between means indicates significant differences at $p < 0.05$ by Duncan's test.

Based on the results of the identification of the drought and leaf scald resistance of 16 sugarcane cultivars, 4 cultivars were selected for follow-up experiments: ROC22 (disease-resistant and drought-tolerant), ZZ13 (disease-resistant and non-drought-tolerant), GT29 (susceptible and non-drought-tolerant), and LC09-15 (susceptible and drought-tolerant).

2.3. Effect of Drought Stress on the Occurrence of Leaf Scald in Sugarcane

To explore the effect of drought stress on the infection of sugarcane by *Xa*, four sugarcane cultivars, ROC22, ZZ13, GT29, and LC09-15, were treated with Xa-FJ1 infection alone and combined stress (Xa-FJ1 infection plus PEG6000 osmotic stress, Xa + PEG6000). The bacterial contents of all the sugarcane cultivars increased significantly after two treatments (Figure 3). At 24 h post-treatment (hpt), the pathogenic bacterium populations in two cultivars (GT29 and LC09-15) susceptible to leaf scald were significantly increased by 143.1% and 39.5% in combined stress compared with Xa-FJ1 infection alone, respectively. Namely, the bacterial populations of GT29 and LC09-15 reached 8507.89 copies/L and 7885.44 copies/L under combined stress, respectively. By contrast, the bacterial populations of both cultivars were 3499.70 copies/L and 5651.30 copies/L under Xa-FJ1 infection alone, respectively. On the other hand, the bacterial populations of two cultivars (ROC22 and ZZ13) resistant to leaf scald did not have significant differences between both treatments at 24 pht. Overall, the proliferation rate of Xa-FJ1 in susceptible cultivars increased significantly under drought stress, but there was no significant difference in the proliferation rate of this pathogenic bacterium in resistant cultivars.

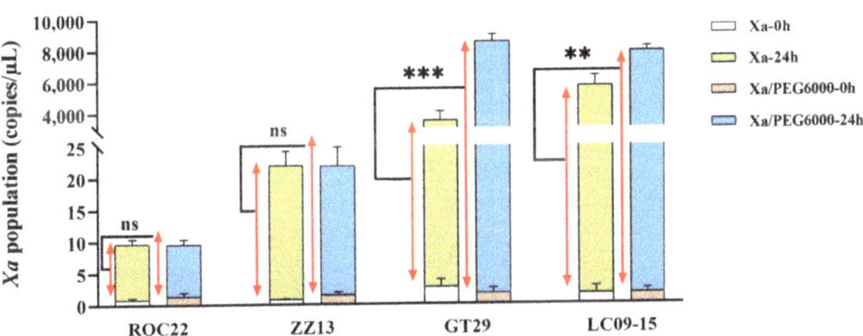

Figure 3. Quantitative PCR (qPCR) detection of bacterial pathogen population in four sugarcane cultivars under *X. albilineans* strain Xa-FJ1 infection or combing with PEG6000 stress. Xa-0h and Xa-24h: 0 and 24 h post-inoculation by Xa-FJ1, respectively; Xa/PEG6000-0h and Xa/PEG6000-24h: 0 and 24 h post-treatment with combined stress, respectively. ns, not significant; **, significant difference at $p < 0.01$; ***, significant difference at $p < 0.001$. Student's *t*-test was used to determine mean differences. The red arrows indicate the bacterial pathogen populations in four sugarcane cultivars at 24 h after *X. albilineans* strain Xa-FJ1 infection alone (Xa-24h) or combined stress (Xa/PEG6000-24h).

2.4. Changes in ROS Contents and ScRBOHD Gene Expression under Xa Infection Plus Drought Stress

The changes in the ROS contents in leaves and the transcriptional expression of *ScRBOHD* (a key gene for ROS synthesis) in four sugarcane cultivars under different treatments are shown in Figure 4. After Xa-FJ1 inoculation alone, the ROS contents in the leaves of ROC22, ZZ13, GT29, and LC09-15 were increased by 1.7-, 1.9-, 1.3-, and 1.4-fold, respectively, compared to the control (0 hpi). Correspondingly, the expression levels of the *ScRBOHD* gene rose by 7.5-, 5.0-, 3.1-, and 2.4-fold at 24 hpt compared to those of the control (0 hpi), respectively. Under combined stress, the ROS contents and *ScRBOHD* expression levels in four sugarcane cultivars were significantly increased, i.e., compared with 0 hpt, the ROS contents and *ScRBOHD* expression levels were enhanced, with increases of 1.2–1.8-fold and 1.6–8.0-fold in four cultivars at 24 hpt, respectively. On the other hand, the ROS contents and *ScRBOHD* expression levels in the two GT29 and LC09-15 cultivars were significantly lower at 24 hpt under combined stress than those under

Xa-FJ1 inoculation alone. Namely, the ROS contents and *ScRBOHD* expression levels in the two GT29 and LC09-15 cultivars were decreased by 9–13% and 27–33% at 24 hpt under combined stress, respectively, compared to *Xa* inoculation alone. Meanwhile, no significant differences in the ROS contents and *ScRBOHD* expression levels were found at 24 hpt between Xa-FJ1 inoculation alone and combined stress in two ROC22 and ZZ13 cultivars.

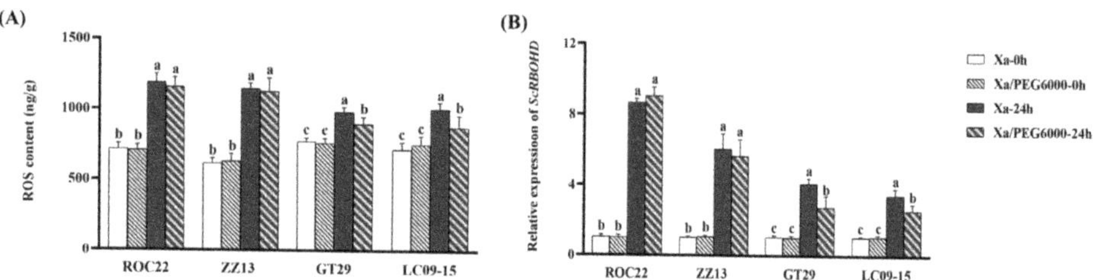

Figure 4. Changes in the ROS contents and transcriptional expression of *ScRBOHD* in sugarcane leaves under *X. albilineans* strain Xa-FJ1 infection along with PEG6000 stress. (**A**) ROS contents determined by an ELISA assay; (**B**) transcriptional expression of *ScRBOHD* determined by real-time quantitative reverse transcription PCR (qRT-PCR). Xa-0h and Xa-24h: 0 and 24 h post-inoculation by Xa-FJ1, respectively; Xa/PEG6000-0h and Xa/PEG6000-24h: 0 and 24 h post-treatment with Xa-FJ1 infection plus PEG6000 stress, respectively. Different letter between means indicates significant differences at $p < 0.05$ by Duncan's test.

2.5. Changes in Antioxidant Enzyme Activity and Gene-Related Expression under Xa Infection Plus Drought Stress

The activities of two key antioxidant enzymes (SOD and CAT) and the transcriptional expression of their related genes were measured in four sugarcane cultivars under different treatments (Figure 5). Under Xa-FJ1 inoculation alone, the SOD activities in four tested cultivars were increased by 9–12% at 24 hpi, while the expression levels of the *ScSOD* gene were enhanced 0.6–1.9-fold at 24 hpt compared with 0 hpi. Under combined stress, the SOD activities and expression levels of *ScSOD* in the four cultivars were increased. Namely, SOD activities were increased 8–9% and *ScSOD* expression levels were increased 1.4–2.0-fold in both cultivars ROC22 and ZZ13 at 24 hpt compared with 0 hpt. Meanwhile, the SOD activities and *ScSOD* expression levels in GT29 and LC09-15 cultivars at 24 hpt were significantly lower than those of Xa-FJ1 inoculation alone. However, no significant difference in the SOD activity and *ScSOD* expression was observed between combined stress and Xa-FJ1 infection alone in the two cultivars ROC22 and ZZ13.

Under *Xa* infection alone, CAT activities decreased, ranging from 11% to 31%, and the expression levels of *ScCAT* decreased from 20% to 44% in the four tested cultivars at 24 hpt compared with 0 hpt. Under combined stress, CAT activities were decreased by 12–32% in three cultivars (ROC22, ZZ13, and GT29), while the expression levels of *ScCAT* in ROC22 and ZZ13 were decreased by 30–50% at 24 hpt compared with 0 hpt. Overall, the CAT activity and expression level of *ScCAT* in the four cultivars were decreased or not significantly changed under *Xa* infection alone or combined stress. No significant difference in CAT activity or *ScCAT* expression level was found between *Xa* infection alone and combined stress in the four cultivars.

The frequency of extreme weather events has increased due to global warming, which has had a significant effect on sugarcane yield and quality [1,14]. Drought has emerged as a predominant factor restricting sugarcane production in China [17]. Numerous studies have highlighted the importance of relative water content, MDA, and chlorophyll fluorescence as pivotal indicators for assessing plant drought resistance [18,19]. Additionally, some parameters such as gauge hydration status, oxidative damage, and photosynthetic efficiency provide comprehensive insights into plant responses to drought stress [7]. In this study, a diverse array of indicators and statistic methods for assessing drought resistance were measured, including these key parameters, to comprehensively evaluate plant responses to drought stress. Some drought-tolerant cultivars such as LC09-15 and GT16-1253 were proposed. In addition, exploring and utilizing disease-resistant germplasm resources has become the most cost-effective and efficient approach for preventing and controlling sugarcane diseases [20,21]. Thus, some cultivars such as ROC22 and ZZ13 resistant to leaf scald were identified. These findings suggested that some sugarcane cultivars like ROC22 are tolerate to drought and leaf scald. To our knowledge, the cultivar ROC22 has even occupied 85% of sugarcane plant areas during the years of 2005–2021, and now it is becoming the main hybrid parent for sugarcane genetic improvement in China.

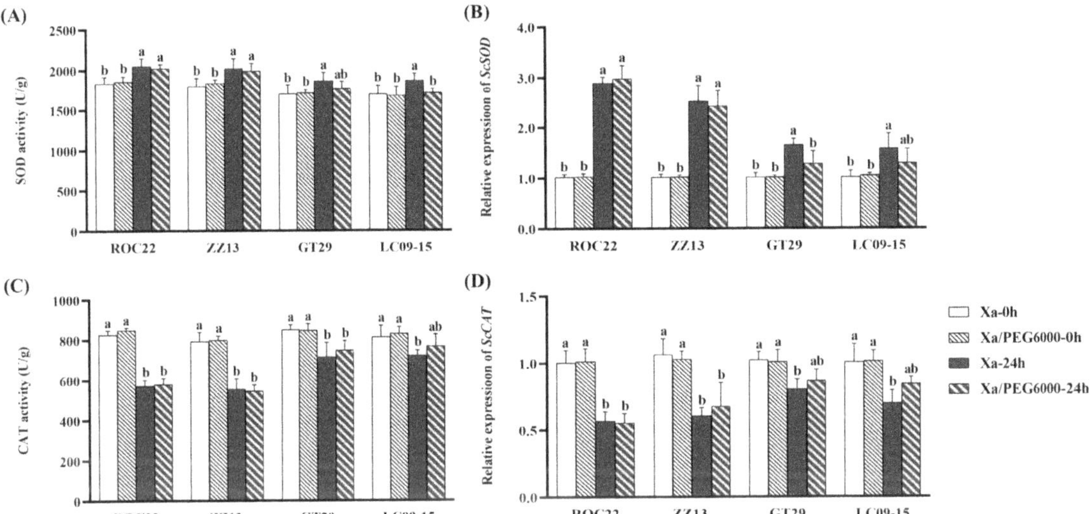

Figure 5. Changes in antioxidant enzyme activity and relative gene expression in sugarcane leaves *X. albilineans* infection along with PEG6000 stress. (**A,C**) Enzyme activities of SOD and CAT, respectively; (**B,D**) transcript levels of *ScSOD* and *ScCAT* determined by qRT-PCR, respectively. Xa-0h and Xa-24h: 0 and 24 h post-inoculation by Xa-FJ1, respectively; Xa/PEG6000-0h and Xa/PEG6000-24h: 0 and 24 h post-treatment with Xa-FJ1 infection plus PEG6000 stress, respectively. Different letter between means indicates significant differences at $p < 0.05$ by Duncan's test. Same letters above bars indicates no significant difference between treatments.3. Discussion.

Abiotic and biotic stressors have together been demonstrated to alter plant defense signaling pathways, as well as the pathogenicity of pathogens, which in turn has a more detrimental effect on agricultural productivity [22,23]. Drought can affect the severity of plant diseases [7]. For instance, drought stress reduces the severity of diseases in *Nicotiana benthamiana* infected by *Sclerotinia sclerotiorum* and *Pseudomonas syringae* pv. *tabaci* [24]. Furthermore, drought-induced stomatal closure restricts the pathogens of *Pseudomonas syringae* and *Melampsora apocyni* infecting *Arabidopsis thaliana* and *Apocynum venetum*, respectively, consequently diminishing disease incidence [25,26]. On the contrary, certain diseases such as charcoal stalk rot in sorghum, smut in cereals, and dry root rot in chickpeas exhibited rapid escalation under drought conditions, which might be caused by the fact that drought exacerbates plant vulnerability by compromising the structural integrity of plant tissues, especially the woody components [6,27,28]. In addition to impacting plants, drought directly hampers the growth and spread of pathogenic bacteria, thereby influencing their pathogenesis. For instance, drought stress significantly restricts the movement of *Verticillium dahliae* conidia, leading to a notable reduction in wilt severity [29]. Plants subjected to severe drought stress are not of benefit to pathogens with a biotrophic phase [7]. Our investigations showed that sugarcane exposed to PEG6000 stress had considerably higher populations of *Xa* in vulnerable cultivars, indicating that drought stress may enhance the incidence of sugarcane leaf scald.

Plants respond to stress by producing a wide range of intricate and sophisticated defense signals [30,31]. Notably, one of the key defense mechanisms of plants is ROS burst [8,10]. The respiratory burst oxidase homolog encoded by the *RBOH* gene is a key enzyme that produces ROS in response to stress signaling [32]. In this study, the levels of ROS production and *ScRBOHD* expression in sugarcane plants, especially in resistant cultivars, showed a significant increase after *Xa* infection. These results suggest that ROS is involved in the process of *Xa* infection, and the intensity of ROS burst is potentially correlated with the tolerance to leaf scald in sugarcane. Previous studies also illustrated that ROS production and *ScRBOHD* expression were involved in sugarcane in response to *Xa* stimuli [33] in rice cultivars resistant to *Magnaporthe oryzae* [34] and in *Gossypium barbadense* resistant to *Verticillium dahliae* [35]. On the other hand, the SOD and CAT are important protective enzymes in the enzymatic scavenging system of ROS, participating in plant defense responses under adverse environmental conditions [8,10]. Our study showed that SOD activity and *ScSOD* gene expression were markedly increased, while CAT activity and *ScCAT* gene expression were decreased in sugarcane, particularly in resistant cultivars, following *Xa* infection. This phenomenon could be attributed to the excessive accumulation of ROS in plants to induce sugarcane possessing stronger defense responses under the infection of this pathogen. Another obversion suggested that this phenomenon could be attributed to the excessive accumulation of salicylic acid in plants under pathogen stress, leading to a reduction in CAT activity [36].

Numerous investigations have demonstrated that a crucial node in the interplay between biotic and abiotic stressors is the ROS burst and scavenging mechanism [22,37]. ROS usually resists a combination of biotic and abiotic stresses by modulating ABA signaling and enhancing disease resistance [3,23]. Some cases revealed that drought imparts tolerance to plants against subsequent pathogen infections by enhanced levels of ROS and higher activity of key antioxidant enzymes [7,24]. For example, the plant of *N. benthamiana* exposure to moderate drought stress resulted in higher ROS production and the induction of some defense genes, decreasing the severity of subsequent infection by *P. syringae* pv. *tabaci* [24]. On the contrary, the rice plant subjected to drought stress increased the susceptibility to *M. oryzae* through downregulating the expression of multiple resistance genes and a reduction in ROS production [38]. Notably, our results showed that the levels of ROS and SOD activity in susceptible cultivars exposed to PEG6000 stress combined with *Xa* infection were markedly higher than those under *Xa* infection alone. A possible reason for increased ROS production might be drought stress. There are many studies that report increased ROS accumulation and oxidative stress under drought stress [39–41].

3. Materials and Methods

3.1. Plant Growth of Sugarcane

Healthy stalks of 16 sugarcane cultivars were collected from a sugarcane nursery (Fuzhou, China), and the single cuttings were immersed in flowing tap water overnight. These cuttings were then immersed in hot water (52 °C) for 3 h. The treated buds were transplanted into a 6 cm^3 planting pot filled with nutrient soil and subsequently placed in an intelligent artificial climate chamber (PLT-RGS-15PF, Ningbo Prandt Instrument Co., Ltd., Ningbo, China) with the following culture conditions: 28 °C, 65% humidity, a light/dark cycle of 16/8 h, and a light intensity of 30,000 Lux. Soil moisture was consistently maintained at 80% through artificial management. After 25 days, sugarcane plants grown until they reached the stage of 3–5 leaves (15–20 cm tall) were used for drought treatment and/or *Xa* inoculation. The soil moisture contents in pots were controlled at 80% before treatment.

3.2. Soil Drought Treatment

A total of 30 buds of sugarcane were used, of which 15 buds were used for soil drought treatment, and the other 15 buds within a controlled environment (80% soil moisture) were used as the control. Three independent repeats were carried out for each experiment. Each experiment was conducted using the Randomized Complete Block Design (RCBD) for allocation. This experiment was carried out in an intelligent artificial climate chamber with the abovementioned conditions. Leaf samples were collected at 0 and 3 days after drought treatment for the subsequent determination of drought resistance indicators.

3.3. Determination of the Drought Indexes

The following formula for calculating the leaf relative water contents (RWCs) of sugarcane was used:

$$RWC = (\text{fresh weight} - \text{dry weight}) \div (\text{turgid weight} - \text{dry weight}) \times 100$$

The maximum quantum yield of PSII photochemistry (Fv/Fm), non-photochemical quenching (qN), and PSII actual photosynthetic efficiency (Y(II)) was measured by the Imaging-PAM Chlorophyll Fluorometer (Shanghai Zealquest Scientific Technology Co., Ltd., China). Sugarcane plants were kept for 30 min in the dark before measurement.

The MDA content of the leaves from treated plants was analyzed using the kit of the biological company, and the specific experimental process was performed according to the manufacturer's manual (Beijing Solarbio Science & Technology Co., Ltd., China). Briefly, we weighed about 0.1 g of leaves and added 1 mL of extract buffer, and then they were ground for homogenization under an ice bath condition. They were centrifuged at $8000 \times g$ at 4 °C for 10 min, and then the supernatant was taken. We added reagents according to the instructions and kept the mixture in a 100 °C water bath for 60 min, and subsequently it was placed in an ice bath to cool. Then it was centrifuged at $10,000 \times g$ at room temperature for 10 min. The absorbance of the supernatant was measured at 532 nm and 600 nm wavelength using a Synergy H1 Multi-Mode Reader (BioTek, Winooski, VT, USA).

3.4. Analysis of Drought-Resistance Grades of Sugarcane Cultivars

The drought-resistance grades of sugarcane cultivars were calculated based on the physiological indicators determined after drought stress using the membership function method in fuzzy mathematics. The calculation method used is as follows.

If the measured index is positively correlated with drought resistance,

$$(X_i) = (X_i - X_{min}) \div (X_{max} - X_{min})$$

if the measured index is negatively correlated with drought resistance,

$$R(X_i) = 1 - (X_i - X_{min}) \div (X_{max} - X_{min})$$

The letters in the formula indicate that X_i = the ratio of the drought resistance index between treatment and control groups; X_{min} and X_{max} stand for minimum and maximum values of X_i in the measured indexes among all the tested cultivars, respectively.

3.5. X. albilineans Inoculation

To assess the disease resistance of the tested sugarcane cultivars, the decapitation method was used following the method of Zhao et al. [33]. The Xa-FJ1 strain of *X. albilineans* was used for inoculation [42]. A set of 35 buds of each sugarcane cultivar were tested, and another set of 35 buds were used as the control. The grown plants were randomly distributed in an intelligent artificial climate chamber with the abovementioned conditions. Three independent experiments were conducted. The disease severity and incidence of the leaf scald was recorded at 0-, 7-, 14-, 21-, and 28 dpi.

3.6. Resistance Assessment of Sugarcane Cultivars against Leaf Scald

The disease index of sugarcane leaf scald disease was identified according to Rott et al. [43] and Zhao et al. [33]. At 28 dpi, the disease index (%) of all the tested sugarcane cultivars was calculated, and resistance grades were grouped according to the criteria of Fu et al. [44].

3.7. Combined Stress Treatments

Four sugarcane cultivars (ROC22, GT29, ZZ13, and LC09-15) were used for combined stress treatment. The buds were placed in a 32 °C incubator for germination for 3 days, and then cuttings with good bud germination were cultivated with clean water under the following conditions: temperature 30 °C, humidity 65%, and a light/dark cycle of 16/8 h. After the plants grew to the 3–5 leaf stage, they were cultured with Hoagland's nutrient solution for 1 week, and then all the plants were divided into two groups for stress treatments. The first group was used for Xa-FJ1 inoculation without PEG6000 stress, while the second group was used for Xa-FJ1 inoculation plus 25% PEG6000 added in the Hoagland nutrient solution. Xa-FJ1 was inoculated using the leaf cutting method [45]. The treated plants were randomly distributed in an intelligent artificial climate chamber with the abovementioned conditions. Plant leaves were collected at 0 h and 24 hpt for the determination of subsequent bacterial contents, physiological and biochemical indexes, and gene expression.

3.8. Determination of Pathogenic Bacterial Contents

Total DNA was extracted from sugarcane leaves with CTAB reagent for quantitative real-time PCR (qPCR) detection according to the specific protocol by Shi et al. [20]. At 0 and 24 hpt, population density determination was performed on three leaf subsamples collected for each treatment. Three technical replicates for each subsample were performed.

3.9. ROS Production and Antioxidant Enzyme Assays

The ROS contents and activities of two antioxidant enzymes (SOD and CAT) were analyzed using the kit from Solarbio Science & Technology Co., Ltd. (Beijing, China) following the manufacturer's manual. Briefly, 0.1 g of leaf samples was weighed and then ground with 1 mL of 10 mM PBS buffer (pH = 7.4), followed by centrifugation at 25 °C at $1200 \times g$ for 20 min, and then the supernatant was used for the subsequent determination of ROS contents by a sandwich ELISA, and the optical density at 450 nm was determined by spectrophotometry. Another set of leaf samples (0.1 g) were ground, followed by centrifugation $(16,770 \times g)$ at 4 °C for 10 min. The supernatant was determined for CAT and SOD activity at visible wavelengths of 240 nm and 560 nm, respectively.

3.10. RNA Extraction and qRT-PCR Analysis

The transcriptional expression of the genes was detected by a real-time quantitative reverse transcription PCR (qRT-PCR) assay with specific primer pairs (Table S1). These genes include the respiratory burst oxidase homologs gene (*ScRBOHD*), which is critical in encoding ROS production in plants, and two genes encoding antioxidant enzymes, the superoxide dismutase gene (*ScSOD*) and the catalase gene (*ScCAT*). Total RNA was extracted from leaf subsamples, and they were reverse-transcribed into cDNA according to the method of Chu et al. [46]. For each treatment, three leaf subsamples were analyzed at each time point. Three technical replicates were carried out for each subsample.

3.11. Statistical Analysis

Variance analysis (ANOVA) was utilized to compare the datasets. Duncan's test (comparison among more than two groups of data) and Student's t-test (comparison between two groups of data) were employed to determine mean differences at $p < 0.05$ or 0.01. Software from IBM (China), called SPSS version 18.0, was used for all the analyses.

4. Conclusions

This study identified drought and leaf scald resistance in 16 recently released sugarcane cultivars, which offers a crucial hint for variety extension. Subsequently, four cultivars contrasting to drought and leaf scald resistance were treated with a combination of PEG6000 stress and *Xa* infection. Drought promoted the incidence of leaf scald disease and *Xa* contents in susceptible cultivars, while there was no significant change in resistant cultivars. The ROS burst and scavenging system was involved in four tested sugarcane cultivars against *Xa* infection. A stronger response of this pathway was observed in resistant cultivars than in susceptible cultivars. However, the response of the ROS production and scavenging system was weakened in sugarcane cultivars under combined stress (*Xa* infection along with PEG6000 stress) compared with *Xa* infection only. Notably, a higher weakening degree existed in susceptible cultivars than in resistant cultivars. Our findings suggest that ROS is a key defense node in sugarcane against *Xa* infection combined with drought stress. This work will lay the foundation for further research on the mechanism altering the prevalence and virulence of *Xa* in sugarcane under drought stress.

Supplementary Materials: The following supporting information can be downloaded at: https://www.mdpi.com/article/10.3390/plants13060862/s1, Table S1. Primer pairs used to analyze *ScRBOHD*, *ScSOD*, and *ScCAT* gene expressions by qRT-PCR assay.

Author Contributions: Conceptualization, S.-J.G.; methodology, Y.-S.W. and J.-Y.Z.; software, Y.-S.W.; validation, S.-J.G., Y.-S.W. and J.-Y.Z.; formal analysis, Y.-S.W. and J.-Y.Z.; investigation, Y.-S.W., M.-T.H. and H.-Y.F.; resources, H.-Y.F. and H.-L.Z.; data curation, Y.-S.W., J.-Y.Z. and H.-L.Z.; writing—original draft preparation, Y.-S.W., T.J. and A.A.; writing—review and editing, S.-J.G. and T.J.; visualization, S.-J.G.; supervision, S.-J.G.; project administration, S.-J.G.; funding acquisition, S.-J.G. All authors have read and agreed to the published version of the manuscript.

Funding: This research was supported by the National Key Research and Development Program of China (grant no. 2023YFD1200700 & 2023YFD1200703) and the earmarked fund for China Agriculture Research System (grant no. CARS-170302).

Data Availability Statement: All data supporting the findings of this study are available within the paper and its supplementary file.

Conflicts of Interest: The authors declare no conflicts of interest.

References

1. Anderegg, W.R.L.; Trugman, A.T.; Badgley, G.; Anderson, C.M.; Bartuska, A.; Ciais, P.; Cullenward, D.; Field, C.B.; Freeman, J.; Goetz, S.J.; et al. Climate-driven risks to the climate mitigation potential of forests. *Science* **2020**, *368*, eaaz7005. [CrossRef]
2. Raymond, C.; Matthews, T.; Horton, R.M. The emergence of heat and humidity too severe for human tolerance. *Sci. Adv.* **2020**, *6*, eaaw1838. [CrossRef] [PubMed]
3. Leisner, C.P.; Potnis, N.; Sanz-Saez, A. Crosstalk and trade-offs: Plant responses to climate change-associated abiotic and biotic stresses. *Plant Cell Environ.* **2023**, *46*, 2946–2963. [CrossRef]
4. Zandalinas, S.I.; Mittler, R. Plant responses to multifactorial stress combination. *New. Phytol.* **2022**, *234*, 1161–1167. [CrossRef] [PubMed]
5. Lamers, J.; van der Meer, T.; Testerink, C. How plants sense and respond to stressful environments. *Plant Physiol.* **2020**, *182*, 1624–1635. [CrossRef]
6. Pandey, P.; Ramegowda, V.; Senthil-Kumar, M. Shared and unique responses of plants to multiple individual stresses and stress combinations: Physiological and molecular mechanisms. *Front. Plant Sci.* **2015**, *6*, 723. [CrossRef]
7. Choudhary, A.; Senthil-Kumar, M. Drought: A context-dependent damper and aggravator of plant diseases. *Plant Cell Environ.* **2024**, *47*, 1–18. [CrossRef] [PubMed]
8. Wu, B.; Qi, F.; Liang, Y. Fuels for ROS signaling in plant immunity. *Trends. Plant Sci.* **2023**, *28*, 1124–1131. [CrossRef]
9. Mittler, R.; Zandalinas, S.I.; Fichman, Y.; Van Breusegem, F. Reactive oxygen species signalling in plant stress responses. *Nat. Rev. Mol. Cell. Biol.* **2022**, *23*, 663–679. [CrossRef]
10. Sahu, P.K.; Jayalakshmi, K.; Tilgam, J.; Gupta, A.; Nagaraju, Y.; Kumar, A.; Hamid, S.; Singh, H.V.; Minkina, T.; Rajput, V.D.; et al. ROS generated from biotic stress: Effects on plants and alleviation by endophytic microbes. *Front. Plant Sci.* **2022**, *13*, 1042936. [CrossRef]
11. Li, M.; Kim, C. Chloroplast ROS and stress signaling. *Plant Commun.* **2022**, *3*, 100264. [CrossRef]

12. Foyer, C.H.; Noctor, G. Redox homeostasis and signaling in a higher-CO_2 world. *Annu. Rev. Plant Biol.* **2020**, *71*, 157–182. [CrossRef]

13. Hasanuzzaman, M.; Bhuyan, M.; Zulfiqar, F.; Raza, A.; Mohsin, S.M.; Mahmud, J.A.; Fujita, M.; Fotopoulos, V. Reactive Oxygen species and antioxidant defense in plants under abiotic stress: Revisiting the crucial role of a universal defense regulator. *Antioxidants* **2020**, *9*, 681. [CrossRef]

14. Javed, T.; Gao, S.J. WRKY transcription factors in plant defense. *Trends. Genet.* **2023**, *39*, 787–801. [CrossRef]

15. Xiao, S.; Wu, Y.; Xu, S.; Jiang, H.; Hu, Q.; Yao, W.; Zhang, M. Field evaluation of *TaDREB2B*-ectopic expression sugarcane (*Saccharum* spp. hybrid) for drought tolerance. *Front. Plant Sci.* **2022**, *13*, 963377. [CrossRef]

16. Ntambo, M.S.; Meng, J.Y.; Rott, P.C.; Henry, R.J.; Zhang, H.L.; Gao, S.J. Comparative transcriptome profiling of resistant and susceptible sugarcane cultivars in response to infection by *Xanthomonas albilineans*. *Int. J. Mol. Sci.* **2019**, *20*, 6138. [CrossRef] [PubMed]

17. Kumar, R.; Sagar, V.; Verma, V.C.; Kumari, M.; Gujjar, R.S.; Goswami, S.K.; Kumar Jha, S.; Pandey, H.; Dubey, A.K.; Srivastava, S.; et al. Drought and salinity stresses induced physio-biochemical changes in sugarcane: An overview of tolerance mechanism and mitigating approaches. *Front. Plant Sci.* **2023**, *14*, 1225234. [CrossRef] [PubMed]

18. Hessini, K.; Wasli, H.; Al-Yasi, H.M.; Ali, E.F.; Issa, A.A.; Hassan, F.A.S.; Siddique, K.H.M. Graded Moisture Deficit Effect on Secondary Metabolites, Antioxidant, and Inhibitory Enzyme Activities in Leaf Extracts of *Rosa damascena* Mill. var. trigentipetala. *Horticulturae* **2022**, *8*, 177. [CrossRef]

19. Mazrou, R.M.; Hassan, F.A.S.; Mansour, M.M.F.; Moussa, M.M. Melatonin Enhanced Drought Stress Tolerance and Productivity of *Pelargonium graveolens* L. (Herit) by Regulating Physiological and Biochemical Responses. *Horticulturae* **2023**, *9*, 1222. [CrossRef]

20. Duarte Dias, V.; Fernandez, E.; Cunha, M.G.; Pieretti, I.; Hincapie, M.; Roumagnac, P.; Comstock, J.C.; Rott, P. Comparison of loop-mediated isothermal amplification, polymerase chain reaction, and selective isolation assays for detection of *Xanthomonas albilineans* from sugarcane. *Trop. Plant Pathol.* **2018**, *43*, 351–359. [CrossRef]

21. Shi, Y.; Zhao, J.Y.; Zhou, J.R.; Ntambo, M.S.; Xu, P.Y.; Rott, P.C.; Gao, S.J. Molecular detection and quantification of *Xanthomonas albilineans* in juice from symptomless sugarcane stalks using a real-time quantitative pcr assay. *Plant Dis.* **2021**, *105*, 3451–3458. [CrossRef]

22. Nejat, N.; Mantri, N. Plant Immune System: Crosstalk between responses to biotic and abiotic stresses the missing link in understanding plant defence. *Curr. Issues. Mol. Biol.* **2017**, *23*, 1–16. [CrossRef] [PubMed]

23. Ramegowda, V.; Da Costa, M.V.J.; Harihar, S.; Karaba, N.N.; Sreeman, S.M. Chapter 17—Abiotic and biotic stress interactions in plants: A cross-tolerance perspective. In *Priming-Mediated Stress and Cross-Stress Tolerance in Crop Plants*; Hossain, M.A., Liu, F., Burritt, D.J., Fujita, M., Huang, B., Eds.; Academic Press: London, UK, 2020; pp. 267–302.

24. Ramegowda, V.; Senthil-Kumar, M.; Ishiga, Y.; Kaundal, A.; Udayakumar, M.; Mysore, K.S. Drought stress acclimation imparts tolerance to *Sclerotinia sclerotiorum* and *Pseudomonas syringae* in *Nicotiana benthamiana*. *Int. J. Mol. Sci.* **2013**, *14*, 9497–9513. [CrossRef] [PubMed]

25. Gao, P.; Duan, T.Y.; Christensen, M.J.; Nan, Z.B.; Liu, Q.T.; Meng, F.J.; Huang, J.F. The occurrence of rust disease, and biochemical and physiological responses on *Apocynum venetum* plants grown at four soil water contents, following inoculation with *Melampsora apocyni*. *Eur. J. Plant Pathol.* **2018**, *150*, 549–563. [CrossRef]

26. Gupta, A.; Dixit, S.K.; Senthil-Kumar, M. Drought stress predominantly endures *Arabidopsis thaliana* to *Pseudomonas syringae* infection. *Front. Plant Sci.* **2016**, *7*, 808. [CrossRef] [PubMed]

27. Irulappan, V.; Kandpal, M.; Saini, K.; Rai, A.; Ranjan, A.; Sinharoy, S.; Senthil-Kumar, M. Drought Stress Exacerbates Fungal Colonization and Endodermal Invasion and Dampens Defense Responses to Increase Dry Root Rot in Chickpea. *Mol. Plant-Microbe Interact.* **2022**, *35*, 583–591. [CrossRef] [PubMed]

28. Hillabrand, R.M.; Hacke, U.G.; Lieffers, V.J. Drought-induced xylem pit membrane damage in aspen and balsam poplar. *Plant Cell Environ.* **2016**, *39*, 2210–2220. [CrossRef] [PubMed]

29. Saadatmand, A.R.; Banihashemi, Z.; Sepaskhah, A.R.; Maftoun, M. Soil Salinity and Water Stress and Their Effect on Susceptibility to Verticillium Wilt Disease, Ion Composition and Growth of Pistachi. *J. Phytopathol.* **2008**, *156*, 287–292. [CrossRef]

30. Verma, V.; Ravindran, P.; Kumar, P.P. Plant hormone-mediated regulation of stress responses. *BMC Plant Biol.* **2016**, *16*, 86. [CrossRef]

31. Choudhury, F.K.; Rivero, R.M.; Blumwald, E.; Mittler, R. Reactive oxygen species, abiotic stress and stress combination. *Plant J.* **2017**, *90*, 856–867. [CrossRef]

32. Kadota, Y.; Shirasu, K.; Zipfel, C. Regulation of the NADPH Oxidase RBOHD During Plant Immunity. *Plant Cell Physiol.* **2015**, *56*, 1472–1480. [CrossRef]

33. Zhao, J.Y.; Chen, J.; Shi, Y.; Fu, H.Y.; Huang, M.T.; Rott, P.C.; Gao, S.J. Sugarcane responses to two strains of *Xanthomonas albilineans* differing in pathogenicity through a differential modulation of salicylic acid and reactive oxygen species. *Front. Plant Sci.* **2022**, *13*, 1087525. [CrossRef]

34. Liu, X.; Zhou, Q.; Guo, Z.; Liu, P.; Shen, L.; Chai, N.; Qian, B.; Cai, Y.; Wang, W.; Yin, Z.; et al. A self-balancing circuit centered on MoOsm1 kinase governs adaptive responses to host-derived ROS in *Magnaporthe oryzae*. *eLife* **2020**, *9*, e61605. [CrossRef]

35. Chang, Y.; Li, B.; Shi, Q.; Geng, R.; Geng, S.; Liu, J.; Zhang, Y.; Cai, Y. Comprehensive analysis of respiratory burst oxidase homologs (Rboh) gene family and function of *GbRboh5/18* on verticillium wilt resistance in *Gossypium barbadense*. *Front. Genet.* **2020**, *11*, 788. [CrossRef]

36. Shim, I.S.; Momose, Y.; Yamamoto, A.; Kim, D.W.; Usui, K. Inhibition of catalase activity by oxidative stress and its relationship to salicylic acid accumulation in plants. *Plant Growth Regul.* **2003**, *39*, 285–292. [CrossRef]
37. Romero-Puertas, M.C.; Terrón-Camero, L.C.; Peláez-Vico, M.; Molina-Moya, E.; Sandalio, L.M. An update on redox signals in plant responses to biotic and abiotic stress crosstalk: Insights from cadmium and fungal pathogen interactions. *J. Exp. Bot.* **2021**, *72*, 5857–5875. [CrossRef]
38. Bidzinski, P.; Ballini, E.; Ducasse, A.; Michel, C.; Zuluaga, P.; Genga, A.; Chiozzotto, R.; Morel, J.B. Transcriptional Basis of Drought-Induced Susceptibility to the Rice Blast Fungus *Magnaporthe oryzae*. *Front. Plant Sci.* **2016**, *7*, 1558. [CrossRef]
39. Cruz, C.D.M.H. Drought stress and reactive oxygen species: Production, scavenging and signaling. *Plant Signal. Behav.* **2008**, *3*, 156–165. [CrossRef]
40. Sachdev, S.; Ansari, S.A.; Ansari, M.I.; Fujita, M.; Hasanuzzaman, M. Abiotic stress and reactive oxygen species: Generation, signaling, and defense mechanisms. *Antioxidants* **2021**, *10*, 277. [CrossRef] [PubMed]
41. Hasanuzzaman, M.; Bhuyan, M.B.; Parvin, K.; Bhuiyan, T.F.; Anee, T.I.; Nahar, K.; Hossen, M.S.; Zulfiqar, F.; Alam, M.M.; Fujita, M. Regulation of ROS metabolism in plants under environmental stress: A review of recent experimental evidence. *Int. J. Mol. Sci.* **2020**, *21*, 8695. [CrossRef] [PubMed]
42. Zhang, H.L.; Ntambo, M.S.; Rott, P.C.; Chen, G.; Chen, L.L.; Huang, M.T.; Gao, S.J. Complete genome sequence reveals evolutionary and comparative genomic features of *Xanthomonas albilineans* causing sugarcane leaf scald. *Microorganisms* **2020**, *8*, 182. [CrossRef]
43. Rott, P.; Mohamed, I.S.; Klett, P.; Soupa, D.; de Saint-Albin, A.; Feldmann, P.; Letourmy, P. Resistance to leaf scald disease is associated with limited colonization of sugarcane and wild relatives by *Xanthomonas albilineans*. *Phytopathology* **1997**, *87*, 1202–1213. [CrossRef]
44. Fu, H.Y.; Zhang, T.; Peng, W.J.; Duan, Y.Y.; Xu, Z.X.; Lin, Y.H.; Gao, S.J. Identification of resistance to leaf scald in newly released sugarcane varieties at seedling stage by artificial inoculation. *Crop J.* **2021**, *47*, 1531–1539. [CrossRef]
45. Lin, L.H.; Ntambo, M.S.; Rott, P.C.; Wang, Q.N.; Lin, Y.H.; Fu, H.Y.; Gao, S.J. Molecular detection and prevalence of *Xanthomonas albilineans*, the causal agent of sugarcane leaf scald, in China. *Crop Prot.* **2018**, *109*, 17–23. [CrossRef]
46. Chu, N.; Zhou, J.R.; Fu, H.Y.; Huang, M.T.; Zhang, H.L.; Gao, S.J. Global gene responses of resistant and susceptible sugarcane cultivars to *Acidovorax avenae* subsp. avenae identified using comparative transcriptome analysis. *Microorganisms* **2019**, *8*, 10. [CrossRef] [PubMed]

Article

Establishment of an Efficient Sugarcane Transformation System via Herbicide-Resistant CP4-EPSPS Gene Selection

Wenzhi Wang [1,2,3,4,†], Talha Javed [1,2,†], Linbo Shen [1,2], Tingting Sun [1,2], Benpeng Yang [1,2] and Shuzhen Zhang [1,2,4,*]

1 National Key Laboratory for Tropical Crop Breeding, Institute of Tropical Bioscience and Biotechnology, Chinese Academy of Tropical Agricultural Sciences, Haikou 571101, China; wangwenzhi@itbb.org.cn (W.W.); talhajaved@itbb.org.cn (T.J.); shenlinbo@itbb.org.cn (L.S.); suntingting@itbb.org.cn (T.S.); yangbenpeng@itbb.org.cn (B.Y.)
2 Sanya Research Institute, Chinese Academy of Tropical Agricultural Sciences, Sanya 571763, China
3 Crop Genomics and Bioinformatics Center and National Key Laboratory of Crop Genetics and Germplasm Enhancement, Nanjing Agricultural University, Nanjing 210095, China
4 Hainan Yazhou Bay Seed Laboratory, Sanya 571763, China
* Correspondence: zhangshuzhen@itbb.org.cn
† These authors contributed equally to this work.

Abstract: Sugarcane (*Saccharum* spp.), a major cash crop that is an important source of sugar and bioethanol, is strongly influenced by the impacts of biotic and abiotic stresses. The intricate polyploid and aneuploid genome of sugarcane has shown various limits for conventional breeding strategies. Nonetheless, biotechnological engineering currently offers the best chance of introducing commercially significant agronomic features. In this study, an efficient *Agrobacterium*-mediated transformation system that uses the herbicide-resistant CP4-EPSPS gene as a selection marker was developed. Notably, all of the plants that were identified by PCR as transformants showed significant herbicide resistance. Additionally, this transformation protocol also highlighted: (i) the high yield of transgenic lines from calli (each gram of calli generated six transgenic lines); (ii) improved selection; and (iii) a higher transformation efficiency. This protocol provides a reliable tool for a routine procedure for the generation of resilient sugarcane plants.

Keywords: sugarcane; *Agrobacterium*-mediated transformation; transgenic technology; herbicide resistance

Citation: Wang, W.; Javed, T.; Shen, L.; Sun, T.; Yang, B.; Zhang, S. Establishment of an Efficient Sugarcane Transformation System via Herbicide-Resistant CP4-EPSPS Gene Selection. *Plants* **2024**, *13*, 852. https://doi.org/10.3390/plants13060852

Academic Editor: Andreas W. Ebert

Received: 16 February 2024
Revised: 6 March 2024
Accepted: 11 March 2024
Published: 15 March 2024

1. Introduction

Sugarcane (*Saccharum* spp.) is one of the most important industrial crops, contributing significantly to economics and food security [1]. Sugarcane is primarily grown as a source of sugar and bioethanol, meeting around 80% and 40% of the global sugar and bioethanol demands, respectively [2]. Climate change adversely affects both the quantity and quality of many crops worldwide. Higher temperatures will exacerbate drought stress and accelerate crop development, as well as increase crop yield variability and the likelihood of pathogen-related yield failures [3]. Additionally, sugarcane productivity is vulnerable to climate change through the direct and indirect effects arising from herbicides [4]. Herbicides may have an impact on the prevalence or severity of infections as well as plant pathogens. Indirect herbicide effects on soil microorganisms that are often hostile to root infections may exacerbate plant diseases. Certain herbicides have the potential to directly harm crop plants, impairing their ability to fend off disease and increasing their susceptibility to infections. When plants sustain sublethal injuries, their root tissues' membrane permeability may change, allowing the cell contents to seep into the root–soil interface [5,6]. A number of the effects of herbicides, including the effects of glyphosate-based herbicides, and herbicide-resistant plants are reviewed by Schütte et al. [6]. Therefore, in the current period of climate change, the creation of sugarcane cultivars that are multi-stress-resistant has become a research hotspot [7–9].

Advances in plant-breeding-based technologies have contributed significantly to sugarcane improvement in recent decades [10–12]. For instance, sugarcane breeders interested in increasing herbicide resistance crossed *Saccharum officinarum* (2n = 8x = 80; x = 10) with a wild and vigorous relative, *S. spontaneum* (2n = 5 − 16x = 40 − 128; x = 8), and then backcrossed the hybrids with *S. officinarum*. The hybridization of both genera resulted in modern cultivars with chromosome numbers ranging from 110 to 130, of which 80% originated from *S. officinarum* and 10–15% from *S. spontaneum*, with about 5–10% being recombinant chromosomes [1]. The complex genome and poly/aneuploidy nature of sugarcane as well as the current climate change scenarios complicate agronomic trait improvement through traditional breeding strategies [1]. Notably, transformation and transgenic techniques are now being applied routinely to diverse plant species, especially sugarcane [13–17]. Furthermore, the application of transgenic technology depends on the creation of an effective embryogenic system. Biotechnologies like cell and tissue culture and genetic transformation techniques play a major role in boosting the yield of sugarcane cultivars. Numerous techniques for plant regeneration in sugarcane have been explored to date, including direct shoot organogenesis [12–14]. Specifically, the *Agrobacterium-tumefaciens*-mediated transformation of commercial sugarcane cultivars is a more desirable approach due to the ease of application and low cost as well as its relatively higher efficiency compared to other techniques [14].

The efficiency of sugarcane transgenesis can be improved by adopting appropriate transgenic selection systems, such as PMI/Mannose, CP4-EPSPS/glyphosate, and bar/Basta selection [13]. For instance, Wang and colleagues reported on the *bar*/Basta sugarcane transformation system. In this system, selection efficiency (SE = PCR-positive shoots/resistant shoots) was close to 100% [13]. Similarly, Baso and colleagues established a *Bar* gene transformation system against a glufosinate-ammonium herbicide. The SP80-3280 sugarcane cultivar was effectively transformed using this technique, which included reduced oxidation and recalcitrance, a high production of embryogenic calli, enhanced selection and shoot regeneration, and the rooting of the altered plants. These enhancements were combined to produce a transformation efficiency of 2.2%. Another study also highlighted that the transgenic sugarcane resistance to the Basta herbicide obtained by *Bar* gene screening had a lower application prospect [14–16]. Conversely, several reports suggested the use of the CP4-EPSPS system for a higher resistance to the Roundup herbicide and vast economic value [14–16]. However, little is known about the CP4-EPSPS transgenic sugarcane transformation system. Therefore, an *Agrobacterium*-mediated transformation system that uses the CP4-EPSPS gene against the Roundup herbicide as a selection marker for sugarcane is developed in this study.

2. Results

2.1. Establishment of Embryogenic Callus

From the immature stalks of sugarcane, explant of about 5 cm in length containing the immature leaf whorl was cut (Figure 1a). After disinfection, the immature top stalk was cut into 1 mm thick transverse sections and transferred to the callus induction medium (Figure 1b). After 25–30 days of culture, mucilaginous calli formed on the transverse sections (Figure 1c). At 40–45 days prior to the herbicide assay and transformation, healthy embryonic calli (light yellow) were selected and cleaned, and the excess culture medium was washed away (Figure 1d).

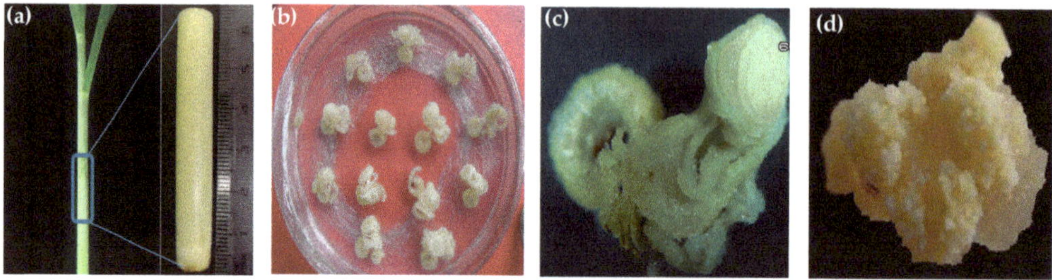

Figure 1. Induction of embryonic callus from immature sugarcane top stalk. (**a**) Preparation of immature top stalks of sugarcane; (**b**) transverse section transferred to callus induction medium; (**c**) mucilaginous calli formed on the transverse sections; (**d**) embryonic calli suitable for transformation after 40–45 days of induction.

2.2. Determination of Minimum Inhibitory Concentration of Roundup

Compact calli were transferred to Petri dishes containing different concentrations (0 mg/L, 10 mg/L, 20 mg/L, 30 mg/L, 40 mg/L, 50 mg/L) of glyphosate for two weeks followed by transfer to regeneration medium containing the abovementioned same concentrations of glyphosate for an additional two weeks (Figure 2). Notably, the regeneration of putative plants decreased with the increase in glyphosate concentration. Plant regeneration in Petri dishes with 0 mg/L and 10 mg/L of glyphosate resulted in a remarkably high rate. The regeneration of the callus started to be reduced significantly in Petri dishes containing 20 mg/L glyphosate. Notably, only a few calli regenerated with fine buds in Petri dishes with 30 mg/L glyphosate. Additionally, no plant regeneration occurred in Petri dishes containing 40 mg/L and 50 mg/L of glyphosate (Figure 2). Overall, the lethal dose of glyphosate for sugarcane callus was 40 mg/L.

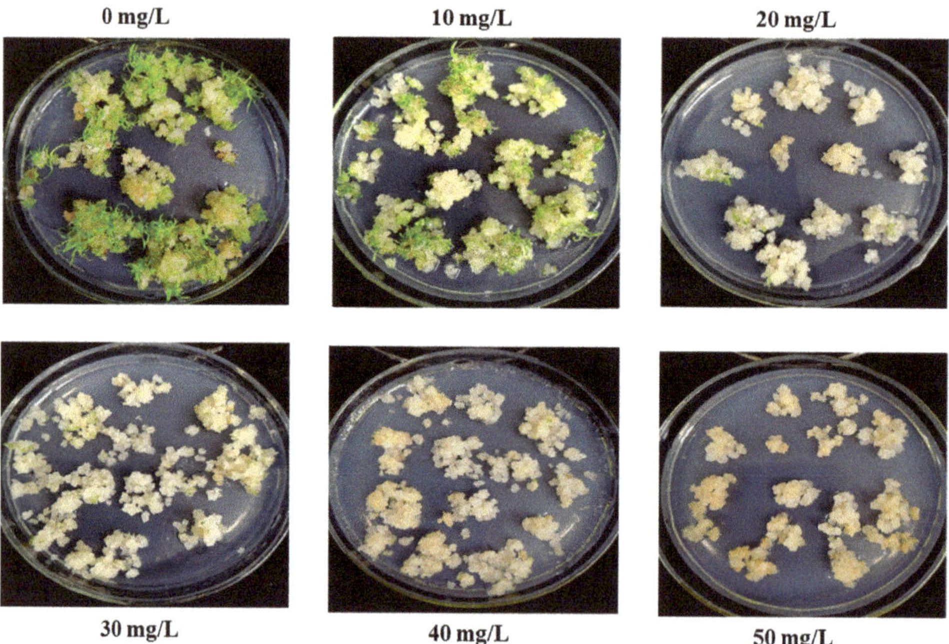

Figure 2. Sugarcane calli maintained on different concentrations (0 mg/L, 10 mg/L, 20 mg/L, 30 mg/L, 40 mg/L, 50 mg/L) of glyphosate.

2.3. Selection of Resistant Shoots

A total of three Petri dishes, each containing 3.5 g of embryogenic calli (air dried), were prepared prior to transformation (Figure 3a). After *Agrobacterium* induction, co-cultivation, and recovery cultivation, the infected calli were transferred to the selection mediums containing 20 mg/L, 30 mg/L, and 40 mg/L of glyphosate, respectively. After 35 days of cultivation in the dark, most of the transformed calli died, and a few calli sprouted which showed significant resistance to glyphosate (Figure 3b). Resistant calli were transferred to a regeneration medium containing the same glyphosate concentrations and cultured for 20 days (Figure 3c). All of the regenerated shoots were transferred to elongation medium and cultured for 20 days followed by transfer to rooting medium (Figure 3d).

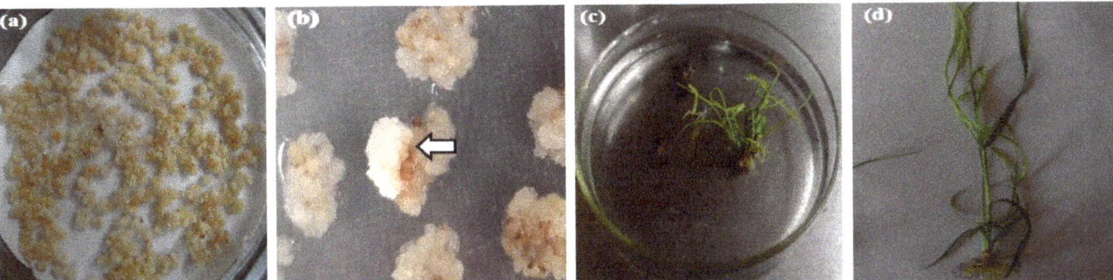

Figure 3. Selection of resistant shoots. (**a**) Air-dried embryonic calli; (**b**) selection of infected calli; (**c**) regeneration of resistant calli; (**d**) root emergence of resistant shoots. Note: arrow indicates resistant calli on the selection medium.

2.4. Green Fluorescent Protein Expression

The infected calli after three days of co-cultivation, resistant calli after 30 days of selection cultivation, and resistant shoots after rooting cultivation were evaluated for GFP expression under a microscope. Bright green fluorescence was observed on the infected calli, indicating a high level of *Agrobacterium* infection. Resistant calli showed bright green fluorescence in comparison with non-resistant calli. Bright green fluorescence on resistant shoots, particularly leaves and roots, indicated an absence of chimera transgenic shoot (Figure 4).

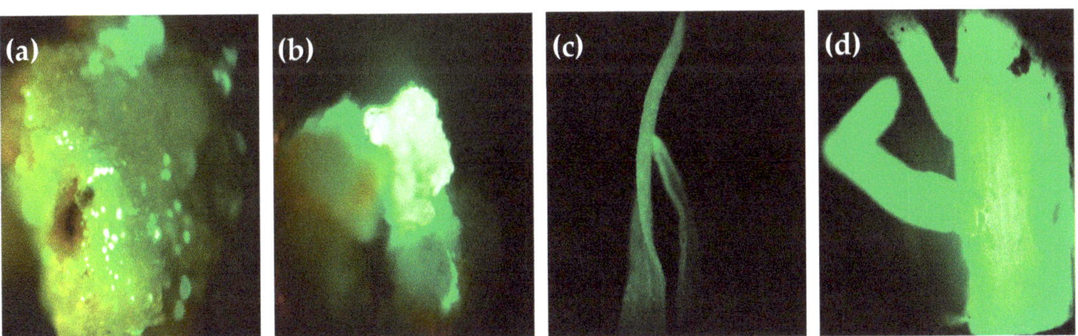

Figure 4. Green fluorescent protein expression in infected calli, resistant calli, and shoots. (**a**) GFP fluorescence of co-cultivated calli; (**b**) GFP fluorescence of resistant calli; (**c**) GFP fluorescence of leaf from resistant shoot; (**d**) GFP fluorescence of root from resistant shoot.

2.5. PCR Assay of Potential Transformed Shoots

After rooting cultivation, all the resistant shoots were evaluated for *CP4-EPSPS* and *GFP* gene integration by PCR. Notably, *CP4-EPSPS* and *GFP* genes were PCR negative for wild-type sugarcane genomic DNA. All the 21 resistant shoots from 40 mg/L glyphosate selection were PCR positive for the *CP4-EPSPS* gene, while 15 shoots among them were PCR positive for the *GFP* gene. *Agrobacterium* absence in the resistant plants was assessed by *VirG* gene PCR detection with primers specific to the *VirG* gene. *Agrobacterium* was not detected in the resistant sugarcane plants from the rooting medium, as evidenced by the lack of amplification in those plants. It is noteworthy that only PCR-positive, resistant plants were transplanted to the greenhouse (Figure 5).

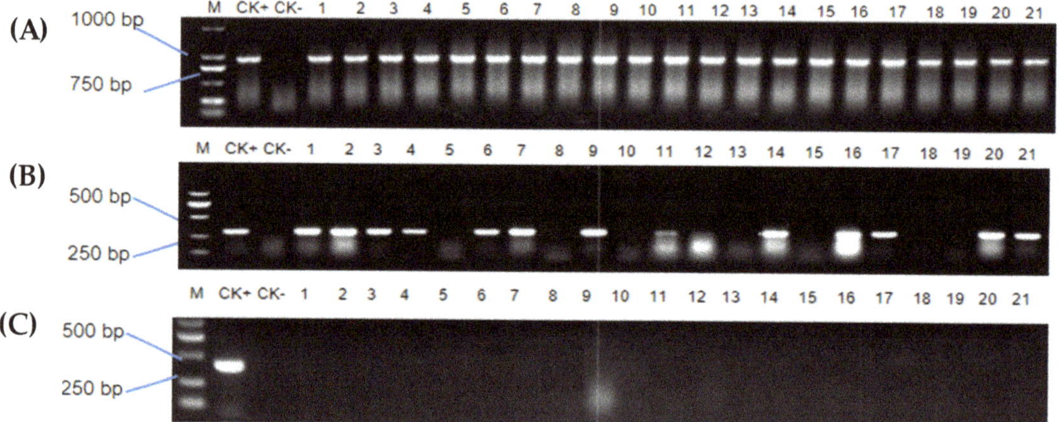

Figure 5. PCR verification of resistant shoots. (**A**) PCR detection of *CP4-EPSPS* gene in resistant shoots; (**B**) PCR detection of GFP gene in resistant shoots; (**C**) identification of *Agrobacterium* contamination from resistant plants via *VirG* gene amplification. M: DNA marker ladder DL2000 (HoldBio. Nanjing, China); CK+: plasmid of plant expression vector; CK−: DNA template from wild-type shoot; 1–21: resistant shoots after tissue culture.

2.6. Transformation Efficiency

The number of resistant shoots from 3.5 g of calli under 20 mg/L, 30 mg/L, and 40 mg/L glyphosate application was 31, 19, and 21, respectively. The highest number of resistant shoots was obtained under a low concentration of glyphosate, 20 mg/L (31 shoots). Notably, 29 shoots were positive for *CP4-EPSPS* with 93.4% selection efficiency (SE). Additionally, SE under 30 mg/L and 40 mg/L glyphosate was 100%. In terms of the transformation efficiency (TE), the number of shoots/weight (gram) of the calli that were PCR positive for the *CP4-EPSPS* gene at the 20 mg/L concentration was 7.4 lines/gram, followed by 4.9 lines/gram at 30 mg/L and 4.2 lines/gram at 40 mg/L. Overall, a low concentration of glyphosate showed lower selection efficiency but higher transformation efficiency (Table 1).

Table 1. Transformation efficiency analysis.

Glyphosate Concentration (mg/L)	Weight of Calli (g)	No. of Resistant Shoots (Line)	CP4-EPSPS Positive (Line)	SE (%)	GFP Positive (Line)	TE (Lines/g)
20	3.5	31	29	93.4	26	7.4
30	3.5	19	19	100	17	4.9
40	3.5	21	21	100	15	4.2

SE (selection efficiency) = shoots/resistant shoots that were PCR positive for the *CP4-EPSPS* gene; TE = number of shoots/weight (gram) of the calli that were PCR positive for the *CP4-EPSPS* gene used for transformation.

2.7. Detection of CP4-EPSPS Gene Expression by Quick Stix Strips

CP4-EPSPS gene expression of 21 transgenic shoots from the 40 mg/L glyphosate selection was tested by Quick Stix Strips (Genesino (Dalian) Biological S&T Development Co. Ltd. Dalian, China). Testing lines appeared on each Stix strip for each of the 21 transgenic shoots. The results were in line with the PCR and showed 100% positive testing of *CP4-EPSPS* gene expression in the 21 transgenic shoots from the 40 mg/L glyphosate selection (Figure 6).

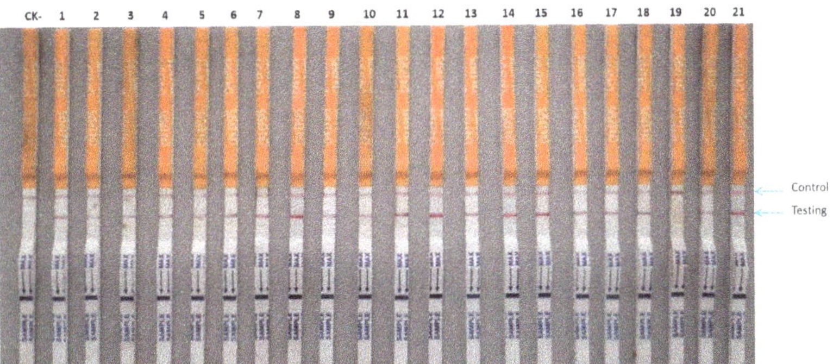

Figure 6. CP4-EPSPS gene expression Quick Stix Strip testing in resistant shoots. CK-: non-transformed shoot; 1–21: resistant shoots.

2.8. Herbicide Sensitivity

The herbicide sensitivity of the transgenic shoots was tested by spraying different concentrations of glyphosate (0%, 0.2%, 0.4%, 0.6%, and 0.8%). After 10 days of spraying different concentrations of glyphosate, all wild-type sugarcane tillers died, while transgenic sugarcane seedlings grew healthily even when treated with up to 0.6% glyphosate. Growth of the transgenic sugarcane seedlings was apparently affected when treated with 0.8% glyphosate (Figure 7).

Figure 7. Herbicide tolerance assay of transgenic and wild-type seedlings.

3. Discussion

Agrobacterium-mediated transformation is the most widely used and genetically stable technique for transferring T-DNA expression constructs into plant genomes. For low-copy expression, transgenic plants that undergo *Agrobacterium*-mediated transformation are mostly genetically stable [17–25]. In the previous research of our lab, two sugarcane transformation selection systems, PMI/Mannose [26] and the *Bar*/Basta selection system [13], were established. Both of these transformation systems had high efficiency due to high-quality sugarcane embryogenic callus induction and a suitable *Agrobacterium*-mediated transformation protocol. High-quality embryogenic callus induction is the key factor for the success of sugarcane genetic transformation. By modifying the 2,4-D concentration in different stages of callus induction, a high quality of sugarcane embryogenic calli was obtained and used for glyphosate killing curve testing and for genetic transformation. *CP4-EPSPS*/glyphosate transformation selection systems have been used in the genetic modification of other crops, such as wheat, cotton, soybean, and corn [27–30]. In this research, by referring to the selection concentration of glyphosate used in the genetic transformation process of maize, the killing curve for the testing of callus cultivation and regeneration cultivation was used, and a candidate concentration of glyphosate for sugarcane genetic transformation was selected. The SE of 20 mg/L, 30 mg/L, and 40 mg/L glyphosate was 93%, 100%, and 100%, which indicates that a higher concentration of glyphosate obtains higher selection efficiency. However, the TE was 7.4 lines/gg, 4.9 lines/g, and 4.2 lines/g, respectively, with transformation with 20 mg/L glyphosate selection obtaining the highest TE. So, 20 mg/L of glyphosate modified in the selection medium was the best concentration for sugarcane transformation using the *CP4-EPSPS* gene as the selection agent. Our research results are consistent with those of previous reports which also obtained a similar trend for TE [13,19].

Using a fluorescence reporter gene to assist in the establishment of a new genetic transformation selection system is helpful. In this research, we observed the fluorescence of GFP reporter genes throughout the transformation process to assay the quality of the embryogenic calli and efficiency of *Agrobacterium* infection protocol. Through the fluorescence from a single site, the formation process of the resistant calli can be observed. A total of 19 and 21 transgenic shoots from 30 mg/L and 40 mg/L glyphosate selection had positive PCR and Quick Stix Strip assay results, indicating that, under higher pressure of glyphosate selection, only the shoots with *CP4-EPSPS* gene integration and expression can survive throughout the selection process. The number of PCR lines positive for *GFP* gene was less than that of the selectable marker genes, mostly because of the fracture of and missing target gene during the process of integration. The fracture and lack of the target gene during transformation also happened in our previous research [13]. In this study, 10–30% of the selected lines lacked the transgenic *GFP* gene, which points to integration of partial transformation cassettes or other unintended changes to the integration constructs. Such effects are a general feature of GE transformation methods, as reviewed by Chu and Agapito [31]. As a consequence, the generated transformants need to be assessed for the integrity of the integrated recombinant constructs, as well as the number and the location of transgenic integrations, before they may be employed in practical agricultural use [31]. Overall, in this study, an *Agrobacterium*-mediated transformation system that uses the herbicide-resistant *CP4-EPSPS* gene was developed. A low concentration of glyphosate showed lower selection efficiency but higher transformation efficiency. Transgenic sugarcane seedlings displayed no sensitivity to 0.6% glyphosate. The enhanced resistance of transgenic sugarcane seedlings to herbicide has great significance for the sugarcane industry.

4. Materials and Methods

4.1. Materials and Media Compositions

In the sugarcane genetic transformation process, all the materials and media compositions for each stage were prepared using the methods described by Wang et al. [13]. Additionally, MS (Murashige and Skoog Salts and Vitamins), sucrose, 2,4-D (2,4-Dichlorophenoxyacetic Acid), 6-BA (6-Benzylaminopurine), NAA (1-naphthylacetic acid), and Timentin were procured from PhytoTechnology Laboratories (PhytoTechnology, Lenexa, KS, USA), while phytoblend was obtained from Caisson Laboratories, Smithfield, UT, USA. Glyphosate (Roundup herbicide) used in this study was obtained from Monsanto, St. Louis, MO, USA (Table 2).

Table 2. Materials and media compositions used in this study.

Medium	MS (g/L)	Sucrose (g/L)	Phytoblend (g/L)	2,4-D (mg/L)	6-BA (mg/L)	NAA (mg/L)	Glyphosate (mg/L)	Timentin (mg/L)	pH
Callus Induction	4.43	30	8	1–3	0–0.5	0	0	0	5.8
Recovery	4.43	30	8	1	0.5	0	0	300	5.8
Callus Screening	4.43	30	8	1	0.5	0	20–40	300	5.8
Regeneration Screening	4.43	30	8	0	1–2	0	10	300	5.8
Elongation Screening	4.43	30	4	0	0	0	10	300	5.8
Rooting Screening	4.43	20	4	0	0	1–2	10	300	5.8

4.2. Plant Material and Callus Induction

Plant material of the sugarcane variety ROC22 was obtained from the Sugarcane Test and Demonstration Base of the Institute of Tropical Bioscience and Biotechnology of the Chinese Academy of Tropical Agriculture Science, Hainan, China. The immature leaf whorl found on the tops of sugarcane tillers served as the raw material for the induction of embryogenic callus, which was started within 24 h of cutting. Immature sugarcane leaf whorl transverse sections were made largely according to Bower and Brich's [25] instructions. Sections that were transverse and 1 mm thick were taken from directly above the meristem and put on callus induction media. Callus cultures were maintained in the dark at 28 °C and sub-cultured on fresh callus induction medium every 2 weeks for a total culture duration of approximately 45 days. According to the actual callus induction status, the adjustment of 2,4-D (1–3 mg/L) and 6-BA (0–0.5 mg/L) was slightly optimized (Table 1). Light-yellow, compact calli were selected and fragmented before glyphosate killing curve testing and *Agrobacterium*-mediated genetic transformation.

4.3. Minimum Inhibitory Concentration of Roundup

The callus induction medium was changed with several concentrations of glyphosate (0 mg/L, 10 mg/L, 20 mg/L, 30 mg/L, 40 mg/L, and 50 mg/L) and inoculated compact calli that were made as previously mentioned for callus induction (Roundup, 41% glyphosate, Monsanto, USA). After being kept at 28 °C in the dark for 14 days, the cultures were transferred to a regeneration medium that had been altered to include the same amounts of glyphosate. They were kept at 30 °C for 20 days while receiving 14 h of light each day. For the purpose of transformation selection, the lowest inhibitory concentration of glyphosate that inhibited the growth of secondary leaves was selected.

4.4. Binary Vector and Agrobacterium Strain

The plant expression vector (*Bar-GFP* containing *GFP* reporter gene and *Bar* screening marker) was constructed in previous research in our lab [13]. The binary vector pCAM-BIA3300 was modified to include the *bar* gene, which is driven by the CaMV 35S promoter, in between the two XhoI sites. The *npt II* gene for bacterial selection was present in the binary vector pCAMBIA3300. The Ubi1 promoter reporter gene green fluorescent protein (GFP) was inserted into pCAMBIA3300 in between the SacI and BamHI restriction sites (Figure 8). Additionally, *CP4-EPSPS* genes with XhoI restriction sites at both C and N sides were synthesized by gene biosynthesis (completed by Nanjing Detai Biological Engineering Co., Ltd., Nanjing, China). The *Bar* gene in the original vector was removed by XhoI single enzyme, and then the linearized binary vector was reconnected with the synthetic *CP4-EPSPS* gene. The recombinant plasmid with forward insertion of *CP4-EPSPS* gene was verified by sequencing. The final binary vector with CP4-EPSPS selection marker was named *CP4-EPSPS-GFP* and mobilized into *Agrobacterium* EHA105 (AC1010, Weidi Co. Ltd., Shanghai, China) via freeze-thawing with liquid nitrogen and verified by PCR analysis. The *Agrobacterium* strain was cultivated in Yeast Extract Peptone (YEP) medium, which contained the necessary antibiotics (rifampicin 10 mg/L and kanamycin 50 mg/L), and was stored in a refrigerator at −80 °C.

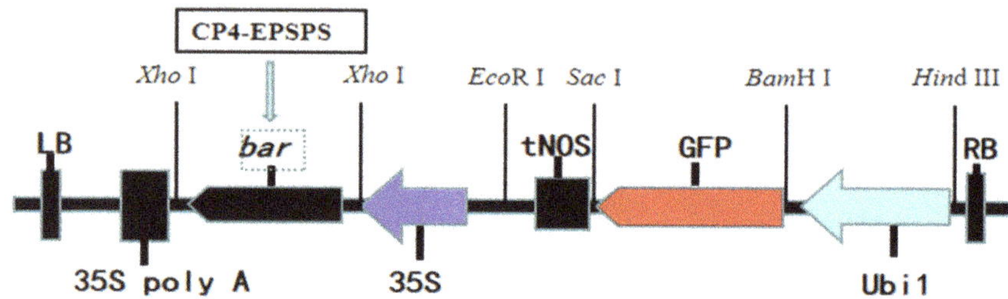

Figure 8. Schematic representation of the binary vector used in sugarcane transformation. Abbreviation: 35S: cauliflower mosaic virus 35S promoter; 35S poly A: cauliflower mosaic virus 35S poly A terminator; tNOS: nopaline synthase terminator; Ubi1: maize Ubi1 promoter.

4.5. Agrobacterium Initiation

The *Agrobacterium* strain containing the *CP4-EPSPS-GFP* plant expression vector was removed from the −80 °C refrigerator and streaked on YEP solid medium (yeast extract 10 g/L, peptone 10 g/L, NaCl 5 g/L, agar 15 g/L, kanamycin 50 mg/L, and rifampicin 10 mg/L) culture at 28 °C for 2–3 days. A single colony was selected and sub-cultured overnight on fresh YEP solid medium at 28 °C. In order to initiate a starter culture of liquid initiating medium (1/5 strength MS basal salt, 30 g/L sucrose, 30 g/L glucose, and 100 uM acetosyringone), bacteria were harvested from the surface of the solid medium. The starter culture was then shaken gently (fewer than 100 revolutions) for two hours at 28 °C in the dark. The bacterial combination was then diluted until it had an optical density of between 0.3 and 0.6 at 600 nm.

4.6. Infection and Co-Cultivation

The compact calli that were ready for transformation were gathered and weighed (3–4 g). Each callus was arranged in a Petri dish, allowed to air dry for one hour on a sterile laboratory bench, and then moved to a 150 mL Erlenmeyer flask. After adding 100 mL of preheated (45 °C) liquid starting media (1/5 strength MS basal salt, 30 g/L sucrose, 30 g/L glucose, 100 uM acetosyringone, pH 5.4) without *Agrobacterium* to cover the calli, all of the calli were heat-shocked in an incubator for five minutes. The liquid blank initiating medium was pipetted out, and 50 mL of initiated *Agrobacterium* mixture was added in and shaken gently for 10 min at 28 °C. The calli/*Agrobacterium* mixture was sonicated for 2 min in an ultrasonic cleaner. The initiated *Agrobacterium* mixture was renewed and vacuumed for 5 min at −0.08 MPa. The calli/*Agrobacterium* mixture was shaken gently for 10 min in the dark. After the *Agrobacterium* suspension was finally pipetted out, the calli were placed on a Petri dish, the excess *Agrobacterium* suspension was wiped dry using filter paper, and the calli were allowed to air dry for about 30 min on the spotless bench. After that, the calli were placed in a fresh, empty Petri dish, covered with paraffin film, and co-cultivated with *Agrobacterium* for two to three days at 22 °C in the dark [13].

4.7. Resistant Plant Screening

After co-cultivation, all of the infected embryogenic calli were transferred to the recovery medium for recovery cultivation in the dark at 28 °C for one week (Table 1). Subsequently, all calli were transferred to a selection medium for 35 days in the dark at 28 °C (each of the 90 mm culture dishes placed around 15 pieces of infected embryogenic calli). After being moved to the regeneration medium, resistant calli were grown for 20 days at 30 °C under 14 h of light per day in an illuminated incubator. After being chosen, the green buds were moved to elongation medium and grown for a further 20 days at 30 °C under 14 h of light every day in the lit incubator. From each bud, a single shoot was transferred to the rooting medium and grown for 30 days at 30 °C under 14 h of light every day in a lit incubator. When the shoots reached a height of around 6–7 cm, the leaves were removed for molecular analysis, moved to pots, and then grown in a greenhouse.

4.8. Visualization of Green Fluorescent Protein Expression

Leaves and roots of resistant shoots were sampled for GFP fluorescence detection using a fluorescence microscope (Carl Zeiss Scope.A1, Oberkochen, Germany). Firstly, the whole infected embryogenic callus particle after co-cultivation was amplified 40 times, the number and distribution of single-cell fluorescence points on the callus were observed, and the initial infection efficiency was analyzed. After callus selection, resistant calli were amplified 10 times for the GFP (510 nm emission filter and 480 nm excitation filter) fluorescence detection followed by the estimation of growth and percentage of resistant calli.

4.9. Total Genomic DNA Extraction and PCR Assay

Total genomic DNA was extracted from the leaf sample of each resistant shoot. The CP4-EPSPS and GFP gene integration was found, and the lack of *Agrobacterium* contamination was verified using a PCR assay. Primers specific to the GFP, VirG, and CP4-EPSPS genes were used in the PCR experiments (Table 3). Total genomic DNA isolated from plants of the wild type was utilized as the negative control, while vector plasmid was utilized as the positive control.

Table 3. List of primers used in this study.

Gene	Product Size (bp)	TM	Primer Sequence	
CP4-EPSPS	969	58 °C	Forward	GGCGACAAGAGCATCAGTCA
			Reverse	CCTCGAGACGTTCATCACGG
GFP	249	58 °C	Forward	CTGGATGAAGTGCCAGTCGG
			Reverse	TGCATGTACCACGAGTCCAA
VirG	332	58 °C	Forward	TTCGTTCCGATGCTCTATGA
			Reverse	AGGTCGTCTTTCTGCTTTCC

The effectiveness of the sugarcane CP4-EPSPS/glyphosate transformation selection system was calculated based on the PCR assay results.

Selection efficiency (SE) = number of shoots/number of resistant shoots that are PCR positive for the CP4-EPSPS gene obtained after rooting selection.

Transformation efficiency (TE) = number of shoots/weight (gram) of the calli that are PCR positive for the CP4-EPSPS gene used for transformation.

4.10. Quick Stix Strips Assay

Resistant shoots were sampled and tested by Quick Stix Strips (Quick Stix Kit for CP4-EPSPS, Genesino (Dalian), Biological S&T Development Co., Ltd., Dalian, China). Approximately 10 mg of leaf tissue was taken from each resistant plant for the testing. Wild-type plant tissue samples served as a negative control.

4.11. Herbicide Sensitivity Assay

Tillers from the resistant shoots (asexual clones) were propagated on the rooting medium for the herbicide tolerance assay. Roundup (41% glyphosate, Monsanto, St. Louis, MO, USA) was diluted into four concentration gradients (0.2%, 0.4%, 0.6%, and 0.8%). A herbicide-free concentration (0%) was formulated as a negative control. The transgenic plants at a height of 10 cm were sprayed with different herbicide concentrations.

5. Conclusions and Future Prospects

In this study, an *Agrobacterium*-mediated transformation system that uses the herbicide-resistant CP4-EPSPS gene was developed. It is noteworthy that an increase in glyphosate concentration resulted in decreased regeneration capacity of the calli. Specifically, a lower glyphosate concentration had the minimum inhibitory impact on calli regeneration. A low concentration of glyphosate showed lower selection efficiency but higher transformation efficiency. Transgenic sugarcane seedlings displayed high resistance to the glyphosate. Interestingly, transgenic lines displayed a significantly decreased glyphosate herbicide sensitivity as compared to wild types. Additionally, a few transgenic plants died even at a higher concentration. Overall, this study demonstrates an effective *Agrobacterium*-mediated transformation system that utilizes the herbicide-resistant CP4-EPSPS gene as a selection marker which can be used for future purposes due to the low cost of glyphosate and efficiency of this system.

Author Contributions: Conceptualization, W.W. and S.Z.; methodology, S.Z.; software, W.W.; validation, T.J., L.S. and S.Z.; formal analysis, W.W.; investigation, W.W.; resources, S.Z.; data curation, T.J.; writing—original draft preparation, W.W. and T.J.; writing—review and editing, T.J., L.S., T.S., B.Y. and S.Z.; supervision, S.Z.; project administration, S.Z.; funding acquisition, S.Z. All authors have read and agreed to the published version of the manuscript.

Funding: This research was funded by Hainan Province Science and Technology Special Fund (ZDYF2023XDNY056), Hainan Yazhou Bay Seed Lab (JBGS+B21HJ0302), and the Central Public-Interest Scientific Institution Basal Research Fund for the Chinese Academy of Tropical Agricultural Sciences (1630052019019).

Data Availability Statement: Data are contained within the article.

Conflicts of Interest: The authors declare no conflicts of interest.

References

1. Javed, T.; Shabbir, R.; Ali, A.; Afzal, I.; Zaheer, U.; Gao, S.-J. Transcription Factors in Plant Stress Responses: Challenges and Potential for Sugarcane Improvement. *Plants* **2020**, *9*, 491. [CrossRef]
2. Aono, H.; Pimenta, G.; Garcia, B.; Correr, H.; Hosaka, K.; Carrasco, M.; Cardoso-Silva, B.; Mancini, C.; Sforça, A.; Santos, B.; et al. The wild sugarcane and sorghum kinomes: Insights into expansion, diversification, and expression patterns. *Front. Plant Sci.* **2021**, *12*, 668623. [CrossRef]
3. Raza, A.; Razzaq, A.; Mehmood, S.S.; Zou, X.; Zhang, X.; Lv, Y.; Xu, J. Impact of climate change on crops adaptation and strategies to tackle its outcome: A review. *Plants* **2019**, *8*, 34. [CrossRef]
4. St. John Warne, M.; Neale, P.A.; Macpherson, M.J. A pesticide decision support tool to guide the selection of less environmentally harmful pesticides for the sugar cane industry. *Environ. Sci. Pollut. Res.* **2023**, *30*, 108036–108050. [CrossRef] [PubMed]
5. Pakdaman, S.B.; Mohammadi, G.E. Weeds, herbicides and plant disease management. In *Sustainable Agriculture Reviews 31: Biocontrol*; Springer: Berlin/Heidelberg, Germany, 2018; pp. 41–178.
6. Schütte, G.; Eckerstorfer, M.; Rastelli, V.; Reichenbecher, W.; Restrepo-Vassalli, S.; Ruohonen-Lehto, M.; Saucy, A.G.W.; Mertens, M. Herbicide resistance and biodiversity: Agronomic and environmental aspects of genetically modified herbicide-resistant plants. *Environ. Sci. Eur.* **2017**, *29*, 5. [CrossRef]
7. Hussain, S.; Khaliq, A.; Mehmood, U.; Tauqeer, Q.; Muhammad, S.; Muhammad, A.I.; Saddam, H. Sugarcane production under changing climate: Effects of environmental vulnerabilities on sugarcane diseases, insects and weeds. In *Climate Change and Agriculture*; Intech Open: Rijeka, Croatia, 2018; pp. 137–151.
8. Sanghera, G.S.; Malhotra, P.K.; Singh, H.; Bhatt, R. Climate change impact in sugarcane agriculture and mitigation strategies. In *Harnessing Plant Biotechnology and Physiology to Stimulate Agricultural Growth*; Agrobios: Jodhpur, India, 2019; pp. 99–115.
9. Luo, T.; Liu, X.; Lakshmanan, P. A combined genomics and phenomics approach is needed to boost breeding in sugarcane. *Plant Phenomics* **2023**, *5*, 0074. [CrossRef] [PubMed]
10. Wang, T.; Fang, J.; Zhang, J. Advances in sugarcane genomics and genetics. *Sugar Tech* **2022**, *24*, 354–368. [CrossRef]
11. Ahmad, M. Plant breeding advancements with "CRISPR-Cas" genome editing technologies will assist future food security. *Front. Plant Sci.* **2023**, *14*, 1133036. [CrossRef]
12. Tolera, B.; Gedebo, A.; Tena, E. Variability, heritability and genetic advance in sugarcane (*Saccharum* spp. hybrid) genotypes. *Cogent Food Agric.* **2023**, *9*, 2194482. [CrossRef]
13. Wang, W.Z.; Yang, B.P.; Feng, C.L.; Wang, J.G.; Xiong, G.R.; Zhao, T.T.; Zhang, S.Z. Efficient sugarcane transformation via bar gene selection. *Trop. Plant Biol.* **2017**, *10*, 77–85. [CrossRef]
14. Basso, M.F.; da Cunha, B.A.D.B.; Ribeiro, A.P.; Martins, P.K.; de Souza, W.R.; de Oliveira, N.G.; Nakayama, T.J.; Augusto das Chagas Noqueli Casari, R.; Santiago, T.R.; Vinecky, F.; et al. Improved genetic transformation of sugarcane (*Saccharum* spp.) embryogenic callus mediated by *Agrobacterium* tumefaciens. *Curr. Protoc. Plant Biol.* **2017**, *2*, 221–239. [CrossRef]
15. Dill, G.M.; Cajacob, C.A.; Padgette, S.R. Glyphosate resistant crops: Adoption, use and future considerations. *Pest. Manag. Sci.* **2008**, *64*, 326–331. [CrossRef]
16. Kumar, K.; Gambhir, G.; Dass, A.; Tripathi, A.K.; Singh, A.; Jha, A.K.; Yadava, P.; Choudhary, M.; Rakshit, S. Genetically modified crops: Current status and future prospects. *Planta* **2020**, *251*, 91. [CrossRef] [PubMed]
17. Zhao, F.Y.; Li, Y.F.; Xu, P. *Agrobacterium*-mediated transformation of cotton (*Gossypium hirsutum* L. cv. Zhongmian 35) using glyphosate as a selectable marker. *Biotechnol. Lett.* **2006**, *28*, 1199–1207. [CrossRef] [PubMed]

18. Vossen, J.H.; van Arkel, G.; Bergervoet, M.; Jo, K.R.; Jacobsen, E.; Visser, R.G. The Solanum demissum R8 late blight resistance gene is an Sw-5 homologue that has been deployed worldwide in late blight resistant varieties. *Theor. Appl. Genet.* **2016**, *129*, 1785–1796. [CrossRef] [PubMed]

19. Wang, W.Z.; Yang, B.P.; Feng, X.Y.; Cao, Z.Y.; Feng, C.L.; Wang, J.G.; Xiong, G.R.; Shen, L.B.; Zeng, J.; Zhao, T.T.; et al. Development and Characterization of Transgenic Sugarcane with Insect Resistance and Herbicide Tolerance. *Front. Plant Sci.* **2017**, *8*, 1535. [CrossRef]

20. Gan, Y.; Fan, Y.; Yang, Y.; Dai, B.; Gao, D.; Wang, X.; Wang, K.; Yao, M.; Wen, H.; Yu, W. Ectopic expression of MNX gene from Arabidopsis thaliana involved in auxin biosynthesis confers male sterility in transgenic cotton (*Gossypium hirsutum* L.) plants. *Mol. Breed.* **2010**, *26*, 77–89. [CrossRef]

21. Lee, K.-W.; Kim, K.-Y.; Kim, K.-H.; Lee, B.-H.; Kim, J.-S.; Lee, S.-H. Development of antibiotic marker-free creeping bentgrass resistance against herbicides. *Acta Biochim. Et Biophys. Sin.* **2010**, *43*, 13–18. [CrossRef]

22. Cui, M.-L.; Liu, C.; Piao, C.-L.; Liu, C.-L. A Stable Agrobacterium rhizogenes-Mediated Transformation of Cotton (*Gossypium hirsutum* L.) and Plant Regeneration From Transformed Hairy Root via Embryogenesis. *Front. Plant Sci.* **2020**, *11*, 604255. [CrossRef]

23. Sharma, K.K.; Pothana, A.; Prasad, K.; Shah, D.; Kaur, J.; Bhatnagar, D.; Chen, Z.; Raruang, Y.; Cary, J.W.; Rajasekaran, K.; et al. Peanuts that keep aflatoxin at bay: A threshold that matters. *Plant Biotechnol. J.* **2017**, *16*, 1024–1033. [CrossRef]

24. Dong, S.; Delucca, P.; Geijskes, R.J.; Ke, J.; Mayo, K.; Mai, P.; Sainz, M.; Caffall, K.; Moser, T.; Yarnall, M.; et al. Advances in agrobacterium-mediated sugarcane transformation and stable transgene expression. *Sugar Tech* **2014**, *16*, 366–371. [CrossRef]

25. Bower, R.; Brich, R.G. Transgenic sugarcane plants via microprojectile bombardment. *Plant J.* **1992**, *2*, 409–416. [CrossRef]

26. Wang, W.; Yang, B.; Cai, W.; Feng, C.; Wang, J.; Xiong, G.; Zhang, S. Establishment of Mannose Selection System in Sugarcane Transformation. *Biotechnol. Bull.* **2015**, *31*, 92–97.

27. Zhou, H.P.; Biest, N.; Chen, G.; Cheng, M.; Feng, X.; Radionenko, M.; Lu, F.; Fry, J.; Hu, T.; Metz, S.; et al. Agrobacterium-mediated large-scale transformation of wheat (*Triticum aestivum* L.) using glyphosate selection. *Plant Cell Rep.* **2003**, *21*, 1010–1019. [CrossRef]

28. Latif, A.; Rao, A.Q.; Khan, M.A.U.; Shahid, N.; Bajwa, K.S.; Ashraf, M.A.; Abbas, M.A.; Azam, M.; Shahid, A.A.; Nasir, I.A.; et al. Herbicide-resistant cotton (*Gossypium hirsutum*) plants: An alternative way of manual weed removal. *BMC Res. Notes* **2015**, *8*, 453. [CrossRef]

29. Bonny, S. Genetically modified glyphosate-tolerant soybean in the USA: Adoption factors, impacts and prospects. A review. *Agron. Sustain. Dev.* **2008**, *28*, 21–32. [CrossRef]

30. Liu, M.-M.; Zhang, X.-J.; Gao, Y.; Shen, Z.-C.; Lin, C.-Y. Molecular characterization and efficacy evaluation of a transgenic corn event for insect resistance and glyphosate tolerance. *J. Zhejiang Univ. Sci. B* **2018**, *19*, 610–619. [CrossRef]

31. Chu, P.; Agapito-Tenfen, S.Z. Z. Unintended Genomic Outcomes in Current and Next Generation GM Techniques: A Systematic Review. *Plants* **2022**, *11*, 2997. [CrossRef]

Review

Genetic Engineering for Enhancing Sugarcane Tolerance to Biotic and Abiotic Stresses

Tanweer Kumar [1,2], Jun-Gang Wang [3], Chao-Hua Xu [1], Xin Lu [1], Jun Mao [1], Xiu-Qin Lin [1], Chun-Yan Kong [1], Chun-Jia Li [1], Xu-Juan Li [1], Chun-Yan Tian [1], Mahmoud H. M. Ebid [1,4], Xin-Long Liu [1] and Hong-Bo Liu [1,*]

[1] National Key Laboratory for Tropical Crop Breeding, Sugarcane Research Institute, Yunnan Academy of Agricultural Sciences, Yunnan Key Laboratory of Sugarcane Genetic Improvement, Kaiyuan 661699, China; tanweerkm_biologist@outlook.com (T.K.)

[2] Sugar Crops Research Institute, Agriculture, Fisheries and Co-Operative Department, Charsadda Road, Mardan 23210, Khyber Pakhtunkhwa, Pakistan

[3] National Key Laboratory for Tropical Crop Breeding, Institute of Tropical Bioscience and Biotechnology, Chinese Academy of Tropical Agricultural Sciences, Sanya 572024, China

[4] Sugar Crops Research Institute, Agricultural Research Center, Giza 12619, Egypt

* Correspondence: liuhongbo1982@126.com

Abstract: Sugarcane, a vital cash crop, contributes significantly to the world's sugar supply and raw materials for biofuel production, playing a significant role in the global sugar industry. However, sustainable productivity is severely hampered by biotic and abiotic stressors. Genetic engineering has been used to transfer useful genes into sugarcane plants to improve desirable traits and has emerged as a basic and applied research method to maintain growth and productivity under different adverse environmental conditions. However, the use of transgenic approaches remains contentious and requires rigorous experimental methods to address biosafety challenges. Clustered regularly interspaced short palindromic repeat (CRISPR) mediated genome editing technology is growing rapidly and may revolutionize sugarcane production. This review aims to explore innovative genetic engineering techniques and their successful application in developing sugarcane cultivars with enhanced resistance to biotic and abiotic stresses to produce superior sugarcane cultivars.

Keywords: sugarcane; biotic and abiotic stress; genetic engineering; transgenic sugarcane; genome editing

Citation: Kumar, T.; Wang, J.-G.; Xu, C.-H.; Lu, X.; Mao, J.; Lin, X.-Q.; Kong, C.-Y.; Li, C.-J.; Li, X.-J.; Tian, C.-Y.; et al. Genetic Engineering for Enhancing Sugarcane Tolerance to Biotic and Abiotic Stresses. *Plants* **2024**, *13*, 1739. https://doi.org/10.3390/plants13131739

Academic Editor: San-Ji Gao

Received: 28 April 2024
Revised: 18 June 2024
Accepted: 18 June 2024
Published: 24 June 2024

1. Introduction

Sugarcane (*Saccharum* spp. hybrids) is a major tropical and sub-tropical crop cultivated in over 121 countries across 27 million hectares of land worldwide. It contributes about 80% of the world's total sugar production and is the most efficient feedstock for bio-ethanol and diesel, accounting for 40% of the world's biofuel production [1]. In addition, sugarcane is the source of valuable products such as paper, acetic acid, plywood, fibers, bio-fertilizer, animal feed, board, and industrial enzymes [2]. Conventional breeding methods for sugarcane aim to develop new hybrid varieties with higher yields and increased sugar contents [3]. However, the high genetic complexity of sugarcane hybrids, and a variable number of chromosomes (ranging from 53 to 143 chromosomes with an estimated genome size of 10 Gb pose significant challenges to the breeders [4]. At the same time, conventional breeding is costly, labor-intensive, and time-consuming, taking up to 10–15 years to release a new elite variety [5]. To acquire more of the intricate nature of the sugarcane genome, numerous initiatives have led to a variety of genome sequencing projects, encompassing parental species and hybrid genotypes [6–13]. More recently, the Chinese Academy of Sciences has initiated a pilot program in collaboration with the National Key Laboratory project, specifically targeting tropical crop breeding. Such scientific collaboration is anticipated to yield significant outcomes and contribute positively to the field of sugarcane genetic improvement in the future.

On the other hand, sugarcane cultivation is adversely affected by biotic and abiotic stresses, including weeds, diseases, insects and pests, drought, salinity, low temperatures, heavy metals, and low soil fertility (Figure 1). More importantly, genetic engineering and genome editing have emerged as powerful tools to address these challenges and increase yield productivity without losses caused by these stresses. Sugarcane is a vegetatively propagative crop and breeding aspects make it an excellent candidate for crop improvement through genetic engineering, to produce green energy. Transgenic technology allows the precise transfer of one or more stacks of genes from unrelated plants or different organisms. Since the 1990s, different genetic transformation techniques have been successfully developed in sugarcane, including electroporation, *Agrobacterium tumefaciens*, the biolistic bombardment method, and so on [14,15]. In the following three decades, *Agrobacterium*-mediated transformation and other new methods have become routine exercises in many research laboratories worldwide. Moreover, genome editing can edit, insert, or replace specific sequences within the genome. A number of genes promoting resistance to biotic and abiotic stresses have been introduced into sugarcane for genetic improvement. Therefore, this review summarizes the advances in genetic improvement of sugarcane and its potential to increase yield productivity, especially under biotic and abiotic stresses.

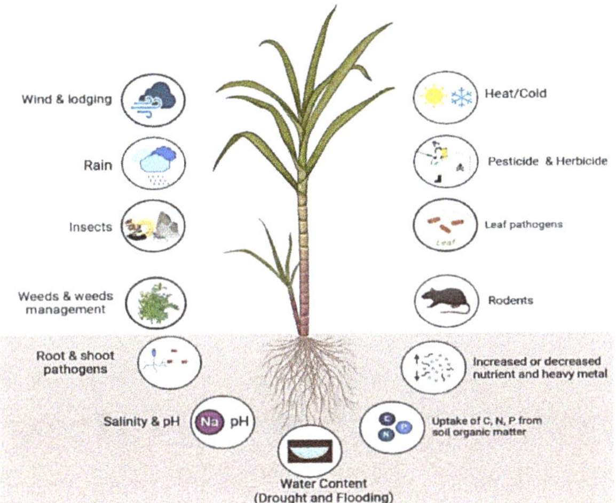

Figure 1. Various abiotic and biotic stress affecting sugarcane plant growth, development, and productivity. (https://biorender.com; accessed on 23 April 2024).

2. Biotic Stress

2.1. The Enhancement of Herbicide Tolerance

Due to a sessile nature, plants must confront various living organisms in the environment (Figure 2). Weed infestation is a serious problem that negatively impacts the yield of sugarcane, resulting in significant economic losses [16,17]. Introducing genes that confer herbicide tolerance into sugarcane can aid in the breeding of transgenic sugarcane tolerant to herbicides, reducing labor intensity and costs while ensuring a higher yield of sugarcane. Traditional breeding for herbicide-tolerant traits is lagging due to the lack of herbicide-tolerant gene pools in wild-sugarcane relative species [18]. An alternative strategy should be adopted for developing crops that are tolerant to broad-spectrum herbicides [19]. The advances in transgenic sugarcane that have been engineered for herbicide tolerance are summarized in Table 1.

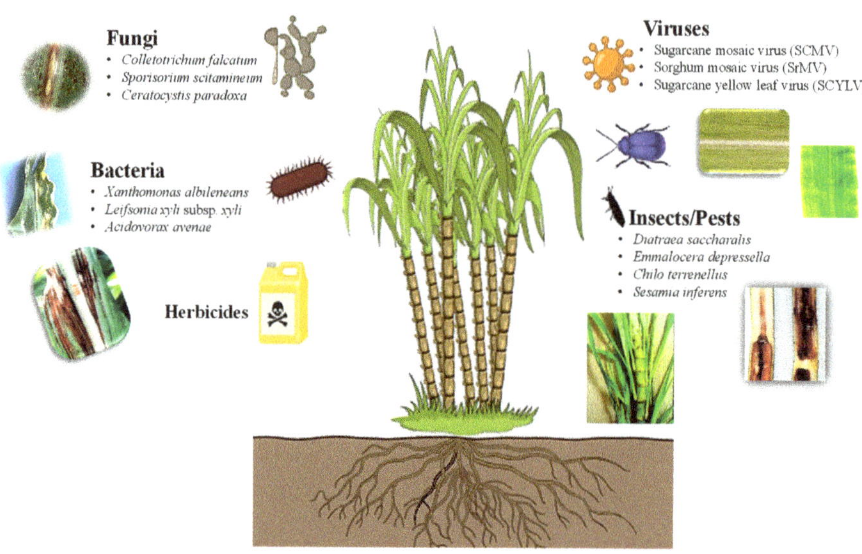

Figure 2. Types of biotic stresses that affect growth, yield, and productivity of sugarcane (https://biorender.com; accessed on 23 April 2024).

Table 1. Genetic modifications for herbicide-tolerant traits in sugarcane.

Variety	Trait Method	Explant	Promoter	Gene	Reference
SP80–180	Particle bombardment	Axillary bud	Ubi-1	*bar*	[20]
ROC22	*Agrobacterium* mediated	Embryogenic calli	Ubi-1	*EPSPS, Cry1Ab*	[21]
CPF-234, CPF-213, HSF-240 & CPF-246	Particle bombardment	Embryogenic calli	CaMV35S	*Glyphosate*	[22]
TUC 03-12	Particle bombardment	Embryogenic calli	Rice Actin	*EPSPS*	[23]
CPF-246	*Agrobacterium* mediated	Embryogenic calli	Ubi-1	*Cry1Ac, Cry2 & GT*	[24]
ROC22	*Agrobacterium* mediated	Embryogenic calli	CaMV35S	*bar*	[25]
ROC22	*Agrobacterium* mediated	Embryogenic calli	Ubi-1	*bar*	[26]
RA87-3	Particle bombardment	Embryogenic calli	Rice Actin	*EPSPS*	[27]
NCo310	Particle bombardment	Meristem		*Pat*	[28]
Ja60-15	*Agrobacterium* mediated	Embryogenic calli	Rice-Ubi	*bar*	[29]
Co92061 and Co671	*Agrobacterium* mediated	Somatic embryos	CaMV35S	*bar*	[30]

The integration of the *bar* gene into the genome of sugarcane using the biolistic transformation method resulted in the production of herbicide-tolerant transgenic sugarcane [20]. Using *Agrobacterium*-mediated transformation to introduce the herbicide-tolerant genes *bar* and *CP4-EPSPS* into the main sugarcane varieties, resulted in transgenic sugarcane exhibiting excellent herbicide tolerance [19,31]. By incorporating the genome of embryonic calli with a glyphosate-tolerant (*GT*) gene, the first study published the genetic transformation of four sugarcane genotypes through the bombardment of embryonic calli with the *GT* gene. These results showed that 88% of the transgenic plants survived on the first application of glyphosate, whereas all the non-transformed plants died. In the second application, increasing the dose of glyphosate resulted in the accumulation of *GT* gene-encoded products that survived [22]. However, transgenic glyphosate-tolerant sugarcane was developed for commercial release through the genetic transformation of cultivar RA87-3. The objective was to obtain glyphosate-tolerant transgenic events in two recently released cultivars, TUC 95-10 and TUC 03-12, by introducing *EPSPS* and *nptII* genes using the microprojectile method. A transgenic event tolerant to the glyphosate herbicide was successfully obtained through the biolistic transformation using the TUC 03-12 genotype as the starting material. The resulting genotype showed greater than 99% similarity with the parental genotype [23,32].

However, sugarcane borers and weeds threaten production, quality, and yield. Genetic modification using stack genes technology (combining two or more genes) was performed by introducing modified cane borer-tolerant (*CEMB-Cry1Ac*) and glyphosate-tolerant (*CEMB-GT*) genes into sugarcane. Transgenic plants showed efficient tolerance to cane borers and tolerance to glyphosate, indicating sustainable tolerance in sugarcane [24]. A transgenic sugarcane cultivar (ROC22) was regenerated using herbicide screening, containing the *bar* gene to obtain sense and antisense strigolactones biosynthesis genes (*ScD27.2*). All transgenic lines tested positive for herbicide tolerance and contained the target gene. Furthermore, PCR detection and 1% Basta (Glufosinate) application on leaves revealed *bar* in all lines [25]. A recent study has presented an efficient transformation system mediated by *Agrobacterium*, which utilizes the herbicide-tolerant *bar* gene as the selectable marker. The transgenic lines were confirmed through PCR and exhibited tolerance to herbicides. Moreover, the protocol demonstrated several advantages, including a higher yield of transgenic lines from calli, improved selection, and increased transformation efficiency [26]; these results showed that a dependable tool now offers a standard procedure for cultivating robust transgenic sugarcane plants.

Additionally, gene editing technology has provided new strategies for breeding herbicide-tolerant sugarcane varieties. CRISPR/Cas9-mediated multi-allelic gene mutation techniques mutate the acetolactate synthase gene (*ALS*) to develop broad-spectrum herbicide-tolerant sugarcane plants [33]. The development of environmentally friendly herbicide-tolerant sugarcane varieties through genetic engineering and gene editing technology will help to reduce the use of harmful herbicides and make sugarcane cultivation more environmentally friendly and sustainable in the future.

2.2. Enhancing Disease Tolerance in Sugarcane

The defense mechanisms of sugarcane are triggered in response to biotic stressors, including different pathogens, such as fungi, bacteria, viruses, and insects [34–38]. These biotic stressors result in serious diseases in sugarcane (Figure 2), such as red rot (*Colletotrichum falcatum*), smut (*Sporisorium scitamineum*), pineapple disease (*Ceratocystis paradoxa*), red stripe (*Acidovorax avenae*), leaf scald (*Xanthomonas albilineans*), grassy shoot disease (*Phytoplasma*), and mosaic virus (ScMV) and yellow leaf virus (ScYLV) infection, as well as sugarcane stem borer (*Diatraea saccharalis*), African sugarcane stalk borer (*Eldana saccharina*), sugarcane weevil (*Sphenophorus levis*), and various pests such as borers, pyrilla, thrips, grasshoppers, sucking pests, and cane grubs (*Melanaphis sacchari*) [16,39,40]. Sugarcane crops are susceptible to over 100 different types of pathogens [41]. Breeding programs play a crucial role in genotype screening for disease resistance. However, sugarcane breeders face significant challenges simultaneously introducing resistance to all pathogens [42].

Therefore, biotechnological strategies have been developed to produce commercial clones with superior agronomic traits; these genetic modifications are summarized in Table 2.

Table 2. List of genetic engineering traits for disease tolerance in sugarcane.

Target Disease	Variety	Trait Method	Explant	Promoter	Gene	Reference
SCMV	SPF-234 & NSG-311	Particle bombardment	Calli	Ubi	*CP*	[1]
SrMV	CP65-357 & CP72-1210	Particle bombardment	Calli	Ubi-1	*CP*	[43]
FDV	Q124	Particle bombardment	Calli	Ubi	Segment 9 of ORF 1	[44]
SCYLV	H62-4671	Particle bombardment	Cell cultures	Ubi	*CP*	[45]
SCMV	Bululawang	*Agrobacterium* mediated	Shoots	CaMV35S	*CP*	[46]
SrMV	ROC22	*Agrobacterium* mediated	Leaf	CaMV35S	*CP*	[47]
SCMV	CP 84-1198 and CP 80-1827	Particle bombardment	Calli	Ubi	*CP*	[48]
SCYLV	CP 92-1666	Particle bombardment	Calli	Ubi	*CP*	[49]
SCMV	Bululawang	PDR & RNAi	Lateral buds		*CP*	[50]
SCSMV	ROC22	*Agrobacterium* mediated	Calli	Ubi 1	*Pac1*	[51]
Red rot	CoJ 83	*Agrobacterium* mediated	Axillary bud	CaMV35S	*β-1,3-glucanase*	[52]
Red rot	S2006SP-93	Particle bombardment	Calli	Ubi	*Chitinase class-II*	[53]
Colletotrichum	SPF-234	Particle bombardment	Calli	Ubi	*SWO, SWT*	[54]
Smut	KRFo93-1	Particle bombardment	Calli	CaMV35S	*BSR1*	[55]
Leaf scald	Q63 & Q87	Particle bombardment	Calli	Ubi	Albicidin detoxifying	[56]

The most prevalent viral diseases affecting sugarcane crops include sugarcane mosaic virus (SCMV), Sorghum mosaic virus (SrMV), and yellow leaf syndrome caused by Sugarcane yellow leaf virus (SCYLV) [41,57,58]. To address this, the preliminary virus capsid protein (CP) was utilized to produce transgenic plants that are immune to viruses. These plants produce a viral protein that co-suppresses and bypasses several stages of the viral life cycle and reduces disease manifestation. The use of untranslated SrMV strain *CP* enabled the cultivation of transgenic sugarcane plants. These plants display complete susceptibility to fully tolerant phenotypes [43]. In another study, transgenic lines with enhanced tolerance to Fiji disease virus (FDV) were obtained using particle bombardment. The transgene encodes a translatable version of FDV segment 9 of 1 open reading frame (ORF) of the virus genome. The tolerance of the transgenic plants was evaluated using a glasshouse experiment. However, the phenotypes of transgenic plants were not entirely consistent with resistance mechanisms based on post-transcriptional gene silencing (PTGS) [44].

Transgenic sugarcane against SCYLV was developed through the biolistic bombardment of cell cultures with an untranslatable *CP* gene. The resistance level exhibited by some transgenic plants was comparable to that of the completely resistant cultivar. These results were based on the virus titer and disease symptom development [45]. A strong innate defensive response against viral infections has been demonstrated using RNA interference (RNAi). In the host defense system, viruses produce suppressors of the host RNA interference (RNAi) pathway. A potential approach to studying viral pathogenesis and managing mosaic disease in sugarcane may involve identifying the microRNAs (miRNAs) encoded by the sugarcane streak mosaic virus (SCSMV) and further requires understanding the host genes that are related to miRNA targets [59]. Moreover, the study examined the resistance of transformed plants to the SCMV, using both the full 927 base pairs (bp) of the *CP-SCMV* gene sequence and truncated sequences of 702 bp. These results demonstrated that plants with the complete gene sequence exhibited a more protective strategy against the virus than those with truncated sequences [46]. Furthermore, the expression of short hairpin RNAs (shRNAs) confers resistance against SCMV infection [1,47]. These findings validated that RNAi technology can confer resistance against SCMV and the ubiquitin promoter is effective in the development of transgenic sugarcane lines [60].

Moreover, a study was conducted to evaluate the agronomic performance and viral resistance of transgenic lines transformed with SCYLV resistance, using antisense expression of a portion of the viral *CP* gene. The parental genotype outperformed the transgenic lines in sugarcane yield. However, the transgenic lines displayed lower SCYLV infection (0–5%) than the parental variety (98%). These variations in yield could be attributed to somaclonal variation during in vitro regeneration of transgenic plants. These findings suggest that transgenic sugarcane exhibits variable agronomic performance and disease resistance [48,49]. These results warrant further investigation to develop more efficient and reliable transgenic sugarcane cultivars. The use of transgenic sugarcane lines expressing the *SCMV-CP* gene led to a higher yield than the wild-type, with all evaluated lines producing significantly more cane and sucrose per hectare than the parental variety. In addition, transgenic lines exhibited a lower incidence of SCMV than the parental variety. These findings indicated that the introduction of transgenic sugarcane expressing *SCMV-CP* could be a viable solution for enhancing agronomic traits and viral disease resistance [61]. Taken together, these studies demonstrated that transgenic plants undergo minor hitherto discernible changes in morphology, physiology, and phytopathology despite low levels of genomic changes. Therefore, it is essential to assess somaclonal variation within transgenic populations to ensure the proper management and evaluation of transgenic sugarcane for field trials [62,63].

A comparative study aimed to evaluate the efficacy of two transgenic sugarcanes, generated through Pathogen-derived resistance (PDR) and RNAi methods using gene-encoding coat protein (CP) of SCMV, against SCMV through artificial viral inoculation. These results indicate that the RNAi approach targeting the CP gene proves more effective in producing resistance against SCMV infection in transgenic sugarcane compared to the PDR approach [50]. These results highlight promising strategies for developing SCMV-resistant sugarcane through genetic engineering. Breeding virus-resistant sugarcane varieties is the main goal of the sugarcane breeding program. Double-stranded RNA-specific ribonuclease (*PAC1*) encoded by the *Pac1* gene from *Schizosaccharomyces pombe* was introduced into a virus-sensitive sugarcane cultivar by *Agrobacterium*-mediated transformation. These results showed that transgenic plants possess significantly milder symptoms and lower viral loads than the wild-type, providing a basic foundation for breeding virus-resistant sugarcane [51].

To combat fungal diseases such as brown rust in sugarcane, glucanase and chitinase genes have been expressed in transgenic sugarcane plants. Transgenic sugarcane resistant to red rot was developed by overexpressing the *β-1,3-glucanase* gene from *Trichoderma* spp. as the primary target of fungal attack. Microscopic examination of parenchyma cells in the stalks of these plants was filled with sucrose, which inhibited the growth of *C. falcatum* hyphae. Transgene overexpression after infection was upregulated and successfully transmitted to the second generation of clones [52]. Moreover, the potential of transgenic sugarcane lines expressing the barley *chitinase class II* gene against *C. falcatum* infection was assessed. Protein extracts from transgenic sugarcane plants inhibited the growth of *C. falcatum* mycelia, and transgenic lines demonstrated strong resistance against *C. falcatum*. Additionally, the mRNA expression of the transgene in *C. falcatum* inoculated transgenic sugarcane lines was outperformed by control parental plants [53]. Two genes, *SUGARWIN1 (SWO)* and *SUGARWIN2 (SWT)*, were overexpressed in sugarcane to improve resistance against *C. falcatum*. Transgenic plants showed significant inhibition of the fungal pathogen infection and proliferation, indicating potential for developing crops with better fungal resistance [54]. Smut disease, caused by *S. scitamineum*, is considered a destructive disease of sugarcane. Rice receptor-like cytoplasmic kinase, known as broad-spectrum resistance 1 (*BSR1*) in sugarcane, confers resistance against smut under greenhouse conditions. *BSR1* overexpressing line displayed normal growth and morphology [55], suggesting that *BSR1* overexpression is effective in conferring broad-spectrum disease resistance in the different crops.

2.3. The Reinforcement of Insect and Pest Resistance

Insects and pests cause significant reduction in the yield and quality of sugarcane using different stem borers (Figure 2), such as Lepidoptera (*D. saccharalis*), root borers (*Emmalocera depressella*), top borers (*Chilo terrenellus*), pink borers (*Sesamia inferens*), and pink stem borers (*Sesamia cretica*) [64,65]. The extensive use of pesticides can harm other life on land such as beneficial insects, vertebrates, humans, and the environment. Pesticides may also pollute different soil zones before reaching deep water levels. Modern biotechnological approaches have made rapid progress in the genetic engineering of sugarcane plants to protect against pests by transferring genes derived from plants, pests, and bacteria [66–68] (Table 3).

Table 3. List of transgenic sugarcane expressing foreign genes for enhanced insect and pest resistance.

Variety	Trait Method	Explant	Promoter	Candidate Gene	Target Pest	Reference
Ja60-5	Electroporation	Calli	CaMV35S	*cry1Ab*	*D. saccharalis*	[15]
TUC95-10/TUC03-12	Particle bombardment	Calli	pCab1	*Bt*	*Diatraea saccharalis.*	[18]
SP80-1842 & SP80-3280	Particle bombardment	Calli	Maize Ubi-1	*skti, sbbi*	*D. saccharalis*	[20]
ROC22	*Agrobacterium* mediated	Calli	Ubi-1	*cry1Ab*	*D. saccharalis*	[21]
YT79-177 & ROC16	Particle bombardment	Calli	Maize Ubi-1	*cry1Ac*	*P. venosatus*	[64]
CP65-357	Paint-sprayer delivery		Maize Ubi-1	*Snowdrop lectin*	*Eoreuma loftini, D. saccharalis*	[69]
ROC25	*Agrobacterium* mediated	Calli	Ubi-1	*Amaranthus viridis & skti*	*D. saccharalis*	[70]
CoC92061 & Co 86032	Particle bombardment	Calli	Maize Ubi-1	*Aprotinin*	*S. excerptalis*	[71]
CPF-246	*Agrobacterium* mediated	Calli	Maize Ubi-1	*Vip3A*	*C. infuscatellus*	[72]
GT54-9(C9)	*Agrobacterium* mediated	Leaves	CaMV 35S	*cry1Ac*	*Sesamia cretica*	[73]
SP 803280	*Agrobacterium* mediated	Calli	35S & FMV	*cry1Ab, cry2Ab*	*D. saccharalis*	[74]
Zhongzhe1 (ZZ1)	Particle bombardment	Direct embryo	Ubi	PPA	*Ceratovacuna lanigera Zehntner*	[75]

Table 3. *Cont.*

Variety	Trait Method	Explant	Promoter	Candidate Gene	Target Pest	Reference
SP80-3280 & 1842	Particle bombardment	Calli	Maize PEPC	*cry1Ab*	*D. saccharalis*	[76]
YT79-177 & ROC16	Particle bombardment	Calli	Maize Ubi-1	*Synthetic-cry1Ac*	*P. venosatus*	[77]
FN81–745 & Badila	*Agrobacterium* mediated	Calli	RSs-1,Ubi-1	*Gna*	*Ceratovacuna lanigera*	[78]
Gui94-119	Particle bombardment	Calli	Ubi	*cry1Ac*	*D. saccharalis*	[79]
SP80-185	Plasmid transformation	Calli	Maize ubi-1	*HIS Cane CPI -1*	*S. levis*	[80]
CoC671	*Agrobacterium* mediated	Leaf roll	CaMV35S	*cry1Aa3*	*C. infuscatellus, C. sacchariphagu & S. excerptalis*	[81]
Co 86032 & CoJ 64	Particle bombardment	Calli	Maize Ubi-1	*cry1Ab*	*C. infuscatellus*	[82]
FN15	Particle bombardment	Calli	CaMV35S	*cry1Ac*	*D. saccharalis*	[83]
LK 92-11	*Agrobacterium* mediated	Calli	CaMV35S	*cry1Ab*	*D. saccharalis*	[84]
SP80-185	Particle bombardment	Calli	Maize ubi-1	*CaneCPI-1*	*S. levis*	[85]
FN15 & ROC22	Particle bombardment	Calli	CaMV35S	*cry1Ac*	*D. saccharalis*	[86]
ROC22	Particle bombardment	Calli	ST-LSI	*cry2A*	*C. sacchariphagus, S. nivella, C. infuscatellus, A. schistaceana & S. inferens*	[87]
Event CTC175-A	*Agrobacterium* mediated	Calli	PEPC	*cry1Ab*	*D. saccharalis*	[88]
Event CTC91087-6	*Agrobacterium* mediated	Calli	Maize ubi-1	*Cry1ac*	*D. saccharalis*	[89]
Bululawang	*Agrobacterium* mediated	Calli	RUBISCO	*CryIAb-CryIAc*	*Scripophaga excerptalis*	[90]

Alternatively, the genetic transformation of several insecticidal proteins, such as lectins and protease inhibitors (PIs), has been used in genetic transformation. Other genes included *avac* (*Amaranthus viridis* L. agglutinin), *skti* (soybean Kunitz trypsin inhibitor), *sbbi* (Bowman–Birk inhibitor), and *gna* (*Galanthus nivalis* agglutinin) [69,70]. Transgenic sugarcane plants expressing either potato proteinase inhibitor II or the snowdrop *lectin* gene showed increased antibiosis in the larvae of the canegrub *Antitrogus consanguineus* under glasshouse conditions. Canegrubs feeding on the transgenic line transformed with the potato gene showed 4.2% weight gain in canegrubs fed on control plants. Similarly, larvae feeding on the roots of transgenic lines transformed with the snowdrop gene showed a 20.6% weight gain of grubs feeding on wild-type plants. In contrast, larvae from the Greyback Cane Beetle (*Dermolepida albohirtum*) fed on transgenic plants expressing snowdrop lectin and proteinase inhibitor decreased in weight compared to larvae fed on nontransformed plants [18]. However, the growth of *D. saccharalis* larvae fed on transgenic sugarcane transformed with *skti* and *sbbi* from soybean was restricted compared to that of larvae from control plants [20]. In vivo bioassay studies of transgenic sugarcane transformed with a synthetic bovine pancreatic trypsin inhibitor (*aprotinin*) gene against the top borer (*Scirpophaga excerptalis*) indicated that larvae fed on transgenic plants exhibited a decrease in larval weight of up to 99.8% [71]. Transgenic sugarcane plants overexpressing sugarcane cysteine peptidase inhibitor 1 (*CaneCPI-1*) were evaluated for their potential resistance through feeding assays with larvae of sugarcane billbug (*S. levis*); significantly less damage by larval attack has been observed in transgenic plants than non-transgenic plants [66]. The trypsin inhibitory activity of PIs from *Erianthus arundinaceus*, a wild-sugarcane relative, was evaluated and showed significant differences among different plant tissues. The highest inhibitory activity was found in meristematic tissue, and PIs isolated from the apical meristem effectively inhibited the mid-gut proteinases of early shoot borer (*C. infuscatellus*) and internode borer (*Chilo sacchariphagus indicus*) [91,92]. These studies have demonstrated the possibility of identifying new and efficient PIs for producing sugarcane plants that are resistant to stem borers. Furthermore, transgenic sugarcane lines displaying the Vip3A protein exhibited resistance to the sugarcane stem borer (*C. infuscatellus*). These results imply that incorporating a single copy of the *Vip3A* gene into transgenic sugarcane lines confers resistance to borers [72]. Consequently, these lines could potentially be used to produce insect-resistant transgenic sugarcane and could also be combined with the *Bacillus thuringiensis* (*Bt*) toxin in gene pyramiding to enhance resistance.

Nevertheless, the most important insecticidal proteins produced by the gram-positive spore-forming bacterium *B. thuringiensis* (*Bt*), which produces proteins during the vegetative phase (*Vip*) and sporulation phase (*Cry*), are toxic to a wide range of insects [73,93]. The development of transgenic *Bt* plants that are resistant to stem borers in sugarcane cultivation has been achieved using specific toxins. Studies have identified various *Bt* genes, including *cry1Ab*, *cry1Aa3*, *cry1Ac*, *s-cry1Ac*, *m-cry1Ac*, *cry2A*, and *vip3A*, that have been deployed (Table 3). More importantly, Bt proteins have been used to control sugarcane borer in maize for the past decade. However, practical use in sugarcane requires the delivery of high doses of protein and harbors the slow evolution of insect resistance. Two *Bacillus* proteins, *Cry1Ab* and *Cry2Ab*, have been expressed in commercial sugarcane varieties, demonstrating efficacy against sugarcane borer in the field. A strategy for trait deployment of tropical crops has also been described [74]. Moreover, the study aimed to develop transgenic sugarcane lines with increased resistance to the sugarcane borer *D. saccharalis*. The *Bt* genes were incorporated into embryogenic sugarcane calli. The transgenic lines with higher levels of *Bt* transcripts were identified and tested for resistance against the pest [18].

To explore the utilization of the *Pinellia pedatisecta agglutinin PPA* in controlling aphids in sugarcane, the resistance of independent transgenic sugarcane lines was assessed. Alterations in stomatal patterns, antioxidant enzymes, chlorophyll content, sugar, and tannins were examined before and after insect infestation. These findings revealed that *PPA* enhanced sugarcane resistance to sugarcane woolly aphids and insect resistance [75]. Taken together, there is an urgent need to fully utilize the recently sequenced sugarcane genome to investigate the interactions between sugarcane and its pathogens, as well as pathogens that are closely related to those affecting sorghum and other grasses sharing similar genomes [6–10,12,13]. However, it is expected that the reference genome for *Saccharum* spp. hybrids is still underway because of its complexity and high polyploidy, which may serve as a tool for searching resistance genes for many common diseases in sugarcane, and functional disease resistance will be maintained when genes are moved across the species using transgenic technology.

3. The Enhancement of Tolerance to Abiotic Stress

Sugarcane cultivation is influenced by a variety of abiotic factors, such as drought, salinity, high temperature, nutrient deficiencies, and heavy metals, which significantly decrease the average yield (Figure 1). To cope with adverse environments, plants must adopt tolerance mechanisms and undergo various morphological, anatomical, physiological, and cellular changes in response to abiotic stresses [94]. The plant response to stress is a multifaceted process that involves various signaling pathways, post-translational modifications, secondary metabolite synthesis, nitrogen metabolism, and the activity of various proteins, including transcription factors, kinases, and transporters. The genes responsible for regulating biotic and abiotic stress responses may function cumulatively or redundantly, activating both common and distinct downstream targets [95]. Adapting to these stresses is also linked to metabolic adaptation, which results in the accumulation of various organic solutes like proline, sugars, betaines, and polyols. To protect from the harmful effects of reactive oxygen species (ROS) (Figure 3), plants utilize antioxidant enzymes such as ascorbate peroxidase APX, SOD, and GR, and non-enzymatic antioxidants like carotenoids, ascorbic acid, and flavonoids [96,97]. Researchers have identified different genes that contribute to abiotic stress tolerance (Table 4) in model/water-tolerant plants, unrelated plants, or completely different organisms, including those involved in the overexpression of transcription factors (TFs) and effector proteins, which play crucial roles in activating genes in response to abiotic stresses [98,99].

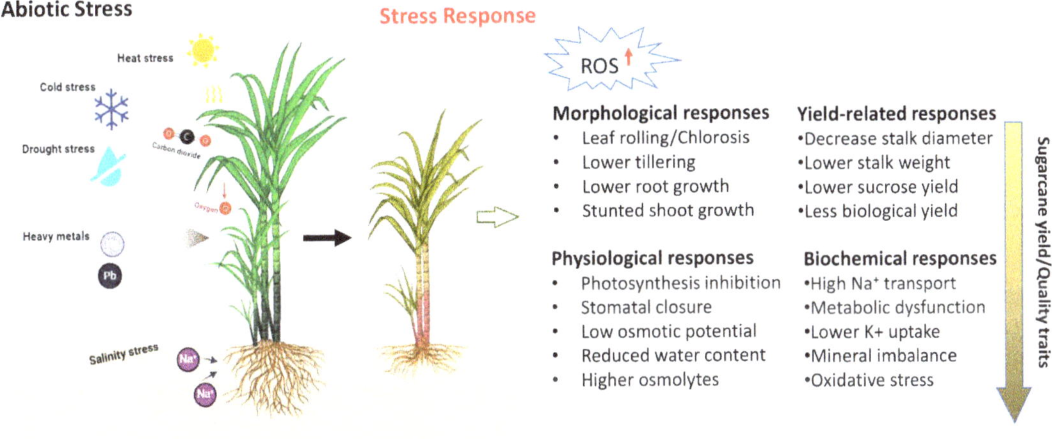

Figure 3. Types of abiotic stresses and stress responses that affect yield productivity of sugarcane (https://biorender.com; accessed on 23 April 2024).

Table 4. Genetic engineering of sugarcane for abiotic stress tolerance.

Variety	Trait Method	Type of Explant	Promoter	Gene	Gene Function	Stress	Reference
ROC22	*Agrobacterium*		CaMV35S	*TERF1*	Gene regulation	Drought	[2]
RB855156	Biolistic	Embryogenic calli	pRab17	*DREB2A CA*	Gene regulation	Drought	[100]
NCo310	Biolistic	Embryogenic calli	UBI	*AtBBX29*	Gene regulation	Drought	[101]
CP-77-400	*Agrobacterium*	Apical buds	CaMV35S	*AVP1*	Osmotic regulation	Drought	[102]
CSSG-668	Biolistic	Embryogenic calli	CaMV35S	*AVP1*	Osmotic regulation	Drought	[103]
RB835089	Biolistic	Embryogeniccalli	UBI	*BI-1*	PCD-regulation	Drought	[104]
RB855156	Biolistic	Nodal buds	ABA-AIPC	*P5CS*	Proline synthesis	Salinity	[105]

Table 4. *Cont.*

Variety	Trait Method	Type of Explant	Promoter	Gene	Gene Function	Stress	Reference
Guitang21	*Agrobacterium*	Embryogenic calli	UBI	*SoP5CS*	Proline synthesis	Drought	[106]
Co 86032	Biolistic		UBI	*EaGly III*	Reduce oxidative stress	Salinity	[107]
Co86032	*Agrobacterium*		UBI	*HSP70*	Cellular stability	Drought/ Salinity	[108]
RB855536	Biolistic	Embryogenic calli	AtCOR15a	*ipt*	Cytokinin	Cold	[109]
ROC22	*Agrobacterium*	Calli	UBI	*SoTUA*	α-tubulin synthesis	Cold	[110]
ROC10	*Agrobacterium*		CaMV35S	*TSase*	Biomolecules stabilization	Drought	[111]
Co86032	*Agrobacterium*/ biobalistic	Embryogenic calli	UBI	*PDH45/DREB2*	Nucleic acids metabolism	Drought/ Salinity	[112]

The COR/DREB (Cold-induced regulated-dehydration-responsive binding element) family of regulatory proteins is the first known set to play a role in the regulation of abiotic stress genes in plants [113]. Studies have shown that the overexpression of *Arabidopsis AtDREB2A* in transgenic sugarcane can enhance drought tolerance under greenhouse conditions. In these experiments, transgenic plants exposed to drought stress exhibited higher relative water content (RWC), carbon assimilation, sugar content, and bud sprouting, without affecting biomass [100]. B-box (BBX) proteins are essential for the control of plant growth and development. Transgenic sugarcane overexpressing the *AtBBX29* gene exhibited improved rates of photosynthesis and higher levels of antioxidants and osmolytes under drought conditions [101]. In another study, the accumulation of osmolytes, such as proline, soluble sugars, and glycine betaine, was observed in sugarcane plants engineered to overexpress the tomato ethylene-responsive factor gene (*TERF*), and these transgenic plants displayed reduced ROS and malondialdehyde (MDA) content [2].

The *Arabidopsis* H^+-pyrophosphatase type I gene (*AVP1*) is involved in the control of apoplastic pH and auxin transport, and overexpression of the *AVP1* gene in sugarcane results in increased RWC and osmotic and turgor potential in the leaves, indicating that improved drought and salinity tolerance. In addition, transgenic sugarcane plants exhibited profuse rooting systems [102,103]. Proapoptotic BAX proteins are responsible for controlling programmed cell death (PCD). Overexpression of the BAX inhibitor gene (*BI-1*) from *A. thaliana* in sugarcane plants increases drought tolerance by reducing the induction of cell-death pathways [104]. The drought-responsive gene *BRK1* from *S. spontaneum* was transformed into sugarcane, and *BRK1* transgenic lines had improved physiological parameters. Interlocking marginal lobes in epidermal leaf cells were observed in all transgenic *BRK1* lines during drought stress compared with the wild-type. These findings suggest that *BRK1* plays a potential role in sugarcane's response to drought stress, promoting leaf epidermal cell morphogenesis and actin polymerization [114].

Plants exposed to salinity stress acquire osmoprotectant molecules, such as proline, which serve not only as a source of nourishment but also as scavengers such as ROS, to maintain cellular activities. By overexpressing the pyrroline-5-carboxylase synthase gene (*P5C5*), salinity-tolerant sugarcane was obtained [105]. Additionally, improved tolerance to water deprivation was achieved through overexpression of a related gene (*SoP5C5*). *P5C5* plays a crucial role in proline production. The ROS-scavenging glyoxalase pathway enzymes, including glyoxalase I (Gly I), glyoxalase II (Gly II), and glyoxalase III (Gly III), help transgenic sugarcane plants to eliminate harmful compounds, such as glyoxylate, under abiotic stress conditions [106]. Furthermore, under salt stress, transgenic sugarcane overexpressing the *EaGlyIII* gene displayed improved RWC, photosynthesis, osmolytes, and ROS-scavenging enzyme activities, resulting in increased biomass and enhancing germination rates [107]. These results indicate that the overexpression of *EaGly III* could be a promising approach for salt and drought tolerance in sugarcane [115]. However, the *ScD27.2* gene was introduced into sugarcane to study its role in drought tolerance. The results showed that interfering with *ScD27.2* expression decreased tolerance to drought stress, indicating its role in sugarcane growth and development [25].

Plant cells respond to stressful conditions by producing a family of stress proteins known as heat shock proteins (HSPs). In *E. arundinaceus*, a sweet cane, *HSP70* overexpression has a strong protective effect when plants are exposed to water and salt stress. Transgenic plants displayed increased photosynthetic efficiency, RWC, stress-induced gene expression, and cell membrane thermostability [108]. Transgenic sugarcane with the drought-tolerant *Ea-DREB2B* gene from *S. arundinaceum*, a wild species of *S. officinarum*, can regulate response to drought stress. In addition, interactions between soil fungi and bacterial communities in the rhizoplane, rhizosphere, and bulk soil were assessed. The results revealed that the host plant genotype plays a crucial role in strengthening plant–fungi interactions and enhancing beneficial fungal function in the root-related area of transgenic sugarcane, allowing response to drought stress. Moreover, changes in the soil environment caused by transgenic sugarcane alter bacterial communities [116,117].

However, limited studies have been conducted on cold-tolerant transgenic sugarcane (Table 4). These plants overexpress the bacterial isopentenyl-transferase gene (*ipt*) using the cold-inducible promoter *AtCOR15* derived from *A. thaliana*. Compared with non-transgenic parent plants exposed to low temperatures, transgenic sugarcane exhibited increased chlorophyll, reduced MDA content, and decreased electrolyte leakage [109]. Transgenic plants overexpressing the tubulin gene (*TUA*), cloned from cold-tolerant sugarcane varieties, enhanced the cold tolerance of cold-susceptible sugarcane varieties, exhibiting increased soluble protein and sugar content, enhanced peroxidase activity, and reduced MDA content compared to non-transgenic plants [110]. Further investigation into the effects of cold stress could have substantial implications for the sugarcane industry, as cold stress negatively affects sugarcane crop yields and quality.

4. Production of New Compounds and Renewable Energy Sources

Genetically modified sugarcane has achieved significant improvement under biotic and abiotic stress conditions. Plant cells, as opposed to animal cells, are cost-effective and safe methods for producing new recombinant (r) proteins. Sugarcane plants have well-developed storage tissue systems, which make them promising candidates to produce high-value compounds and pharmaceuticals. Recent advancements in bio-pharming have enabled the production of a wide range of essential products, such as high biomass production, rapid growth, and efficient carbon fixation, making sugarcane more attractive for novel compound applications [41,57,118]. The utilization of sugarcane for molecular farming has gained significant momentum owing to its minimal transgene dispersal and high amount of extractable juice with low protein content. The culm of sugarcane comprises 70% dry weight and is predominantly composed of parenchyma cells, a desirable target for r-protein production. Within plant cells, vacuoles constitute a large portion of the cell volume, contributing 60% of the cane weight [57,119]. Thus, targeting r-proteins in the vacuole can result in high-protein yields and relatively simple purification.

Initial attempts were made to express proteins in the cytoplasm and purify from the leaves, resulting in low yields due to the presence of numerous other proteins that may interfere with downstream processing [120,121]. However, targeting proteins to vacuoles in the stem parenchyma may lead to higher levels of recombinant proteins. Previous studies have shown that newly identified vacuolar-targeting determinants (VT) of 78-bp or 18-bp in combination with strong constitutive promoter (Port ubi882) have proven to be effective in the production of large quantities of r-proteins. However, the presence of vacuolar proteolytic enzymes and the acidic nature of lytic vacuoles must be considered when selecting recombinant candidate proteins. It is estimated that 4.8–7.2 kg of r-protein can be produced per acre (40,000 canes per acre) with a purity of 50% after affinity chromatography [122]. This yield is comparable to commercially available r-proteins. The abundance of extractable juice with low protein content and well-established juice processing technology has motivated researchers to focus on the culm of sugarcane. Thus, culms have drawn attention due to their significant biomass output and potential for the large-scale production of recombinant proteins. Sugarcane could be an effective platform to produce recombinant proteins, and further research could significantly impact the biotechnology sector. Taken together, targeting r-proteins to the lytic vacuoles of sugarcane shows promising results in the production of recombinant proteins. Sugarcane can extract a large volume of juice coupled with newly identified vacuolar-targeting determinants and strong constitutive promoter, making it an efficient platform for molecular farming.

Trehalose is a naturally occurring non-reducing disaccharide that plays an important role in regulating carbon metabolism and photosynthesis. To investigate the effects of trehalose synthesis on sucrose accumulation in sugarcane, overexpression of *Escherichia. coli, otsA* (trehalose-6-phosphate synthase TPS), *otsB* (trehalose-6-phosphate phosphatase; TPP), and silenced native *TPS* expression altered the effect of trehalose synthesis. These findings suggest that manipulating Tre6P/trehalose metabolism could be a potential strategy for modifying the sugar profile of sugarcane stems [123]. Sucrose phosphate synthase (*SPS*) is a vital enzyme that plays a key role in regulating sucrose content in sugarcane by controlling the synthesis of sucrose. Enhancing *SPS* activity may be a useful approach to increasing sugarcane yield [124]. The effects of overexpressing the *SoSPS1* gene on sucrose accumulation and carbon partitioning in transgenic sugarcane were studied. Overexpression of *SoSPS1* resulted in increased levels of sucrose and enhanced enzymatic activity in the leaves and stalks, leading to improved plant growth [125]. Transgenic sugarcane with enhanced sucrose yield was developed by the over-expression (*SoSPS1*) gene. Furthermore, the nutritional and mineral composition of transgenic lines and non-transgenic sugarcane were analyzed after 11 months in field conditions. Protein and potassium content was higher in stems of transgenic lines. However, wild-type and transgenic sugarcane were found to be substantially equivalent in terms of nutritional and mineral compositions [126]. The development of new sucrose isomers, trehalulose and isomaltulose, is facilitated by the insertion of sucrose isomerase genes. The transgene expression in the sugarcane culms exhibited significantly greater expression levels than in leaf tissues. The overall sugar estimation conducted in internodes of the transgenic sugarcane revealed increased sugar concentrations in mature sugarcane culms. The transgenic sugarcane lines demonstrated the highest sugar recovery of 14.9%, compared to 8.5% in the control lines [127]. However, consistent attempts to modify sugar metabolism to increase sugar yield remain challenging for molecular farming.

Another field of study is bioplastics such as polyhydroxyalkanoates (PHAs), which are biodegradable. Previous studies have shown that these polymers can be produced from sugarcane, including polyhydroxybutyrate (PHB) [128,129]. Utilization of sugarcane as a renewable bioenergy crop is a new emerging task. The selection of superior multipurpose cultivars for agro-industries is based on genetic and biotechnological strategies. To meet the demand for renewable energy sources, sugarcane has received numerous genes promoting resistance to biotic and abiotic stresses [41,57,118]. Nonetheless, considerable emphasis has been placed on improving the production of bioenergy and biofuel. Various approaches have been devised to modify lignin, facilitate saccharification, and enhance second-generation bioethanol [130,131]. Recently developed transgenic sugarcane, termed oilcane, has been genetically engineered to direct carbon flux towards biosynthesis and accumulation of energy-rich triacylglyceride (TAG) molecules in vegetative tissues [132]. This oilcane was employed to produce five high-value bioproducts: hydroxymethylfurfural (HMF), furfural, acetic acid, fermentable sugars, and vegetative lipids [132–134]. Genetic and biotechnological strategies have been employed to improve sugarcane productivity, including the development of transgenic sugarcane that directs carbon flux toward the biosynthesis and accumulation of energy-rich molecules. These advances in sugarcane biotechnology hold great potential for the development of future energy sources for biofuel production. Conversely, the success of these developments depends on the careful selection of techniques employed for sugarcane transformation, choice of promoters and marker genes, target tissue/explants, and tissue culture system (Tables 1–4). Further research and development in sugarcane biotechnology will be crucial for meeting the increasing demand for renewable energy sources and sustainable agro-industries. However, little progress has been made in cultivating sugarcane for commercialization through genetic engineering in a few countries (Table 5).

Table 5. List of Sugarcane genetically modified (GM) event names approved for food/feed and commercialization.

Developer	Event Name	Trait Method	Gene	Gene Source	Product	Function	Country/Year
Centro de Tecnologia Canavieira (CTC)	CTB141175/01-A	Microparticle bombardment	*cry1Ab*	*B. thuringiensis* subsp. kurstaki	Cry1Ab delta-endotoxin	lepidopteron insects	Brazil 2017, Canada & United States 2018
	CTC-92015-7	*Agrobacterium* mediated	*cry1Ac*	*B. thuringiensis* subsp. Kurstaki strain HD73	Cry1Ac delta-endotoxin	lepidopteron insects	Brazil 2022
			nptII	*E.coli* Tn5 transposon	neomycin phosphotransferase II	neomycin & kanamycin antibiotics	Brazil 2022
	CTC75064-3	*Agrobacterium* mediated	*cry1Ac*	*B.thuringiensis* subsp. Kurstaki strain HD73	Cry1Ac delta-endotoxin	lepidopteron insects	Brazil 2020, Canada 2022
			nptII	*E.coli* Tn5 transposon	neomycin phosphotransferase II	neomycin & kanamycin antibiotics	Brazil 2020, Canada 2022
	CTC91087-6		*cry1Ac*	*B. thuringiensis* subsp. Kurstaki strain HD73	Cry1Ac delta-endotoxin	lepidopteron insects	Brazil 2018 & United States 2020
	CTC93209	*Agrobacterium* mediated	*cry1Ac*	*B. thuringiensis* subsp. Kurstaki strain HD73	Cry1Ac delta-endotoxin	lepidopteron insects	Brazil 2019
	CTC95019-5	*Agrobacterium* mediated	*cry1Ac*	*B. thuringiensis* subsp. Kurstaki strain HD73	Cry1Ac delta-endotoxin	lepidopteron insects	Brazil 2021
			nptII	*E. coli* Tn5 transposon	neomycin phospho transferase II	neomycin & kanamycin antibiotics	Brazil 2021
PT Perkebunan Nusantara XI (Persero)	NXI-1T	*Agrobacterium* mediated	*EcBetA*	*E. coli*	choline dehydrogenase	Osmoprotectant, glycine betaine	Indonesia 2011
			nptII	*E. coli* Tn5 transposon	neomycin phospho transferase II	neomycin & kanamycin	Indonesia 2011
			aph4 hpt)	*E. coli*	hygromycin-B phosphotransferase	hygromycin B	Indonesia 2011
PT Perkebunan Nusantara XI (Persero)	NXI-4T	*Agrobacterium* mediated	*RmBetA*	*Rhizobium meliloti*	choline dehydrogenase	osmoprotectantglycine betaine)	Indonesia 2013
PT Perkebunan Nusantara XI (Persero)	NXI-6T	*Agrobacterium* mediated	*RmBetA*	*Rhizobium meliloti*	choline dehydrogenase	Osmoprotectant, glycine betaine	Indonesia 2013
Estación Experimental Agroindustrial Obispo Colombres (EEAOC)	TUC-873RH-7	Microparticle bombardment	*cp4 epsps (aroA:CP4)*	*A. tumefaciens* strain CP4	(EPSPS) enzyme	glyphosate herbicide	Argentina 2015
			nptII	*E. coli* Tn5 transposon	neomycin phosphotransferase II	neomycin & kanamycin antibiotics	Argentina 2015

Table 5. *Cont.*

Developer	Event Name	Trait Method	Gene	Gene Source	Product	Function	Country/Year
Monsanto Company & Bayer Crop Science	MON87427 × MON95379 × MON87411	Conventional breeding-cross hybridization-transgenic donor(s)	*cry1Da_7*	*B. thuringiensis*	crystalline protein prototoxin Cry1Da_7	*S. italica* promoter and *O. sativa* gos2 terminator-Rice actin 15 gene-intron	Brazil 2021
			cry1B.868	*B. thuringiensis*	crystalline protein prototoxin Cry1B.868	*S. italica* promoter and *O. sativa* gos2 terminator-Rice actin 15 gene-intron	Brazil 2021
			cry3Bb1	*B. thuringiensis* subsp. kumamotoensis	Cry3Bb1 delta endotoxin	Tolerance to coleopteran insects & corn rootworm	Brazil 2021

Source 2024. Crop Biotech Update April. Available online at: https://www.isaaa.org/gmapprovaldatabase/crop/default.asp? CropID=27&Crop=Sugarcane (accessed on 24 April 2024).

5. Status and Concerns about GM Sugarcane

Since its inception, the debate over genetically modified organisms (GMOs) has continued, with a particular focus on the release of herbicide-tolerant sugarcane and its potential impact on the sugar export market. Several studies have been conducted on sugarcane using various genetic transformation techniques. For example, the safety of transgenic sugarcane with the *AVP1* gene [103] was assessed through animal feeding and genotoxicity assays. Acute and subchronic toxicity studies were conducted in rats, and no significant differences were observed among the different treatments [135]. Moreover, genotoxicity assessment revealed that GM sugarcane was neither genotoxic nor cytotoxic to rats. These results suggest that GM sugarcane is non-toxic to experimental animals and is suitable for commercial release. A field study conducted on transgenic sugarcane expressing insect and glyphosate tolerance genes showed poor agronomic and industrial traits compared to non-transformed plants [21]. Some studies have reported successful field trials on disease-resistant transgenic sugarcane varieties. For instance, two cultivars resistant to SCMV showed large variations in yield and disease resistance in the field condition [48]. In another trial, transgenic sugarcane lines transformed for SCYLV resistance showed reduced performance compared to the parental genotype, despite their viral resistance [49]. Conversely, a positive result was observed in a field experiment conducted on transgenic sugarcane lines expressing the *CP* gene of SCMV, with greater yield and lower SCMV disease incidence at four different experiment locations in China across two successive growing seasons [61]. However, only a few officially approved varieties (ISAAA 2024: Table 5) have been released.

A transgenic trait expressing the herbicide-resistant sugarcane *CP4-EPSPS* gene was developed in Argentina. In Argentina, the commercial approval/deregulation of transgenic events requires continuous effort. Extensive research has been conducted on health and environmental issues to evaluate the transgenic event impact on agricultural systems and food safety [27,32,136]. Later, the Ministry of Agriculture, Livestock, and Fisheries denied grants for final approval and commercialization. This may concern the export of sugars produced from such transgenic crops. Nonetheless, this process has led to the development of new regulations for the cultivation of sugarcane. Argentina's regulatory system for assessing transgenic crops has served as a model in many other countries. In 2013, a significant change was made in the requirements for evaluating vegetatively propagated and highly polyploid genome crops such as sugarcane. This change allowed for fast-track evaluation of new events containing gene constructs equal or similar to those that have already been approved. This resolution has greatly accelerated the approval of new sugarcane varieties with the same or similar gene constructs regardless of their position in the genome. The introgression of a transgene through backcrossing is impossible. Thus, two possible ways to produce transgenic sugarcane varieties have emerged: forward breeding and the direct transformation of elite varieties. Both methods are time-consuming and often result in outdated technology by the time the transgenic variety is ready for commercial release. This policy has allowed the production of new varieties with desirable traits in a timely and cost-effective manner and has been adopted by regulatory agencies in Canada, the United States, and Brazil (Table 5). However, regulatory authorities in various countries play crucial roles in monitoring and overseeing the research and development of GMOs to ensure that their introduction does not pose any threat to human health or the environment. Moreover, due to deregulation, the cultivation and use of transgenic varieties are no longer subject to regulatory restrictions. Therefore, it is necessary to establish stewardship programs to safeguard these varieties and neighboring crops. Implementing current strategies would not only facilitate the successful reintroduction of GM sugarcane but also establish standards for introducing new transgenic traits into sugarcane.

Moreover, the first drought-tolerant transgenic sugarcane expressing the bacterial choline dehydrogenase gene, the osmoregulator glycine betaine, plays a crucial role in plant defenses against cellular dehydration. Persero (PT Perkebunan Nusantara XI) was commercially released in Indonesia (ISAAA 2024; Table 5). Two genetically modified sugarcane varieties produced by Embrapa Agro-energy were approved by the Brazilian regulatory authority, the National Biosafety Technical Commission (CTNBio). Brazilian science has developed the first sugarcane variety with CRISPR/Cas9 altered, Cana Flex I and II, which showed improved cell-wall digestibility. Currently, plants with altered genomes are controlled as transgenic in certain nations but are not regarded as GM in others. Recently, Brazilian authorization has been encouraging as it will significantly reduce the expenses and work required for commercial releases. It is anticipated that more nations will soon assist in the deregulation of varieties created using transgene technology.

6. Genome Editing for Sugarcane Improvement

Sugarcane is characterized by a huge genome size (~10 GB), high degree of complexity, auto-allopolyploid, limited genetic diversity, and extensive recombination between the two subgenomes. In addition, it is highly heterozygous in nature, taking a long generation time to propagate vegetatively and preserve its genetic makeup. However, alternative strategies such as genetic transformation and the application of different omics technologies have been widely employed for the genetic improvement of sugarcane [137]. Nonetheless, sugarcane can be edited using the CRISPR/Cas9 system, even at high levels of ploidy. CRISPR/Cas9 is a highly effective and dependable tool for targeted mutagenesis (Figure 4). To date, four classes of nuclease-mediated genome editing methods/technologies have been developed; meganucleases, zinc finger nucleases (ZFN), transcription activator-like effector nucleases (TALENs), and the clustered regularly interspaced short palindromic repeats (CRISPR)/Cas9 system [138,139].

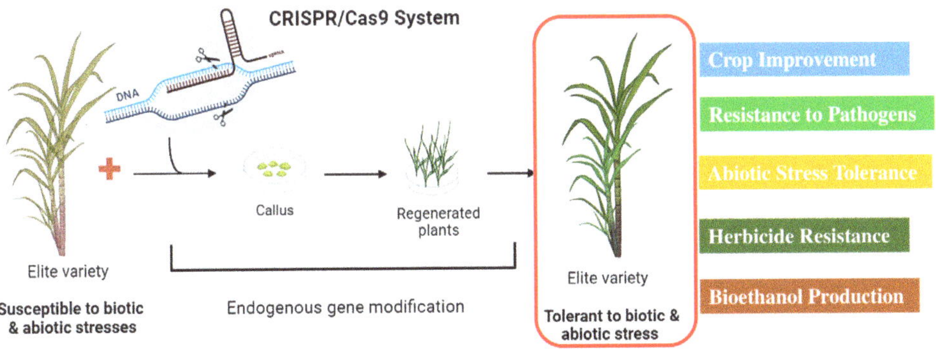

Figure 4. A general description of the genome editing in sugarcane plants to produce elite varieties against biotic and abiotic stress tolerance (https://biorender.com; accessed on 23 April 2024).

Targeted mutagenesis using TALENs was first demonstrated in the sugarcane gene responsible for lignin biosynthesis, caffeic acid O-methyltransferase (*COMT*). The gene was identified, and TALEN-mediated mutations were introduced into the *COMT* gene to assess effects on agronomic performance, lignin content, cell wall composition, and saccharification efficiency. These results showed a 19.7% reduction in lignin and a 44% increase in saccharification of cell wall-bound sugars, without influencing biomass, disease resistance, and logging under field conditions. These modifications significantly increased the production of sugarcane-derived biofuels [130,131]. Furthermore, targeted co-mutagenesis of more than 100 alleles of the *COMT* gene improves saccharification and enhances bioethanol yield from biomass without affecting agronomic performance [138,140]. However, optimizing SgRNA expression cassettes, vectors, and delivery systems is necessary for co-editing multiple alleles. Although TALEN and ZFN-based genome engineering of crops have proven useful, their application for precise gene editing has been limited because of the design and complex interactions of zinc fingers and DNA [138,141]. Of note, sugarcane is a vegetative propagating crop once mutagenesis occurs, and can be stably transmitted to progeny.

The CRISPR/Cas9 system is a groundbreaking technology that emerged as the most crucial tool for crop improvement following genetic transformation [142–144]. Among three types of CRISPR/Cas9 mechanisms, Types I, II, and III were classified based on the presence of cas 3, cas 9, and cas 10 genes, respectively; Type II is preferred for genome editing. Using this technology, plants classified under the non-GMO category can be produced, which boosts public acceptance as the target genome does not contain any foreign DNA molecules [145]. Insect pests and diseases are the major causes of reduced sugarcane yield; also, ratoon crops should be considered. Red rot is a major disease that results in significant yield loss in sugarcane. For example, the antifungal diene compound (AFD) is a potent antifungal agent produced in sugarcane that helps to alter dormancy in *Colletotrichum* spp. CRISPR/Cas9 technology can be utilized to modify genes involved in AFD biosynthesis, providing sugarcane resistance to red rot caused by *Colletotrichum* spp. [146].

Additionally, smut disease caused by *S. scitamineum* can be managed by modifying the genes involved in *β-1,3-glucanase* biosynthesis using CRISPR/Cas9. Deletion of *Sspep1* plays a crucial role in various biological activities of *S. scitamineum*, such as mating, defense mechanisms, and virulence, and is helpful in smut disease resistance [147]. Furthermore, deletion of the *Ssubc2* gene encoding kinase regulator in *S. scitamineum* plays a potential role in the growth and proliferation of fungi in sugarcane, resulting in disruption of mating and a decrease in the virulence potential of the fungus [148,149]. Deletion of the *SsRSS1*, which regulates salicylic acid, can result in a reduction in fungal virulence. For instance, R genes play a crucial role in determining the capacity of plants to resist insect pests and diseases, while susceptible genes (*S* genes) make plants sensitive to biotic stress [150]. The adoption of CRISPR/Cas9-mediated genome editing in sugarcane offers several benefits including multiplex capacity, adaptability, and ease of design. This approach is considered more advantageous than the TALEN [151].

CRISPR/Cas9 technology has the potential to improve the expression of the non-specific lipid transfer (*ScNsLTP*) gene in sugarcane, thereby providing tolerance to drought and cold stress [152]. The transcripts of *ScGluD2*, a novel member of subfamily D beta-1,3-glucanase, partially increased after 12 h of salt stress and were significantly upregulated from 6 to 24 h under abscisic acid (ABA), H_2O_2, and $CdCl_2$. These results suggest that ABA may act as a signaling molecule that helps to regulate oxidative stress and plays a critical role in stress-induced *ScGluD2* transcripts [153]. Interestingly, *ScGluD2* is a stress-responsive gene in sugarcane expressed under defense response against smut, salt, and heavy metal stress [154]. Using CRISPR/Cas9 technology, *ScGluD2* expression can be enhanced to provide tolerance against heavy metal and saline stress in sugarcane.

The use of specific co-editing techniques involving multiple alleles and CRISPR/Cas9 technology can facilitate the rapid development of herbicide tolerance in sugarcane. Analysis of 146 individually engineered plants from five separate experiments in sugarcane revealed targeted nucleotide substitutions, resulting in alterations in W574L and S653I in the acetolactate synthase (*ALS*) gene in 11 lines. In addition to single targeted amino acid substitutions, W574L and S653I were present in 25 and 18 lines, respectively. Co-editing of three *ALS* alleles that confer herbicide tolerance has been established [33], and such techniques can be used to convert inferior to superior alleles through precise substitution mutations.

The application of TALEN-based technology enabled precise editing of conserved locations on the gene, resulting in the development of *COMT*-engineered lines that exhibited a 19.7% reduction in lignin content and a 43.8% improvement in saccharification. The lignin composition of *COMT* mutants increases from 29 to 32% [130]. Considering this fact, and the complex and polyploid nature of the sugarcane genome, precise mutation using CRISPR/Cas9 technology could serve as a valuable tool for reducing lignin content. The utilization of a multigene expression/suppression strategy resulted in the hyper-accumulation of triacylglycerol (TAG) and total fatty acids (FA) in the leaves, as two-fold higher than observed in unmodified energy cane [155,156]. The strongest link between TAG accumulation, *ZmDGAT1* expression, and *SDP1* repression was also observed. Moreover, genetically modified sugarcane upregulates TAG biosynthesis genes and downregulates TAG catabolism genes to direct carbon flux toward oil synthesis and accumulation in the leaf and stem tissues. Additionally, under field conditions, constitutive expression of *WRI 1*, along with lipogenic factors *DGAT1-2*, *OLE1*, and *PXA1*, resulted in hyper-accumulation of TAG and reduced biomass yield [157,158].

7. Future Perspectives and Conclusions

Sugarcane is a valuable cash crop that not only provides sugar but also serves as a renewable energy source. The application of genetic engineering and genome editing tools can reduce losses caused by biotic and abiotic stresses. Several transgenic sugarcane traits outperformed under greenhouse and field conditions. However, the commercial release of transgenic sugarcane varieties lags behind that of other crops, such as soybean, corn, and cotton. Nonetheless, some countries, including Argentina, recently allowed the rapid release of second varieties of sugarcane, with genetic modifications similar to previously released varieties. This action is expected to encourage the development of new transgenic varieties in other countries, such as India, China, Thailand, and the United States. With the continuous increase in the world population, excellent transgenic sugarcane varieties such as those with insect resistance, disease resistance, and drought tolerance will attract increasing attention from various countries, thus promoting sucrose production to meet people's demands. Recent marketing approvals for transgenic sugarcane in Indonesia and Brazil are expected to accelerate the development of new transgenic sugarcane varieties that can provide effective solutions to the challenges faced by crops in different agro-ecological conditions. These improved sugarcane varieties are most urgently needed in Yunnan province, China, because of its huge mountains and extremely complex ecological environment compared to elsewhere in the world. At the same time, advanced biotechnological methods can be used to produce r-proteins, biomolecules, and industrial and pharmaceutical products from the stem vacuoles of sugarcane. Sugarcane can also serve as a platform to produce edible vaccines and other useful products, making plant-based production cost-effective. The successful implementation of sugarcane-based vaccine production could significantly reduce global demand; with the potential to fuel the industry, stimulate economic growth, and reduce carbon emissions, sugarcane is poised to be a game changer.

The CRISPR-Cas9 gene editing method has shown promising results in rice and wheat. However, identifying appropriate sequences for modification and predicting the potential consequences on the genome is a challenging task in sugarcane. However, genome editing in sugarcane has been hindered by transgene silencing at pre–post transcriptional levels. To address this, efficient promoters can be used to regulate specific genome-editing tools, such as Cas9. Like other genome-editing nucleases, Cas9 may also have off-target effects, resulting in unwanted mutations. The gRNA–Cas9 complexes, in addition to cutting target

DNA, can also cleave off-target DNA sequences. Overall, several challenges still need to be addressed. The use of different Cas9 variants and other CRISPR-associated nucleases has the potential to be a powerful tool for successful genome editing in sugarcane in the near future.

Furthermore, to effectively utilize CRISPR/Cas9 technology for defending biotic and abiotic stress in sugarcane, a comprehensive understanding of the genomic sequences is indispensable. Fortunately, recent advances in genome sequencing technology and the availability of the reference genome, along with other transcriptome resources, can significantly enhance sugarcane genetic improvement. The CRISPR-Cas9 technology can efficiently edit multiple copies of sugarcane genes after knowing haplotypes clearly, including those with a high number of homologs and homeologs (orthologous and paralogous). Optimizing genome editing reagents and their delivery can facilitate the co-editing of numerous alleles, thereby increasing editing efficiency. Recent achievements in targeting multiple copies of genes in sugarcane have demonstrated that genome editing is possible for polyploid crops. Thus, this opens new avenues for future research to develop new elite varieties with enhanced tolerance to biotic and abiotic stressors. The utilization of gene editing technology is expected to increase soon due to its potential to significantly reduce the time and costs associated with commercial deregulation.

Author Contributions: T.K. and H.-B.L. are responsible for conceptualizing the relevant literature, critical writing, and editing and revising the manuscript. J.-G.W. wrote/revised the biotic stress section; C.-H.X., X.L., J.M., X.-Q.L., C.-Y.K., C.-J.L., X.-J.L., C.-Y.T., M.H.M.E. and X.-L.L. put forward some ideas and suggestions, repair this paper; H.-B.L. participated in revising and updating this manuscript. All authors have read and agreed to the published version of the manuscript.

Funding: This work was supported by grants from the Joint Special Project of Basic Agricultural Research of Yunnan Province (202101BD070001-025, 2017FG001-069); NSFC (31760411); The key research plan of Yunnan Province (202203AK140029); The Project of National Key Laboratory for Tropical Crop Breeding (NKLTCB-YAAS-2024-S03); the National Key R&D Program of China (2018YFD1000503); Yunnan Intelligence Union Program (202303AM140017); The Yunnan Seed Laboratory (202205AR070001-13); Yunnan Science and Technology Talent and platform program (202205AM070001); Yunnan Haizhi Station for Sugarcane Research Institute of Yunnan Academy of Agricultural Sciences; Project of Sanya Science and Technology Innovation (No. 2022KJCX17).

Acknowledgments: The authors express their sincere gratitude to the Sugarcane Research Institute of the Yunnan Academy of Agricultural Sciences, China, for their invaluable support. We also extend our thankful gratitude to the Talented Young Scientist Program (TYSP), funded by the Chinese government.

Conflicts of Interest: All the authors declare that they have no conflicts of interest.

References

1. Aslam, U.; Tabassum, B.; Nasir, I.A.; Khan, A.; Husnain, T. A Virus-Derived Short Hairpin RNA Confers Resistance against Sugarcane Mosaic Virus in Transgenic Sugarcane. *Transgenic Res.* **2018**, *27*, 203–210. [CrossRef] [PubMed]
2. Rahman, M.A.; Wu, W.; Yan, Y.; Bhuiyan, S.A. Overexpression of *TERF1* in Sugarcane Improves Tolerance to Drought Stress. *Crop Pasture Sci.* **2021**, *72*, 268–279. [CrossRef]
3. Grandis, A.; Fortirer, J.S.; Navarro, B.V.; de Oliveira, L.P.; Buckeridge, M.S. Biotechnologies to Improve Sugarcane Productivity in a Climate Change Scenario. *BioEnergy Res.* **2023**, *17*, 1–26. [CrossRef]
4. Piperidis, N.; D'Hont, A. Sugarcane Genome Architecture Decrypted with Chromosome-Specific Oligo Probes. *Plant J.* **2020**, *103*, 2039–2051. [CrossRef] [PubMed]
5. Vieira, M.L.C.; Almeida, C.B.; Oliveira, C.A.; Tacuatiá, L.O.; Munhoz, C.F.; Cauz-Santos, L.A.; Pinto, L.R.; Monteiro-Vitorello, C.B.; Xavier, M.A.; Forni-Martins, E.R. Revisiting Meiosis in Sugarcane: Chromosomal Irregularities and the Prevalence of Bivalent Configurations. *Front. Genet.* **2018**, *9*, 213. [CrossRef] [PubMed]
6. Zhang, J.; Zhang, X.; Tang, H.; Zhang, Q.; Hua, X.; Ma, X.; Zhu, F.; Jones, T.; Zhu, X.; Bowers, J.; et al. Allele-Defined Genome of the Autopolyploid Sugarcane *Saccharum spontaneum* L. *Nat. Genet.* **2018**, *50*, 1565–1573. [CrossRef] [PubMed]
7. Garsmeur, O.; Droc, G.; Antonise, R.; Grimwood, J.; Potier, B.; Aitken, K.; Jenkins, J.; Martin, G.; Charron, C.; Hervouet, C.; et al. A Mosaic Monoploid Reference Sequence for the Highly Complex Genome of Sugarcane. *Nat. Commun.* **2018**, *9*, 2638. [CrossRef] [PubMed]

8.   Souza, G.M.; Van Sluys, M.-A.; Lembke, C.G.; Lee, H.; Margarido, G.R.A.; Hotta, C.T.; Gaiarsa, J.W.; Diniz, A.L.; Oliveira, M.d.M.; Ferreira, S.d.S.; et al. Assembly of the 373k Gene Space of the Polyploid Sugarcane Genome Reveals Reservoirs of Functional Diversity in the World's Leading Biomass Crop. *Gigascience* **2019**, *8*, giz129. [CrossRef]

9.   Zhang, Q.; Qi, Y.; Pan, H.; Tang, H.; Wang, G.; Hua, X.; Wang, Y.; Lin, L.; Li, Z.; Li, Y.; et al. Genomic Insights into the Recent Chromosome Reduction of Autopolyploid Sugarcane *Saccharum spontaneum*. *Nat. Genet.* **2022**, *54*, 885–896. [CrossRef]

10. Wang, T.; Wang, B.; Hua, X.; Tang, H.; Zhang, Z.; Gao, R.; Qi, Y.; Zhang, Q.; Wang, G.; Yu, Z.; et al. A Complete Gap-Free Diploid Genome in *Saccharum* Complex and the Genomic Footprints of Evolution in the Highly Polyploid *Saccharum* Genus. *Nat. Plants* **2023**, *9*, 554–571. [CrossRef]

11. Kui, L.; Majeed, A.; Wang, X.; Yang, Z.; Chen, J.; He, L.; Di, Y.; Li, X.; Qian, Z.; Jiao, Y.; et al. A Chromosome-Level Genome Assembly for *Erianthus fulvus* Provides Insights into Its Biofuel Potential and Facilitates Breeding for Improvement of Sugarcane. *Plant Commun.* **2023**, *4*, 100562. [CrossRef]

12. Bao, Y.; Zhang, Q.; Huang, J.; Zhang, S.; Yao, W.; Yu, Z.; Deng, Z.; Yu, J.; Kong, W.; Yu, X.; et al. A Chromosomal-Scale Genome Assembly of Modern Cultivated Hybrid Sugarcane Provides Insights into Origination and Evolution. *Nat. Commun.* **2024**, *15*, 3041. [CrossRef]

13. Healey, A.L.; Garsmeur, O.; Lovell, J.T.; Shengquiang, S.; Sreedasyam, A.; Jenkins, J.; Plott, C.B.; Piperidis, N.; Pompidor, N.; Llaca, V.; et al. The Complex Polyploid Genome Architecture of Sugarcane. *Nature* **2024**, *628*, 804–810. [CrossRef]

14. Bower, R.; Birch, R.G. Transgenic Sugarcane Plants via Microprojectile Bombardment. *Plant J.* **1992**, *2*, 409–416. [CrossRef]

15. Arencibia, A.D.; Carmona, E.R.; Tellez, P.; Chan, M.-T.; Yu, S.-M.; Trujillo, L.E.; Oramas, P. An Efficient Protocol for Sugarcane (*Saccharum* spp. L.) Transformation Mediated by *Agrobacterium tumefaciens*. *Transgenic Res.* **1998**, *7*, 213–222. [CrossRef]

16. Budeguer, F.; Enrique, R.; Perera, M.F.; Racedo, J.; Castagnaro, A.P.; Noguera, A.S.; Welin, B. Genetic Transformation of Sugarcane, Current Status and Future Prospects. *Front. Plant Sci.* **2021**, *12*, 768609. [CrossRef]

17. Li, A.-M.; Chen, Z.-L.; Liao, F.; Zhao, Y.; Qin, C.-X.; Wang, M.; Pan, Y.-Q.; Wei, S.-L.; Huang, D.-L. Sugarcane Borers: Species, Distribution, Damage and Management Options. *J. Pest. Sci.* **2024**, 1–31. [CrossRef]

18. Budeguer, F.; Racedo, J.; Enrique, R.; Perera, M.F.; Ostengo, S.; Noguera, A.S. Transgenic Sugarcane for the Sustainable Management of the Sugarcane Borer *Diatraea saccharalis*. *Sugar Ind.* **2024**, *149*, 133–138. [CrossRef]

19. Mulwa, R.M.S.; Mwanza, L.M. Review-Biotechnology Approaches to Developing Herbicide Tolerance/Selectivity in Crops. *Afr. J. Biotechnol.* **2006**, *5*, 396–404.

20. Falco, M.C.; Tulmann Neto, A.; Ulian, E.C. Transformation and Expression of a Gene for Herbicide Resistance in a Brazilian Sugarcane. *Plant Cell Rep.* **2000**, *19*, 1188–1194. [CrossRef]

21. Wang, W.Z.; Yang, B.P.; Feng, X.Y.; Cao, Z.Y.; Feng, C.L.; Wang, J.G.; Xiong, G.R.; Shen, L.B.; Zeng, J.; Zhao, T.T.; et al. Development and Characterization of Transgenic Sugarcane with Insect Resistance and Herbicide Tolerance. *Front. Plant Sci.* **2017**, *8*, 1535. [CrossRef]

22. Nasir, I.; Tabassum, B.; Qamar, Z.; Javed, M.; Tariq, M.; Farooq, A.; Butt, S.; Qayyum, A.; Husnain, T. Herbicide-Tolerant Sugarcane (*Saccharum officinarum* L.) Plants: An Unconventional Method of Weed Removal. *Turk. J. Biol.* **2014**, *38*, 439–449. [CrossRef]

23. Racedo, J.; Noguera, A.S.; Castagnaro, A.P.; Perera, M.F. Biotechnological Strategies Adopted for Sugarcane Disease Management in Tucumán, Argentina. *Plants* **2023**, *12*, 3994. [CrossRef]

24. Qamar, Z.; Nasir, I.A.; Abouhaidar, M.G.; Hefferon, K.L.; Rao, A.Q.; Latif, A.; Ali, Q.; Anwar, S.; Rashid, B.; Shahid, A.A. Novel Approaches to Circumvent the Devastating Effects of Pests on Sugarcane. *Sci. Rep.* **2021**, *11*, 12428. [CrossRef] [PubMed]

25. Zan, F.; Wu, Z.; Wang, W.; Hu, X.; Feng, L.; Liu, X.; Liu, J.; Zhao, L.; Wu, C.; Zhang, S.; et al. Strigolactones in Sugarcane Growth and Development. *Agronomy* **2023**, *13*, 1086. [CrossRef]

26. Wang, W.; Javed, T.; Shen, L.; Sun, T.; Yang, B.; Zhang, S. Establishment of an Efficient Sugarcane Transformation System via Herbicide-Resistant *CP4-EPSPS* Gene Selection. *Plants* **2024**, *13*, 852. [CrossRef] [PubMed]

27. Noguera, A.; Enrique, R.; Ostengo, S.; Perera, M.F.; Racedo, J.; Costilla, D.; Zossi, S.; Cuenya, M.I.; Paula, M.; Welin, B.; et al. Development of the Transgenic Sugarcane Event TUC 87-3RG Resistant to Glyphosate. *Proc. Int. Soc. Sugar Cane Technol.* **2019**, *30*, 493–501.

28. Leibbrandt, N.B.; Snyman, S.J. Stability of Gene Expression and Agronomic Performance of a Transgenic Herbicide-Resistant Sugarcane Line in South Africa. *Crop Sci.* **2003**, *43*, 671–677. [CrossRef]

29. Enríquez-Obregón, G.A.; Vázquez-Padrón, R.I.; Prieto-Samsonov, D.L.; De la Riva, G.A.; Selman-Housein, G. Herbicide-Resistant Sugarcane (*Saccharum officinarum* L.) Plants by *Agrobacterium*-Mediated Transformation. *Planta* **1998**, *206*, 20–27. [CrossRef]

30. Manickavasagam, M.; Ganapathi, A.; Anbazhagan, V.R.; Sudhakar, B.; Selvaraj, N.; Vasudevan, A.; Kasthurirengan, S. *Agrobacterium*-Mediated Genetic Transformation and Development of Herbicide-Resistant Sugarcane (*Saccharum* Species Hybrids) Using Axillary Buds. *Plant Cell Rep.* **2004**, *23*, 134–143. [CrossRef]

31. Wang, W.Z.; Yang, B.P.; Feng, C.L.; Wang, J.G.; Xiong, G.R.; Zhao, T.T.; Zhang, S.Z. Efficient Sugarcane Transformation via Bar Gene Selection. *Trop. Plant Biol.* **2017**, *10*, 77–85. [CrossRef]

32. Ostengo, S.; Serino, G.; Perera, M.F.; Racedo, J.; Mamaní González, S.Y.; Yáñez Cornejo, F.; Cuenya, M.I. Sugarcane Breeding, Germplasm Development and Supporting Genetic Research in Argentina. *Sugar Tech* **2022**, *24*, 166–180. [CrossRef]

33. Oz, M.T.; Altpeter, A.; Karan, R.; Merotto, A.; Altpeter, F. CRISPR/Cas9-mediated Multi-Allelic Gene Targeting in Sugarcane Confers Herbicide Tolerance. *Front. Genome Ed.* **2021**, *3*, 673566. [CrossRef] [PubMed]

34. Souza, T.P.; Dias, R.O.; Silva-Filho, M.C. Defense-Related Proteins Involved in Sugarcane Responses to Biotic Stress. *Genet. Mol. Biol.* **2017**, *40*, 360–372. [CrossRef]

35. Chu, N.; Zhou, J.-R.; Rott, P.C.; Li, J.; Fu, H.-Y.; Huang, M.-T.; Zhang, H.-L.; Gao, S.-J. *ScPR1* Plays a Positive Role in the Regulation of Resistance to Diverse Stresses in Sugarcane (*Saccharum* spp.) and *Arabidopsis thaliana*. *Ind. Crops Prod.* **2022**, *180*, 114736. [CrossRef]

36. Zhou, J.-R.; Sun, H.-D.; Ali, A.; Rott, P.C.; Javed, T.; Fu, H.-Y.; Gao, S.-J. Quantitative Proteomic Analysis of the Sugarcane Defense Responses Incited by *Acidovorax avenae* subsp. *avenae Causing Red Stripe*. *Ind. Crops Prod.* **2021**, *162*, 113275. [CrossRef]

37. Javed, T.; Gao, S.-J. WRKY Transcription Factors in Plant Defense. *Trends Genet.* **2023**, *39*, 787–801. [CrossRef] [PubMed]

38. Javed, T.; Shabbir, R.; Ali, A.; Afzal, I.; Zaheer, U.; Gao, S.-J. Transcription Factors in Plant Stress Responses: Challenges and Potential for Sugarcane Improvement. *Plants* **2020**, *9*, 491. [CrossRef] [PubMed]

39. Wang, Y.; Zhang, J.; Wang, R.; Hou, Y.; Fu, H.; Xie, Y.; Gao, S.; Wang, J. Unveiling Sugarcane Defense Response to *Mythimna separata* Herbivory by a Combination of Transcriptome and Metabolic Analyses. *Pest. Manag. Sci.* **2021**, *77*, 4799–4809. [CrossRef]

40. Hu, Z.-T.; Ntambo, M.S.; Zhao, J.-Y.; Javed, T.; Shi, Y.; Fu, H.-Y.; Huang, M.-T.; Gao, S.-J. Genetic Divergence and Population Structure of *Xanthomonas albilineans* Strains Infecting *Saccharum* spp. Hybrid and *Saccharum officinarum*. *Plants* **2023**, *12*, 1937. [CrossRef]

41. Verma, K.K.; Song, X.-P.; Budeguer, F.; Nikpay, A.; Enrique, R.; Singh, M.; Zhang, B.-Q.; Wu, J.-M.; Li, Y.-R. Genetic Engineering: An Efficient Approach to Mitigating Biotic and Abiotic Stresses in Sugarcane Cultivation. *Plant Signal. Behav.* **2022**, *17*, 2108253. [CrossRef]

42. Cursi, D.E.; Castillo, R.O.; Tarumoto, Y.; Umeda, M.; Tippayawat, A.; Ponragdee, W.; Racedo, J.; Perera, M.F.; Hoffmann, H.P.; Carneiro, M.S. Origin, Genetic Diversity, Conservation, and Traditional and Molecular Breeding Approaches in Sugarcane. In *Cash Crops: Genetic Diversity, Erosion, Conservation and Utilization*; Priyadarshan, P.M., Jain, S.M., Eds.; Springer International Publishing: Cham, Switzerland, 2022; pp. 83–116. ISBN 978-3-030-74926-2.

43. Ingelbrecht, I.L.; Irvine, J.E.; Mirkov, T.E. Posttranscriptional Gene Silencing in Transgenic Sugarcane. Dissection of Homology-Dependent Virus Resistance in a Monocot That Has a Complex Polyploid Genome. *Plant Physiol.* **1999**, *119*, 1187–1198. [CrossRef]

44. McQualter, R.B.; Dale, J.L.; Harding, R.M.; McMahon, J.A.; Smith, G.R. Production and Evaluation of Transgenic Sugarcane Containing a Fiji Disease Virus (FDV) Genome Segment S9-Derived Synthetic Resistance Gene. *Aust. J. Agric. Res.* **2004**, *55*, 139–145. [CrossRef]

45. Zhu, Y.J.; McCafferty, H.; Osterman, G.; Lim, S.; Agbayani, R.; Lehrer, A.; Schenck, S.; Komor, E. Genetic Transformation with Untranslatable Coat Protein Gene of Sugarcane Yellow Leaf Virus Reduces Virus Titers in Sugarcane. *Transgenic Res.* **2011**, *20*, 503–512. [CrossRef]

46. Apriasti, R.; Widyaningrum, S.; Hidayati, W.N.; Sawitri, W.D.; Darsono, N.; Hase, T.; Sugiharto, B. Full Sequence of the Coat Protein Gene Is Required for the Induction of Pathogen-Derived Resistance against Sugarcane Mosaic Virus in Transgenic Sugarcane. *Mol. Biol. Rep.* **2018**, *45*, 2749–2758. [CrossRef]

47. Guo, J.; Gao, S.; Lin, Q.; Wang, H.; Que, Y.; Xu, L. Transgenic Sugarcane Resistant to *Sorghum Mosaic Virus* Based on Coat Protein Gene Silencing by RNA Interference. *BioMed Res. Int.* **2015**, *2015*, e861907. [CrossRef]

48. Gilbert, R.A.; Gallo-Meagher, M.; Comstock, J.C.; Miller, J.D.; Jain, M.; Abouzid, A. Agronomic Evaluation of Sugarcane Lines Transformed for Resistance to Sugarcane Mosaic Virus Strain E. *Crop Sci.* **2005**, *45*, 2060–2067. [CrossRef]

49. Gilbert, R.A.; Glynn, N.C.; Comstock, J.C.; Davis, M.J. Agronomic Performance and Genetic Characterization of Sugarcane Transformed for Resistance to Sugarcane Yellow Leaf Virus. *Field Crops Res.* **2009**, *111*, 39–46. [CrossRef]

50. Hidayati, W.N.; Apriasti, R.; Addy, H.S.; Sugiharto, B. Distinguishing Resistances of Transgenic Sugarcane Generated from RNA Interference and Pathogen-derived Resistance Approaches to Combating Sugarcane Mosaic Virus. *Indones. J. Biotechnol.* **2021**, *26*, 107–114. [CrossRef]

51. Wang, W.; Wang, J.; Feng, X.; Shen, L.; Feng, C.; Zhao, T.; Xiao, H.; Li, S.; Zhang, S. Breeding of Virus-Resistant Transgenic Sugarcane by the Integration of the *Pac1* Gene. *Front. Sustain. Food Syst.* **2022**, *6*, 925839. [CrossRef]

52. Nayyar, S.; Sharma, B.K.; Kaur, A.; Kalia, A.; Sanghera, G.S.; Thind, K.S.; Yadav, I.S.; Sandhu, J.S. Red Rot Resistant Transgenic Sugarcane Developed through Expression of *β-1,3-glucanase* Gene. *PLoS ONE* **2017**, *12*, e0179723. [CrossRef]

53. Tariq, M.; Khan, A.; Tabassum, B.; Toufiq, N.; Bhatti, M.U.; Riaz, S.; Nasir, I.A.; Husnain, T. Antifungal Activity of Chitinase II against *Colletotrichum falcatum* Went. Causing Red Rot Disease in Transgenic Sugarcane. *Turk. J. Biol.* **2018**, *42*, 45–53. [CrossRef] [PubMed]

54. Parvaiz, A.; Mustafa, G.; Khan, M.S.; Ali, M.A. Over-Expression of Endogenous *SUGARWIN* Genes Exalted Tolerance against *Colletotrichum* Infection in Sugarcane. *Plants* **2021**, *10*, 869. [CrossRef]

55. Maeda, S.; Ackley, W.; Yokotani, N.; Sasaki, K.; Ohtsubo, N.; Oda, K.; Mori, M. Enhanced Resistance to Fungal and Bacterial Diseases due to Overexpression of *BSR1*, a Rice *RLCK*, in Sugarcane, Tomato, and Torenia. *Int. J. Mol. Sci.* **2023**, *24*, 3644. [CrossRef] [PubMed]

56. Zhang, L.; Xu, J.; Birch, R.G. Engineered Detoxification Confers Resistance against a Pathogenic Bacterium. *Nat. Biotechnol.* **1999**, *17*, 1021–1024. [CrossRef]

57. Babu, K.H.; Devarumath, R.M.; Thorat, A.S.; Nalavade, V.M.; Saindane, M.; Appunu, C.; Suprasanna, P. Sugarcane Transgenics: Developments and Opportunities. In *Genetically Modified Crops: Current Status, Prospects and Challenges Volume 1*; Kavi Kishor, P.B., Rajam, M.V., Pullaiah, T., Eds.; Springer: Singapore, 2021; pp. 241–265. ISBN 9789811558979.

58. Sawitri, W.D.; Harmoko, R.; Sugiharto, B. Induction of Resistance against Sugarcane Mosaic Virus by Pathogen-Derived Resistance and RNA Interference Methods in Transgenic Sugarcane. *AIP Conf. Proc.* **2024**, *3080*, 020002. [CrossRef]

59. Viswanathan, C.; Anburaj, J.; Prabu, G. Identification and Validation of Sugarcane Streak Mosaic Virus-Encoded microRNAs and Their Targets in Sugarcane. *Plant Cell Rep.* **2014**, *33*, 265–276. [CrossRef]

60. Widyaningrum, S.; Pujiasih, D.R.; Sholeha, W.; Harmoko, R.; Sugiharto, B. Induction of Resistance Tolerance to Sugarcane Mosaic Virus by RNA Interference Targeting Coat Protein Gene Silencing in Transgenic Sugarcane. *Mol. Biol. Rep.* **2021**, *48*, 3047–3054. [CrossRef]

61. Yao, W.; Ruan, M.; Qin, L.; Yang, C.; Chen, R.; Chen, B.; Zhang, M. Field Performance of Transgenic Sugarcane Lines Resistant to Sugarcane Mosaic Virus. *Front. Plant Sci.* **2017**, *8*, 104. [CrossRef]

62. Noguera, A.; Enrique, R.; Perera, M.F.; Ostengo, S.; Racedo, J.; Costilla, D.; Zossi, S.; Cuenya, M.I.; Filippone, M.P.; Welin, B.; et al. Genetic Characterization and Field Evaluation to Recover Parental Phenotype in Transgenic Sugarcane: A Step toward Commercial Release. *Mol. Breed.* **2015**, *35*, 115. [CrossRef]

63. Nerkar, G.; Thorat, A.; Sheelavantmath, S.; Kassa, H.B.; Devarumath, R. Genetic Transformation of Sugarcane and Field Performance of Transgenic Sugarcane. In *Biotechnologies of Crop Improvement, Volume 2: Transgenic Approaches*; Gosal, S.S., Wani, S.H., Eds.; Springer International Publishing: Cham, Switzerland, 2018; pp. 207–226. ISBN 978-3-319-90650-8.

64. Weng, L.-X.; Deng, H.-H.; Xu, J.-L.; Li, Q.; Zhang, Y.-Q.; Jiang, Z.-D.; Li, Q.-W.; Chen, J.-W.; Zhang, L.-H. Transgenic Sugarcane Plants Expressing High Levels of Modified *cry1Ac* Provide Effective Control against Stem Borers in Field Trials. *Transgenic Res.* **2011**, *20*, 759–772. [CrossRef]

65. Iqbal, A.; Khan, R.S.; Khan, M.A.; Gul, K.; Jalil, F.; Shah, D.A.; Rahman, H.; Ahmed, T. Genetic Engineering Approaches for Enhanced Insect Pest Resistance in Sugarcane. *Mol. Biotechnol.* **2021**, *63*, 557–568. [CrossRef] [PubMed]

66. Schneider, V.K.; Soares-Costa, A.; Chakravarthi, M.; Ribeiro, C.; Chabregas, S.M.; Falco, M.C.; Henrique-Silva, F. Transgenic Sugarcane Overexpressing *CaneCPI*-1 Negatively Affects the Growth and Development of the Sugarcane Weevil *Sphenophorus levis*. *Plant Cell Rep.* **2017**, *36*, 193–201. [CrossRef] [PubMed]

67. Gill, R.; Malhotra, P.K.; Gosal, S.S. Direct Plant Regeneration from Cultured Young Leaf Segments of Sugarcane. *Plant Cell Tissue Organ Cult.* **2006**, *84*, 227–231. [CrossRef]

68. Gosal, S.S.; Wani, S.H. Plant Genetic Transformation and Transgenic Crops: Methods and Applications. In *Biotechnologies of Crop Improvement, Volume 2: Transgenic Approaches*; Gosal, S.S., Wani, S.H., Eds.; Springer International Publishing: Cham, Switzerland, 2018; pp. 1–23. ISBN 978-3-319-90650-8.

69. Sétamou, M.; Bernal, J.S.; Legaspi, J.C.; Mirkov, T.E.; Legaspi, B.C. Evaluation of *Lectin*-Expressing Transgenic Sugarcane against Stalkborers (Lepidoptera: Pyralidae): Effects on Life History Parameters. *J. Econ. Entomol.* **2002**, *95*, 469–477. [CrossRef] [PubMed]

70. Deng, Z.-N.; Wei, Y.-W.; Lü, W.-L.; Li, Y.-R. Fusion Insect-Resistant Gene Mediated by Matrix Attachment Region (MAR) Sequence in Transgenic Sugarcane. *Sugar Tech* **2008**, *10*, 87–90. [CrossRef]

71. Christy, L.A.; Arvinth, S.; Saravanakumar, M.; Kanchana, M.; Mukunthan, N.; Srikanth, J.; Thomas, G.; Subramonian, N. Engineering Sugarcane Cultivars with Bovine Pancreatic Trypsin Inhibitor (*Aprotinin*) Gene for Protection against Top Borer (*Scirpophaga excerptalis* Walker). *Plant Cell Rep.* **2009**, *28*, 175–184. [CrossRef]

72. Riaz, S.; Nasir, I.A.; Bhatti, M.U.; Adeyinka, O.S.; Toufiq, N.; Yousaf, I.; Tabassum, B. Resistance to *Chilo infuscatellus* (Lepidoptera: Pyraloidea) in Transgenic Lines of Sugarcane Expressing *Bacillus thuringiensis* Derived Vip3A Protein. *Mol. Biol. Rep.* **2020**, *47*, 2649–2658. [CrossRef] [PubMed]

73. Dessoky, E.S.; Ismail, R.M.; Elarabi, N.I.; Abdelhadi, A.A.; Abdallah, N.A. Improvement of Sugarcane for Borer Resistance Using *Agrobacterium* Mediated Transformation of *cry1Ac* Gene. *GM Crops Food* **2021**, *12*, 47–56. [CrossRef] [PubMed]

74. Cristofoletti, P.T.; Kemper, E.L.; Capella, A.N.; Carmago, S.R.; Cazoto, J.L.; Ferrari, F.; Galvan, T.L.; Gauer, L.; Monge, G.A.; Nishikawa, M.A.; et al. Development of Transgenic Sugarcane Resistant to Sugarcane Borer. *Trop. Plant Biol.* **2018**, *11*, 17–30. [CrossRef]

75. Zhao, M.; Zhou, Y.; Su, L.; Li, G.; Huang, Z.; Huang, D.; Wu, W.; Zhao, Y. Expression of *Pinellia pedatisecta* Agglutinin *PPA* Gene in Transgenic Sugarcane Led to Stomata Patterning Change and Resistance to Sugarcane Woolly Aphid, *Ceratovacuna lanigera* Zehntner. *Int. J. Mol. Sci.* **2022**, *23*, 7195. [CrossRef]

76. Braga, D.P.V.; Arrigoni, E.D.B.; Silva-Filho, M.C.; Ulian, E.C. Expression of the *Cry1Ab* Protein in Genetically Modified Sugarcane for the Control of *Diatraea saccharalis* (Lepidoptera: Crambidae). *J. New Seeds* **2003**, *5*, 209–221. [CrossRef]

77. Weng, L.-X.; Deng, H.; Xu, J.-L.; Li, Q.; Wang, L.-H.; Jiang, Z.; Zhang, H.B.; Li, Q.; Zhang, L.-H. Regeneration of Sugarcane Elite Breeding Lines and Engineering of Stem Borer Resistance. *Pest. Manag. Sci.* **2006**, *62*, 178–187. [CrossRef]

78. Zhangsun, D.; Luo, S.; Chen, R.; Tang, K. Improved *Agrobacterium*-Mediated Genetic Transformation of *GNA* Transgenic Sugarcane. *Biologia* **2007**, *62*, 386–393. [CrossRef]

79. Xu, J.-S.; Gao, S.; Xu, L.; Chen, R. Construction of Expression Vector of *CryIA(c)* Gene and Its Transformation in Sugarcane. *Sugar Tech* **2008**, *10*, 269–273. [CrossRef]

80. Ribeiro, C.W.; Soares-Costa, A.; Falco, M.C.; Chabregas, S.M.; Ulian, E.C.; Cotrin, S.S.; Carmona, A.K.; Santana, L.A.; Oliva, M.L.V.; Henrique-Silva, F. Production of a His-Tagged Canecystatin in Transgenic Sugarcane and Subsequent Purification. *Biotechnol. Prog.* **2008**, *24*, 1060–1066. [CrossRef]

81. Kalunke, R.M.; Kolge, A.M.; Babu, K.H.; Prasad, D.T. *Agrobacterium* Mediated Transformation of Sugarcane for Borer Resistance Using *Cry 1Aa3* Gene and One-Step Regeneration of Transgenic Plants. *Sugar Tech* **2009**, *11*, 355–359. [CrossRef]

82. Arvinth, S.; Arun, S.; Selvakesavan, R.K.; Srikanth, J.; Mukunthan, N.; Ananda Kumar, P.; Premachandran, M.N.; Subramonian, N. Genetic Transformation and Pyramiding of Aprotinin-Expressing Sugarcane with *cry1Ab* for Shoot Borer (*Chilo infuscatellus*) Resistance. *Plant Cell Rep.* **2010**, *29*, 383–395. [CrossRef]

83. Gao, S.; Yang, Y.; Wang, C.; Guo, J.; Zhou, D.; Wu, Q.; Su, Y.; Xu, L.; Que, Y. Transgenic Sugarcane with a *cry1Ac* Gene Exhibited Better Phenotypic Traits and Enhanced Resistance against Sugarcane Borer. *PLoS ONE* **2016**, *11*, e0153929. [CrossRef]

84. Islam, N.; Laksana, C.; Chanprame, S. *Agrobacterium*-mediated Transformation and Expression of *Bt* Gene in Transgenic Sugarcane. *J. Int. Soc. Southeast Asian Agric. Sci.* **2016**, *22*, 84–95.

85. Shibao, P.Y.T.; Santos-Júnior, C.D.; Santiago, A.C.; Mohan, C.; Miguel, M.C.; Toyama, D.; Vieira, M.A.S.; Narayanan, S.; Figueira, A.; Carmona, A.K.; et al. Sugarcane Cystatins: From Discovery to Biotechnological Applications. *Int. J. Biol. Macromol.* **2021**, *167*, 676–686. [CrossRef] [PubMed]

86. Zhou, D.; Liu, X.; Gao, S.; Guo, J.; Su, Y.; Ling, H.; Wang, C.; Li, Z.; Xu, L.; Que, Y. Foreign *cry1Ac* Gene Integration and Endogenous Borer Stress-Related Genes Synergistically Improve Insect Resistance in Sugarcane. *BMC Plant Biol.* **2018**, *18*, 342. [CrossRef] [PubMed]

87. Gao, S.; Yang, Y.; Xu, L.; Guo, J.; Su, Y.; Wu, Q.; Wang, C.; Que, Y. Particle Bombardment of the *cry2A* Gene Cassette Induces Stem Borer Resistance in Sugarcane. *Int. J. Mol. Sci.* **2018**, *19*, 1692. [CrossRef]

88. Cheavegatti-Gianotto, A.; Gentile, A.; Oldemburgo, D.A.; Merheb, G.d.A.; Sereno, M.L.; Lirette, R.P.; Ferrseira, T.H.S.; de Oliveira, W.S. Lack of Detection of *Bt* Sugarcane *Cry1Ab* and *nptII* DNA and Proteins in Sugarcane Processing Products Including Raw Sugar. *Front. Bioeng. Biotechnol.* **2018**, *6*, 24. [CrossRef] [PubMed]

89. Gianotto, A.C.; Rocha, M.S.; Cutri, L.; Lopes, F.C.; Dal'Acqua, W.; Hjelle, J.J.; Lirette, R.P.; Oliveira, W.S.; Sereno, M.L. The Insect-Protected CTC91087-6 Sugarcane Event Expresses *Cry1Ac* Protein Preferentially in Leaves and Presents Compositional Equivalence to Conventional Sugarcane. *GM Crops Food* **2019**, *10*, 208–219. [CrossRef] [PubMed]

90. Koerniati, S.; Sukmadjaja, D.; Samudra, I.M. C Synthetic Gene of *Cry1Ab-Cry1Ac* Fusion to Generate Resistant Sugarcane to Shoot or Stem Borer. *IOP Conf. Ser. Earth Environ. Sci.* **2020**, *418*, 012069. [CrossRef]

91. Punithavalli, M.; Jebamalaimary, A. Inhibitory Activities of Proteinase Inhibitors on Developmental Characteristics of Sugarcane *Chilo infuscatellus* (Snellen). *Phytoparasitica* **2019**, *47*, 43–53. [CrossRef]

92. Punithavalli, M. Spatial Distribution of Proteinase Inhibitors among Diverse Groups of Sugarcane and Their Interaction with Sugarcane Borers. *Indian J. Entomol.* **2022**, *84*, 693–696. [CrossRef]

93. Salgado, L.D.; Wilson, B.E.; Villegas, J.M.; Richard, R.T.; Penn, H.J. Resistance to the Sugarcane Borer (Lepidoptera: Crambidae) in Louisiana Sugarcane Cultivars. *Environ. Entomol.* **2022**, *51*, 196–203. [CrossRef] [PubMed]

94. Dixit, S.; Sivalingam, P.N.; Baskaran, R.K.M.; Senthil-Kumar, M.; Ghosh, P.K. Plant Responses to Concurrent Abiotic and Biotic Stress: Unravelling Physiological and Morphological Mechanisms. *Plant Physiol. Rep.* **2024**, *29*, 6–17. [CrossRef]

95. Zhang, H.; Zhu, J.; Gong, Z.; Zhu, J.-K. Abiotic Stress Responses in Plants. *Nat. Rev. Genet.* **2022**, *23*, 104–119. [CrossRef]

96. Wei, Y.-S.; Zhao, J.-Y.; Javed, T.; Ali, A.; Huang, M.-T.; Fu, H.-Y.; Zhang, H.-L.; Gao, S.-J. Insights into Reactive Oxygen Species Production-Scavenging System Involved in Sugarcane Response to *Xanthomonas albilineans* Infection under Drought Stress. *Plants* **2024**, *13*, 862. [CrossRef] [PubMed]

97. Mishra, N.; Jiang, C.; Chen, L.; Paul, A.; Chatterjee, A.; Shen, G. Achieving Abiotic Stress Tolerance in Plants through Antioxidative Defense Mechanisms. *Front. Plant Sci.* **2023**, *14*, 1110622. [CrossRef] [PubMed]

98. Tardieu, F. Any Trait or Trait-Related Allele Can Confer Drought Tolerance: Just Design the Right Drought Scenario. *J. Exp. Bot.* **2012**, *63*, 25–31. [CrossRef] [PubMed]

99. Cominelli, E.; Conti, L.; Tonelli, C.; Galbiati, M. Challenges and Perspectives to Improve Crop Drought and Salinity Tolerance. *New Biotechnol.* **2013**, *30*, 355–361. [CrossRef] [PubMed]

100. Reis, R.R.; Andrade Dias Brito da Cunha, B.; Martins, P.K.; Martins, M.T.B.; Alekcevetch, J.C.; Chalfun-Júnior, A.; Andrade, A.C.; Ribeiro, A.P.; Qin, F.; Mizoi, J.; et al. Induced Over-Expression of *AtDREB2A CA* Improves Drought Tolerance in Sugarcane. *Plant Sci.* **2014**, *221–222*, 59–68. [CrossRef] [PubMed]

101. Mbambalala, N.; Panda, S.K.; van der Vyver, C. Overexpression of *AtBBX29* Improves Drought Tolerance by Maintaining Photosynthesis and Enhancing the Antioxidant and Osmolyte Capacity of Sugarcane Plants. *Plant Mol. Biol. Rep.* **2021**, *39*, 419–433. [CrossRef]

102. Kumar, T.; Uzma; Khan, M.R.; Abbas, Z.; Ali, G.M. Genetic Improvement of Sugarcane for Drought and Salinity Stress Tolerance Using *Arabidopsis* Vacuolar Pyrophosphatase (*AVP1*) Gene. *Mol. Biotechnol.* **2014**, *56*, 199–209. [CrossRef] [PubMed]

103. RAZA, G.; Ali, K.; Ashraf, M.; Mansoor, S.; Javid, M.; Asad, S. Overexpression of an *H⁺-PPase* Gene from *Arabidopsis* in Sugarcane Improves drought Tolerance, Plant Growth, and Photosynthetic Responses. *Turk. J. Biol.* **2016**, *40*, 109–119. [CrossRef]

104. Ramiro, D.A.; Melotto-Passarin, D.M.; Barbosa, M.d.A.; dos Santos, F.; Gomez, S.G.P.; Massola Júnior, N.S.; Lam, E.; Carrer, H. Expression of Arabidopsis Bax Inhibitor-1 in Transgenic Sugarcane Confers Drought Tolerance. *Plant Biotechnol. J.* **2016**, *14*, 1826–1837. [CrossRef]

105. Guerzoni, J.T.S.; Belintani, N.G.; Moreira, R.M.P.; Hoshino, A.A.; Domingues, D.S.; Filho, J.C.B.; Vieira, L.G.E. Stress-Induced Δ1-Pyrroline-5-Carboxylate Synthetase (*P5CS*) Gene Confers Tolerance to Salt Stress in Transgenic Sugarcane. *Acta Physiol. Plant.* **2014**, *36*, 2309–2319. [CrossRef]

106. Li, J.; Phan, T.-T.; Li, Y.-R.; Xing, Y.-X.; Yang, L.-T. Isolation, Transformation and Overexpression of Sugarcane *SoP5CS* Gene for Drought Tolerance Improvement. *Sugar Tech* **2018**, *20*, 464–473. [CrossRef]

107. Mohanan, M.V.; Pushpanathan, A.; Padmanabhan, S.; Sasikumar, T.; Jayanarayanan, A.N.; Selvarajan, D.; Ramalingam, S.; Ram, B.; Chinnaswamy, A. Overexpression of Glyoxalase III Gene in Transgenic Sugarcane Confers Enhanced Performance under Salinity Stress. *J. Plant Res.* **2021**, *134*, 1083–1094. [CrossRef] [PubMed]

108. Augustine, S.M.; Cherian, A.V.; Syamaladevi, D.P.; Subramonian, N. *Erianthus arundinaceus* HSP70 (*EaHSP70*) Acts as a Key Regulator in the Formation of Anisotropic Interdigitation in Sugarcane (*Saccharum* spp. Hybrid) in Response to Drought Stress. *Plant Cell Physiol.* **2015**, *56*, 2368–2380. [CrossRef] [PubMed]

109. Belintani, N.G.; Guerzoni, J.T.S.; Moreira, R.M.P.; Vieira, L.G.E. Improving Low-Temperature Tolerance in Sugarcane by Expressing the *Ipt* Gene under a Cold Inducible Promoter. *Biol. Plant* **2012**, *56*, 71–77. [CrossRef]

110. Chen, J.-Y.; Khan, Q.; Sun, B.; Tang, L.-H.; Yang, L.-T.; Zhang, B.-Q.; Xiu, X.-Y.; Dong, D.-F.; Li, Y.-R. Overexpression of Sugarcane *SoTUA* Gene Enhances Cold Tolerance in Transgenic Sugarcane. *Agron. J.* **2021**, *113*, 4993–5005. [CrossRef]

111. Zhang, S.-Z.; Yang, B.-P.; Feng, C.-L.; Chen, R.-K.; Luo, J.-P.; Cai, W.-W.; Liu, F.-H. Expression of the *Grifola frondosa* Trehalose Synthase Gene and Improvement of Drought-Tolerance in Sugarcane (*Saccharum officinarum* L.). *J. Integr. Plant Biol.* **2006**, *48*, 453–459. [CrossRef]

112. Augustine, S.M.; Ashwin Narayan, J.; Syamaladevi, D.P.; Appunu, C.; Chakravarthi, M.; Ravichandran, V.; Tuteja, N.; Subramonian, N. Overexpression of *EaDREB2* and Pyramiding of *EaDREB2* with the Pea DNA Helicase Gene (*PDH45*) Enhance Drought and Salinity Tolerance in Sugarcane (*Saccharum* spp. Hybrid). *Plant Cell Rep.* **2015**, *34*, 247–263. [CrossRef]

113. Moran, J.F.; Becana, M.; Iturbe-Ormaetxe, I.; Frechilla, S.; Klucas, R.V.; Aparicio-Tejo, P. Drought Induces Oxidative Stress in Pea Plants. *Planta* **1994**, *194*, 346–352. [CrossRef]

114. Narayan, J.A.; Manoj, V.M.; Nerkar, G.; Chakravarthi, M.; Dharshini, S.; Subramonian, N.; Premachandran, M.N.; Valarmathi, R.; Kumar, R.A.; Gomathi, R.; et al. Transgenic Sugarcane with Higher Levels of *BRK1* Showed Improved Drought Tolerance. *Plant Cell Rep.* **2023**, *42*, 1611–1628. [CrossRef]

115. Mohanan, M.V.; Thelakat Sasikumar, S.P.; Jayanarayanan, A.N.; Selvarajan, D.; Ramanathan, V.; Shivalingamurthy, S.G.; Raju, G.; Govind, H.; Chinnaswamy, A. Transgenic Sugarcane Overexpressing *Glyoxalase III* Improved Germination and Biomass Production at Formative Stage under Salinity and Water-Deficit Stress Conditions. *3 Biotech* **2024**, *14*, 52. [CrossRef]

116. Zhao, X.; Jiang, Y.; Liu, Q.; Yang, H.; Wang, Z.; Zhang, M. Effects of Drought-Tolerant *Ea-DREB2B* Transgenic Sugarcane on Bacterial Communities in Soil. *Front. Microbiol.* **2020**, *11*, 704. [CrossRef] [PubMed]

117. Zhao, X.; Liu, Q.; Xie, S.; Jiang, Y.; Yang, H.; Wang, Z.; Zhang, M. Response of Soil Fungal Community to Drought-Resistant *Ea-DREB2B* Transgenic Sugarcane. *Front. Microbiol.* **2020**, *11*, 562775. [CrossRef]

118. Mall, A.K.; Manimekalai, R.; Misra, V.; Pandey, H.; Srivastava, S.; Sharma, A. CRISPR/Cas-mediated Genome Editing for Sugarcane Improvement. *Sugar Tech* **2024**, 1–13. [CrossRef]

119. Appunu, C.; Ram, B.; Subramonian, N. Sugarcane: An Efficient Platform for Molecular Farming. In *Sugarcane Biotechnology: Challenges and Prospects*; Mohan, C., Ed.; Springer International Publishing: Cham, Switzerland, 2017; pp. 87–110, ISBN 978-3-319-58946-6.

120. Fischer, R.; Stoger, E.; Schillberg, S.; Christou, P.; Twyman, R.M. Plant-Based Production of Biopharmaceuticals. *Curr. Opin. Plant Biol.* **2004**, *7*, 152–158. [CrossRef]

121. Jackson, M.A.; Nutt, K.A.; Hassall, R.; Rae, A.L. Comparative Efficiency of Subcellular Targeting Signals for Expression of a Toxic Protein in Sugarcane. *Funct. Plant Biol.* **2010**, *37*, 785–793. [CrossRef]

122. Palaniswamy, H.; Syamaladevi, D.P.; Mohan, C.; Philip, A.; Petchiyappan, A.; Narayanan, S. Vacuolar Targeting of R-Proteins in Sugarcane Leads to Higher Levels of Purifiable Commercially Equivalent Recombinant Proteins in Cane Juice. *Plant Biotechnol. J.* **2016**, *14*, 791–807. [CrossRef]

123. Gabriel, C.; Fernhout, J.; Fichtner, F.; Feil, R.; Lunn, J.E.; Kossmann, J.; Lloyd, J.R.; van der Vyver, C. Genetic Manipulation of Trehalose-6-phosphate Synthase Results in Changes in the Soluble Sugar Profile in Transgenic Sugarcane Stems. *Plant Direct* **2021**, *5*, e358. [CrossRef] [PubMed]

124. Liu, H.; Lin, X.; Li, X.; Luo, Z.; Lu, X.; You, Q.; Yang, X.; Xu, C.; Liu, X.; Liu, J.; et al. Haplotype Variations of *Sucrose Phosphate Synthase* Gene among Sugarcane Accessions with Different Sucrose Content. *BMC Genom.* **2023**, *24*, 42. [CrossRef]

125. Anur, R.M.; Mufithah, N.; Sawitri, W.D.; Sakakibara, H.; Sugiharto, B. Overexpression of *Sucrose Phosphate Synthase* Enhanced Sucrose Content and Biomass Production in Transgenic Sugarcane. *Plants* **2020**, *9*, 200. [CrossRef]

126. Suherman; Wijayanto, S.I.; Anur, R.M.; Neliana, I.R.; Dewanti, P.; Sugiharto, B. Field Evaluation on Growth and Productivity of the Transgenic Sugarcane Lines Overexpressing Sucrose-Phosphate Synthase. *Sugar Technol.* **2022**, *24*, 1689–1698. [CrossRef]

127. Awan, M.F.; Ali, S.; Iqbal, M.S.; Sharif, M.N.; Ali, Q.; Nasir, I.A. Enhancement of Healthful Novel Sugar Contents in Genetically Engineered Sugarcane Juice Integrated with Molecularly Characterized *ThSyGII* (CEMB-SIG2). *Sci. Rep.* **2022**, *12*, 18621. [CrossRef] [PubMed]

128. Petrasovits, L.A.; Purnell, M.P.; Nielsen, L.K.; Brumbley, S.M. Production of Polyhydroxybutyrate in Sugarcane. *Plant Biotechnol. J.* **2007**, *5*, 162–172. [CrossRef] [PubMed]

129. Petrasovits, L.A.; Zhao, L.; McQualter, R.B.; Snell, K.D.; Somleva, M.N.; Patterson, N.A.; Nielsen, L.K.; Brumbley, S.M. Enhanced Polyhydroxybutyrate Production in Transgenic Sugarcane. *Plant Biotechnol. J.* **2012**, *10*, 569–578. [CrossRef] [PubMed]

130. Jung, J.H.; Altpeter, F. TALEN Mediated Targeted Mutagenesis of the Caffeic Acid O-Methyltransferase in Highly Polyploid Sugarcane Improves Cell Wall Composition for Production of Bioethanol. *Plant Mol. Biol.* **2016**, *92*, 131–142. [CrossRef] [PubMed]

131. Kannan, B.; Jung, J.H.; Moxley, G.W.; Lee, S.-M.; Altpeter, F. TALEN-Mediated Targeted Mutagenesis of More than 100 *COMT* Copies/Alleles in Highly Polyploid Sugarcane Improves Saccharification Efficiency without Compromising Biomass Yield. *Plant Biotechnol. J.* **2018**, *16*, 856–866. [CrossRef] [PubMed]

132. Jia, Y.; Maitra, S.; Singh, V. Chemical-Free Production of Multiple High-Value Bioproducts from Metabolically Engineered Transgenic Sugarcane 'Oilcane' Bagasse and Their Recovery Using Nanofiltration. *Bioresour. Technol.* **2023**, *371*, 128630. [CrossRef] [PubMed]

133. Zale, J.; Jung, J.H.; Kim, J.Y.; Pathak, B.; Karan, R.; Liu, H.; Chen, X.; Wu, H.; Candreva, J.; Zhai, Z.; et al. Metabolic Engineering of Sugarcane to Accumulate Energy-Dense Triacylglycerols in Vegetative Biomass. *Plant Biotechnol. J.* **2016**, *14*, 661–669. [CrossRef] [PubMed]

134. Parajuli, S.; Kannan, B.; Karan, R.; Sanahuja, G.; Liu, H.; Garcia-Ruiz, E.; Kumar, D.; Singh, V.; Zhao, H.; Long, S.; et al. Towards Oilcane: Engineering Hyperaccumulation of Triacylglycerol into Sugarcane Stems. *GCB Bioenergy* **2020**, *12*, 476–490. [CrossRef]

135. Bhatti, F.; Asad, S.; Khan, Q.M.; Mobeen, A.; Iqbal, M.J.; Asif, M. Risk Assessment of Genetically Modified Sugarcane Expressing *AVP1* Gene. *Food Chem. Toxicol.* **2019**, *130*, 267–275. [CrossRef]

136. Perera, M.F.; Ovejero, S.N.; Racedo, J.; Noguera, A.S.; Cuenya, M.I.; Castagnaro, A.P. TRAP Markers Allow the Identification of Transgenic Lines That Are Genetically Close to Their Parental Genotype. *Sugar Tech* **2020**, *22*, 750–755. [CrossRef]

137. Schaart, J.G.; van de Wiel, C.C.M.; Smulders, M.J.M. Genome Editing of Polyploid Crops: Prospects, Achievements and Bottlenecks. *Transgenic Res.* **2021**, *30*, 337–351. [CrossRef] [PubMed]

138. Eid, A.; Mohan, C.; Sanchez, S.; Wang, D.; Altpeter, F. Multiallelic, Targeted Mutagenesis of Magnesium Chelatase with CRISPR/Cas9 Provides a Rapidly Scorable Phenotype in Highly Polyploid Sugarcane. *Front. Genome Ed.* **2021**, *3*, 654996. [CrossRef] [PubMed]

139. Shabbir, R.; Javed, T.; Afzal, I.; Sabagh, A.E.; Ali, A.; Vicente, O.; Chen, P. Modern Biotechnologies: Innovative and Sustainable Approaches for the Improvement of Sugarcane Tolerance to Environmental Stresses. *Agronomy* **2021**, *11*, 1042. [CrossRef]

140. Ko, J.K.; Jung, J.H.; Altpeter, F.; Kannan, B.; Kim, H.E.; Kim, K.H.; Alper, H.S.; Um, Y.; Lee, S.-M. Largely Enhanced Bioethanol Production through the Combined Use of Lignin-Modified Sugarcane and Xylose Fermenting Yeast Strain. *Bioresour. Technol.* **2018**, *256*, 312–320. [CrossRef] [PubMed]

141. Sander, J.D.; Dahlborg, E.J.; Goodwin, M.J.; Cade, L.; Zhang, F.; Cifuentes, D.; Curtin, S.J.; Blackburn, J.S.; Thibodeau-Beganny, S.; Qi, Y.; et al. Selection-Free Zinc-Finger-Nuclease Engineering by Context-Dependent Assembly (CoDA). *Nat. Methods* **2011**, *8*, 67–69. [CrossRef] [PubMed]

142. Kumar, T.; Bao, A.-K.; Bao, Z.; Wang, F.; Gao, L.; Wang, S.-M. The Progress of Genetic Improvement in Alfalfa (*Medicago sativa* L.). *Czech J. Genet. Plant Breed.* **2018**, *54*, 41–51. [CrossRef]

143. Tsanova, T.; Stefanova, L.; Topalova, L.; Atanasov, A.; Pantchev, I. DNA-Free Gene Editing in Plants: A Brief Overview. *Biotechnol. Biotechnol. Equip.* **2021**, *35*, 131–138. [CrossRef]

144. Javaid, D.; Ganie, S.Y.; Hajam, Y.A.; Reshi, M.S. CRISPR/Cas9 System: A Reliable and Facile Genome Editing Tool in Modern Biology. *Mol. Biol. Rep.* **2022**, *49*, 12133–12150. [CrossRef] [PubMed]

145. Wolter, F.; Puchta, H. Knocking out Consumer Concerns and Regulator's Rules: Efficient Use of CRISPR/Cas Ribonucleoprotein Complexes for Genome Editing in Cereals. *Genome Biol.* **2017**, *18*, 43. [CrossRef]

146. Singh, R.K.; Banerjee, N.; Khan, M.S.; Yadav, S.; Kumar, S.; Duttamajumder, S.K.; Lal, R.J.; Patel, J.D.; Guo, H.; Zhang, D.; et al. Identification of Putative Candidate Genes for Red Rot Resistance in Sugarcane (*Saccharum* Species Hybrid) Using LD-Based Association Mapping. *Mol. Genet. Genom.* **2016**, *291*, 1363–1377. [CrossRef]

147. Lu, S.; Zhang, H.; Guo, F.; Yang, Y.; Shen, X.; Chen, B. *SsUbc2*, a Determinant of Pathogenicity, Functions as a Key Coordinator Controlling Global Transcriptomic Reprogramming during Mating in Sugarcane Smut Fungus. *Front. Microbiol.* **2022**, *13*, 954767. [CrossRef]

148. Zhang, H.; Yang, Y.; Guo, F.; Shen, X.; Lu, S.; Chen, B. *SsRSS1* Mediates Salicylic Acid Tolerance and Contributes to Virulence in Sugarcane Smut Fungus. *J. Integr. Agric.* **2023**, *22*, 2126–2137. [CrossRef]

149. Zhang, J.; Xing, J.; Mi, Q.; Yang, W.; Xiang, H.; Xu, L.; Zeng, W.; Wang, J.; Deng, L.; Jiang, J.; et al. Highly Efficient Transgene-Free Genome Editing in Tobacco Using an Optimized CRISPR/Cas9 System, pOREU3TR. *Plant Sci.* **2023**, *326*, 111523. [CrossRef]

150. Tyagi, S.; Kesiraju, K.; Saakre, M.; Rathinam, M.; Raman, V.; Pattanayak, D.; Sreevathsa, R. Genome Editing for Resistance to Insect Pests: An Emerging Tool for Crop Improvement. *ACS Omega* **2020**, *5*, 20674–20683. [CrossRef]

151. Eid, A.; Mahfouz, M.M. Genome Editing: The Road of CRISPR/Cas9 from Bench to Clinic. *Exp. Mol. Med.* **2016**, *48*, e265. [CrossRef] [PubMed]

152. Chen, Y.; Ma, J.; Zhang, X.; Yang, Y.; Zhou, D.; Yu, Q.; Que, Y.; Xu, L.; Guo, J. A Novel Non-Specific Lipid Transfer Protein Gene from Sugarcane (*NsLTPs*), Obviously Responded to Abiotic Stresses and Signaling Molecules of SA and MeJA. *Sugar Tech* **2017**, *19*, 17–25. [CrossRef]

153. Su, Y.; Wang, Z.; Liu, F.; Li, Z.; Peng, Q.; Guo, J.; Xu, L.; Que, Y. Isolation and Characterization of *ScGluD2*, a New Sugarcane *Beta-1,3-glucanase* D Family Gene Induced by *Sporisorium scitamineum*, ABA, H_2O_2, NaCl, and $CdCl_2$ Stresses. *Front. Plant Sci.* **2016**, *7*, 1348. [CrossRef] [PubMed]

154. Wu, Q.; Pan, Y.-B.; Su, Y.; Zou, W.; Xu, F.; Sun, T.; Grisham, M.P.; Yang, S.; Xu, L.; Que, Y. WGCNA Identifies a Comprehensive and Dynamic Gene Co-Expression Network That Associates with Smut Resistance in Sugarcane. *Int. J. Mol. Sci.* **2022**, *23*, 10770. [CrossRef] [PubMed]

155. Luo, G.; Cao, V.D.; Kannan, B.; Liu, H.; Shanklin, J.; Altpeter, F. Metabolic Engineering of Energycane to Hyperaccumulate Lipids in Vegetative Biomass. *BMC Biotechnol.* **2022**, *22*, 24. [CrossRef]
156. Laksana, C.; Sophiphun, O.; Chanprame, S. Lignin Reduction in Sugarcane by Performing CRISPR/Cas9 Site-Direct Mutation of *SoLIM* Transcription Factor. *Plant Sci.* **2024**, *340*, 111987. [CrossRef]
157. Kannan, B.; Liu, H.; Shanklin, J.; Altpeter, F. Towards Oilcane: Preliminary Field Evaluation of Metabolically Engineered Sugarcane with Hyper-Accumulation of Triacylglycerol in Vegetative Tissues. *Mol. Breed.* **2022**, *42*, 64. [CrossRef] [PubMed]
158. Li, C.; Iqbal, M.A. Leveraging the Sugarcane CRISPR/Cas9 Technique for Genetic Improvement of Non-Cultivated Grasses. Front. *Plant Sci.* **2024**, *15*, 1369416. [CrossRef]

MDPI AG
Grosspeteranlage 5
4052 Basel
Switzerland
Tel.: +41 61 683 77 34

Plants Editorial Office
E-mail: plants@mdpi.com
www.mdpi.com/journal/plants